CALCIUM AND CELL FUNCTION

Volume VI

Molecular Biology

An International Series of Monographs and Textbooks

Editors

BERNARD HORECKER
Graduate School of Medical Sciences
Cornell University Medical College
New York, New York

JULIUS MARMUR
Department of Biochemistry
Albert Einstein College of Medicine
Yeshiva University
Bronx, New York

NATHAN O. KAPLAN
Department of Chemistry
University of California
at San Diego
La Jolla, California

HAROLD A. SCHERAGA
Department of Chemistry
Cornell University
Ithaca, New York

Recently published titles

TERRANCE LEIGHTON AND WILLIAM F. LOOMIS, JR. (Editors). The Molecular Genetics of Development: An Introduction to Recent Research on Experimental Systems. 1980

ROBERT B. FREEDMAN AND HILARY C. HAWKINS (Editors). The Enzymology of Post-Translational Modification of Proteins, Volume 1, 1980. Volume 2, 1985

WAI YIU CHEUNG (Editor). Calcium and Cell Function, Volume I: Calmodulin, 1980. Volume II, 1983. Volume III, 1982. Volume IV, 1983. Volume V, 1984. Volume VI, 1986

OLEG JARDETZKY AND G. C. K. ROBERTS. NMR in Molecular Biology. 1981

DAVID A. DUBNAU (Editor). The Molecular Biology of the Bacilli, Volume I: *Bacillus subtilis,* 1982. Volume II, 1985

GORDON G. HAMMES. Enzyme Catalysis and Regulation. 1982

GUNTER KAHL AND JOSEF S. SCHELL (Editors). Molecular Biology of Plant Tumors. 1982

P. R. CAREY. Biochemical Applications of Raman and Resonance Raman Spectroscopies. 1982

OSAMU HAYAISHI AND KUNIHIRO UEDA (Editors). ADP-Ribosylation Reactions: Biology and Medicine. 1982

G. O. ASPINALL (Editor). The Polysaccharides, Volume 1, 1982. Volume 2, 1983. Volume 3, 1985

CHARIS GHELIS AND JEANNINE YON. Protein Folding. 1982

ALFRED STRACHER (Editor). Muscle and Nonmuscle Motility, Volume 1, 1983. Volume 2, 1983

In preparation

IRVIN E. LIENER, NATHAN SHARON, AND IRWIN J. GOLDSTEIN (Editors). The Lectins: Properties, Functions, and Applications in Biology and Medicine. 1986

CALCIUM AND CELL FUNCTION

Volume VI

Edited by

WAI YIU CHEUNG

Department of Biochemistry
St. Jude Children's Research Hospital
and University of Tennessee Center for the Health Sciences
Memphis, Tennessee

1986

ACADEMIC PRESS, INC.

Harcourt Brace Jovanovich, Publishers
Orlando   San Diego   New York   Austin
London   Montreal   Sydney   Tokyo   Toronto

ACADEMIC PRESS, INC.
Orlando, Florida 32887

United Kingdom Edition published by
ACADEMIC PRESS INC. (LONDON) LTD.
24–28 Oval Road, London NW1 7DX

Library of Congress Cataloging in Publication Data
(Revised for vol. 6)
Main entry under title:

Calcium and cell function.

(Molecular biology, an international series of
monographs and textbooks)
Vols. 2-   have no special title.
Includes bibliographies and indexes.
Contents: v. 1. Calmodulin.
1. Calcium—Physiological effect—Collected works.
2. Calcium—Metabolism—Collected works.   3. Cell
physiology—Collected works.   I. Cheung, Wai Yiu.
II. Series.   [DNLM: 1. Calcium.   2. Calcium—Binding
proteins.   QU55 C144]
QP535.C2C26   612'.3924   80-985
ISBN 0–12–171406–3 (v. 6)

Contents

Chapter 3 Calmodulin-Dependent Protein Phosphatase
E. Ann Tallant and Wai Yiu Cheung

Chapter 4 Biophysical Studies of Calmodulin
Sture Forsén, Hans J. Vogel, and Torbjörn Drakenberg

Chapter 5 Regulation of Ca^{2+}-Dependent Proteinase of Human Erythrocytes
Sandro Pontremoli and Edon Melloni

Chapter 6 Toxicological Implications of Perturbation of Ca²⁺ Homeostasis in Hepatocytes
Sten Orrenius and Giorgio Bellomo

Chapter 7 The Role of Calcium in Meiosis
Gene A. Morrill and Adele B. Kostellow

Chapter 8 Calcium and the Control of Insulin Secretion
Bo Hellman and Erik Gylfe

Chapter 9   Roles of Calcium in Photosynthesis
Clanton C. Black, Jr. and Jerry J. Brand

Contributors

Numbers in parentheses indicate the pages on which the authors' contributions begin.

Sonia Ruth Anderson (1), Department of Biochemistry and Biophysics, Oregon State University, Corvallis, Oregon 97331

Giorgio Bellomo (185), Medical Clinic, University of Pavia, 27100 Pavia, Italy

Clanton C. Black, Jr. (327), Biochemistry Department, University of Georgia, Athens, Georgia 30602

Jerry J. Brand (327), Department of Botany, University of Texas, Austin, Texas 78712

Wai Yiu Cheung (71), Department of Biochemistry, St. Jude Children's Research Hospital, and University of Tennessee Center for the Health Sciences, Memphis, Tennessee 38101

Torbjörn Drakenberg (113), Physical Chemistry, Chemical Centre, Lund University, S-221 00 Lund, Sweden

Sture Forsén (113), Physical Chemistry, Chemical Centre, Lund University, S-221 00 Lund, Sweden

Erik Gylfe (253), Department of Medical Cell Biology, University of Uppsala, S-751 23 Uppsala, Sweden

Bo Hellman (253), Department of Medical Cell Biology, University of Uppsala, S-751 23 Uppsala, Sweden

Randall L. Kincaid (43), Laboratory of Cellular Metabolism, National Heart, Lung and Blood Institute, National Institutes of Health, Bethesda, Maryland 20205

Adele B. Kostellow (209), Department of Physiology and Biophysics, Albert Einstein College of Medicine, Bronx, New York 10461

Dean A. Malencik (1), Department of Biochemistry and Biophysics, Oregon State University, Corvallis, Oregon 97331

Edon Melloni (159), Institute of Biological Chemistry, University of Genoa, 16132 Genoa, Italy

Gene A. Morrill (209), Department of Physiology and Biophysics, Albert Einstein College of Medicine, Bronx, New York 10461

Sten Orrenius (185), Department of Toxicology, Karolinska Institute, S-104 01 Stockholm, Sweden

Sandro Pontremoli (159), Institute of Biological Chemistry, University of Genoa, 16132 Genoa, Italy

E. Ann Tallant[1] (71), Department of Biochemistry, St. Jude Children's Research Hospital, and University of Tennessee Center for the Health Sciences, Memphis, Tennessee 38101

Martha Vaughan (43), Laboratory of Cellular Metabolism, National Heart, Lung and Blood Institute, National Institutes of Health, Bethesda, Maryland 20205

Hans J. Vogel[2] (113), Physical Chemistry, Chemical Centre, Lund University, S-221 00 Lund, Sweden

[1]Present address: Department of Pharmacology, University of Alabama at Birmingham, Birmingham, Alabama 35294.
[2]Present address: Division of Biochemistry, Department of Chemistry, University of Calgary, Calgary, Alberta, Canada T2N 4N1.

Preface

This volume deals with several topics that are of current interest. In the first chapter, Anderson and Malencik describe the various peptides that recognize calmodulin in a calcium-dependent fashion. Chapter 2 by Kincaid and Vaughan updates the molecular and regulatory properties of calmodulin-dependent phosphodiesterase, a topic covered earlier in Volume I. In Chapter 3, Tallant and Cheung review a calmodulin-dependent protein phosphatase, the most recent addition to the list of calmodulin-dependent enzymes. Forsén et al. in Chapter 4 summarize their studies on the biophysical characterization of calmodulin. In Chapter 5, Pontremoli and Melloni present their recent studies on a calcium-dependent protease from human erythrocytes. As shown by many investigators in this treatise, calcium plays a dominant role in cellular regulation, but an uncontrolled level of calcium often precipitates toxic consequences, a problem addressed by Orrenius and Bellomo in Chapter 6. The role of calcium in meiosis has been studied extensively by many investigators. Morrill and Kostellow bring us up-to-date on this important topic in Chapter 7. The calcium ion has been intimately involved with the secretory process; Hellman and Gylfe review this subject in Chapter 8 with particular emphasis on the secretion of insulin. Many cellular processes in plants are known to be regulated by calcium. In the final chapter, Black and Brand discuss its role in photosynthesis, a process that captures the solar energy and sustains all living systems.

It has been my pleasure to work with the many contributors of this volume. I take this opportunity to thank them for their splendid efforts.

Wai Yiu Cheung

Contents of Previous Volumes

Volume II

Volume IV

Volume V

Chapter 1

Peptides Recognizing Calmodulin

SONIA RUTH ANDERSON
DEAN A. MALENCIK

Department of Biochemistry and Biophysics
Oregon State University
Corvallis, Oregon

1

CALCIUM AND CELL FUNCTION, VOL. VI

I. INTRODUCTION

Calmodulin is the major known intracellular receptor for calcium in plants and animals. It is a vital part of the intricate system for cellular communication and coordination evolved by higher organisms. The integration of physiological processes in vertebrates is facilitated by an array of extracellular messenger molecules together with a small number of intracellular second messengers, such as calcium and the cyclic nucleotides. The nervous system stimulates membrane depolarization and the release of calcium recognized by calmodulin and related intracellular proteins (cf. reviews by Kretsinger, 1980; Cheung, 1980; Means, 1981; Klee and Vanaman, 1982). Cell surface receptors for hormones such as adrenocorticotropin (ACTH), glucagon, and epinephrine transfer information across the cell membrane, with cAMP acting as the second messenger (cf. review by Krebs and Beavo, 1979). Calcium and cAMP often act together in mediating cellular response. The enzymes catalyzing cAMP synthesis and degradation, adenylate cyclase and cyclic nucleotide phosphodiesterase, are both activated by the calcium–calmodulin complex (cf. review by Cheung, 1980). The activities of phosphorylase kinase and smooth muscle myosin light chain kinase, strictly calmodulin-dependent enzymes, are stimulated and inhibited, respectively, after phosphorylation by the cAMP-dependent protein kinase (cf. review by Malencik and Fischer, 1982; Conti and Adelstein, 1981). The close interconnection between the nervous and endocrine systems is shown further by the occurrence of peripheral peptide hormones including somatostatin, angiotensin II, insulin, glucagon, and members of the gastrin–cholecystokinin group in the brain, where they may function as neurotransmitters (cf. reviews by Snyder and Innis, 1979; Hokfelt *et al.*, 1980; Krieger, 1983).

The mutual recognition sites in calmodulin and the affected enzymes are yet to be characterized in terms of primary and higher levels of structure. In fact, no calmodulin-dependent enzyme has been completely sequenced up to the present time. Various approaches have been made to the topography of the protein binding site in calmodulin. The binding of phenothiazine drugs, widely used calmodulin antagonists, correlates with hydrophobicity (Norman *et al.*, 1979) and charge (Weiss *et al.*, 1982, 1985; Prozialeck and Weiss, 1982). Presumed fluorescent probes of hydrophobic binding sites in proteins—8-anilino-1-naphthalenesulfonate, *N*-phenyl-1-naphthylamine, and 9-anthroylcholine—exhibit calcium-dependent binding by calmodulin (LaPorte *et al.*, 1980; Tanaka and Hidaka, 1980). Troponin I, the inhibitory subunit of troponin, has been widely studied as a model calmodulin-binding protein (Grand *et al.*, 1979; Keller *et al.*, 1982). But with 184 amino acid residues, it provides few clues regarding local interactions. Experiments with model peptides have contributed information on diverse aspects of protein structure. For example, proteolytically derived fragments—20-residue S-peptide and 104-residue S-protein—were used to probe structure–function relationships in ribonuclease A (cf. review by Richards and Wyckoff (1971). The specific association of these fragments, with accompanying restoration of catalytic activity, was even suggested as a model for hormone–receptor interaction (Finn and Hofmann, 1973). More recently, synthetic peptide substrates helped to establish a class of recognition sequences surrounding the serine and threonine residues which can be phosphorylated by the cAMP-dependent protein kinase (cf. review by Carlson *et al.*, 1979). Although similar types of investigation might shed light on the protein binding specificity of calmodulin, an initial survey of several small peptides revealed no calcium-dependent complex formation. The study also showed that, in spite of the apparent lack of specificity, highly basic polypeptides such as polyarginine and polylysine inhibit calmodulin-dependent cyclic nucleotide phosphodiesterase (Itano *et al.*, 1980).

In 1980, Weiss *et al.* discovered that ACTH and β-endorphin inhibit the purified cyclic nucleotide phosphodiesterase and showed that the peptides compete with the enzyme for calmodulin. Their report was followed up with *in vitro* binding measurements, performed in our laboratory and others, demonstrating efficient calcium-dependent binding of a number of biologically active peptides by calmodulin: ACTH, β-endorphin, glucagon, substance P (Malencik and Anderson, 1982); the dynorphins (Sellinger-Barnette and Weiss, 1982, 1984; Malencik and Anderson, 1983a, 1984); secretin, gastric inhibitory peptide (GIP), vasoactive intestinal peptide (VIP) (Malencik and Anderson, 1983a); melittin (Comte *et al.*, 1983; Barnette *et al.*, 1983); and the mastoparans (Barnette *et al.*, 1983; Malencik and Anderson, 1983b, 1984). All told, about 50 small peptides, including several belonging to common gene families, have been examined for calmodulin binding. Although the above peptide–calmodulin associations do not necessarily occur *in vivo*, they are nonetheless interesting and important, providing information on molecules that are more or less direct partners in cellular regulation. The structural and functional significance of these interactions, together with a description of the methods used to detect peptide binding *in vitro*, are the subjects of this review.

II. METHODS FOR THE DETERMINATION OF PEPTIDE–CALMODULIN INTERACTION *IN VITRO*

We have summarized the principal methods used to detect peptide binding by calmodulin,[1] both for understanding of this article and for future applications.

A. Enzyme Inhibition

ACTH, β-endorphin, melittin, and other specific peptides inhibit the calmodulin-dependent activity of purified cyclic nucleotide phosphodiesterase (Weiss *et*

---

[1] Porcine brain or bovine brain calmodulin was used throughout this work.

al., 1980; Sellinger-Barnette and Weiss, 1982, 1984; Barnette and Weiss, 1983; Barnette *et al.*, 1983; Comte *et al.*, 1983) and myosin light chain kinase (Malencik *et al.*, 1982a; Katoh *et al.*, 1982). Inhibition constants, reflecting the affinities of the peptides for calmodulin, are in principle extractable from catalytic activity measurements performed at varying peptide concentrations. However, the accumulation of quantitative information from these experiments is difficult for several reasons. First, the assays for these enzymes are fixed-time rather than continuous. Additionally, cyclic nucleotide phosphodiesterase has considerable baseline activity in the absence of calmodulin, while the myosin light chains alone interact moderately with a number of the peptides (see Section VI). In view of these problems, we have developed several more efficient and sensitive methods for the determination of peptide binding. They are described in detail in the following three sections.

The results of the enzyme inhibition experiments are often expressed by values of IC_{50}. Weiss's group reports the concentration of peptide required to inhibit calmodulin-dependent cyclic nucleotide phosphodiesterase activity by 50% in solutions initially containing sufficient calmodulin to produce 50% of the maximum stimulation. When the relationship is competitive, the value of IC_{50} obtained under these conditions is determined by the dissociation constants of the peptide–calmodulin complex (K_d), of the enzyme–calmodulin complex (K_E), and by the total calmodulin concentration according to the equation

$$IC_{50} = \frac{(K_E + 3K_d)([CaM]_{total} + K_E)}{K_E} \quad \frac{2}{} \quad - \frac{K_E}{3} - K_d$$

Hence the value of IC_{50} is larger than the true dissociation constant. The relative difference between the two quantities is greatest when $K_d \ll K_E$ and $[CaM]_{total} \gg K_E$. The effect of the substrate is included in the term K_E.

B. 9-Anthroylcholine: Probe for the Competitive Displacement of Smooth Muscle Myosin Light Chain Kinase from Calmodulin

Turkey gizzard myosin light chain kinase, an absolutely calmodulin-dependent enzyme (cf. reviews by Adelstein and Klee, 1980; Hartshorne and Siemankowski, 1981), contains an interaction site for the fluorescent dye 9-anthroylcholine (9AC) which may correspond to the ATP binding site (Malencik *et al.*, 1982a). Association of the enzyme with calmodulin increases its affinity for 9AC, with the dissociation constant decreasing from 20 to 6.4 μM. The resulting change in fluorescence is useful in stoichiometric titrations of the enzyme with calmodulin (Fig. 1). The addition of excess EDTA to the enzyme–calmodulin complex reverses the enhancement, giving the same fluorescence obtained with the enzyme and 9AC alone. In passing, we note that rabbit skeletal muscle myosin light chain kinase does not bind 9AC appreciably.

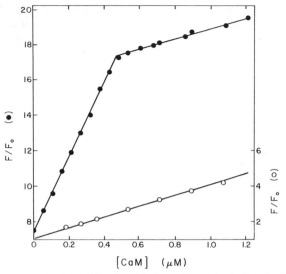

Fig. 1. Fluorescence titrations of 9-anthroylcholine with porcine brain calmodulin performed in the absence (○) and presence (●) of turkey gizzard myosin light chain kinase (0.50 μM). F/F_0 is the ratio of the observed fluorescence intensity to the intensity of 9AC in the absence of protein. The two linear stages obtained in the presence of the enzyme indicate that the dissociation constant of the enzyme–calmodulin complex is at least 100-fold smaller than the enzyme concentration. Conditions: 5.0 μM 9AC, 1.0 mM $CaCl_2$, 1.0 mM dithiothreitol, 0.20 N KCl, and 10.0 mM 3-(N-morpholino)propanesulfonic acid, pH 7.3, 25.0°. Excitation, 360 nm; emission, 460 nm.

We have developed a competitive displacement technique based on the binding of 9-anthroylcholine by turkey gizzard myosin kinase. First, we determined that calmodulin-binding proteins and peptides usually displace 9AC from calmodulin to about the same extent found by LaPorte *et al.* (1980) with troponin I. Next, we showed that they do not interact with 9AC themselves and that they have no effect on its binding by the enzyme in the absence of calmodulin. Addition of one of the competing proteins or peptides (P) to the enzyme–calmodulin complex results in the following net equilibrium:

$$MLCK \cdot CaM + P \rightleftarrows MLCK + P \cdot CaM$$

Since the concentration of added calmodulin equals that of the enzyme, the fraction of unbound calmodulin is negligible. Thus the fluorescence obtained on the addition of P should range between the values obtained with the enzyme–calmodulin complex and with the enzyme alone. Our experimental results are expressed as the fraction of the difference (ΔF) between these reference values.

$$\Delta F = (F_{MLCK \cdot CaM} - F_{obs})/(F_{MLCK \cdot CaM} - F_{MLCK})$$

The distribution of calmodulin between the two complexes and the ratio of the dissociation constants for the competing equilibria can be evaluated even when the individual constants are beyond the range of direct determination.

$$MLCK·CaM \xrightleftharpoons{K_{MLCK·CaM}} MLCK + CaM$$

$$P·CaM \xrightleftharpoons{K_{P·CaM}} P + CaM$$

$$\frac{K_{MLCK·CaM}}{K_{P·CaM}} = \frac{\Delta F^2 [MLCK]_0}{(1-\Delta F)([P]_0 - \Delta F[MLCK]_0)}$$

where $[MLCK]_0$ and $[P]_0$ represent the total concentrations of the enzyme and competing protein or peptide, respectively. This equation is simplified whenever $[P]_0 \gg \Delta F[MLCK]_0$.

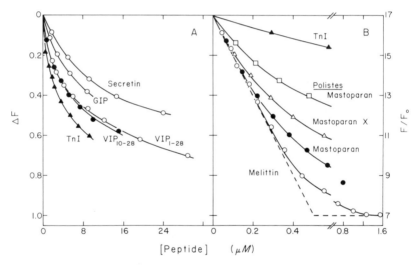

Fig. 2. Dissociation of the turkey gizzard myosin light chain kinase–calmodulin complex by calmodulin-binding peptides. The left-hand ordinate shows the fractional changes in 9-anthroylcholine fluorescence (ΔF), while the right-hand ordinate shows the corresponding relative fluorescence intensities (F/F_0). Panel A depicts the results obtained with varying concentrations of intermediate-affinity peptides, and panel B, those obtained with high-affinity peptides. Titrations with rabbit skeletal troponin I are included for comparison. The peptides have negligible effect on the fluorescence obtained with the enzyme in the absence of calmodulin. The fluorescence approaches this baseline value when excess concentrations of mastoparan or melittin are added to the enzyme–calmodulin complex. Conditions: 0.50 μM MLCK–CaM, 5.0 μM 9AC, 1.0 mM CaCl₂, 1.0 mM dithiothreitol, 0.15M (panel A) or 0.20 M, (panel B) KCl, and 45 mM 3-(N-morpholino)propanesulfonic acid, pH 7.3, 25.0°. Excitation, 360 nm; emission, 460 nm. Panel A reprinted with permission from Malencik, D. A. and Anderson, S. R., *Biochemistry* **22,** 1995–2001. Copyright (1983) American Chemical Society. Panel B from Malencik and Anderson (1983b).

$$\frac{K_{\text{MLCK·CaM}}}{K_{\text{P·CaM}}} = \frac{\Delta F^2}{(1-\Delta F)} \frac{[\text{MLCK}]_0}{[\text{P}]_0}$$

The ratio $K_{\text{MLCK·CaM}}/K_{\text{P·CaM}}$ is a reasonable representation of the relative affinities for calmodulin: The predicted value for $K_{\text{MLCK·CaM}}$ in the presence of 5 μM 9-anthroylcholine is 71% of that expected in its absence.[2] The comparatively weak binding of 9AC by calmodulin, with $K_{\text{av}} = \sim440$ μM (LaPorte et al., 1980), exerts minimal effect. This method has advantages over enzyme inhibition in that continuous titrations spanning a range of concentrations can be performed on a single sample. Perturbation of the equilibrium by 9AC is likely to be no greater than that caused by the substrates present in enzyme assays.

Titration of the smooth muscle myosin light chain kinase–calmodulin complex with the more familiar calmodulin-binding proteins and peptides produces the expected changes in 9-anthroylcholine fluorescence (Fig. 2A). Rabbit skeletal muscle troponin I, corresponding to a dissociation constant of 20–60 nM (cf. review by Keller et al., 1982), gives a value of 0.048 for $K_{\text{MLCK·CaM}}/K_{\text{TnI·CaM}}$. Calculation of $K_{\text{MLCK·CaM}}$ gives figures, 1–3 nM, comparable to the K_{m} of the enzyme for calmodulin. Secretin and β-endorphin form complexes with dansylcalmodulin having K_{d}'s of 0.14 and 1.9 μM, respectively. Competition of these peptides with myosin light chain kinase yields corresponding ratios of 0.010 and 0.0008. Since appreciable dissociation requires values of $[\text{P}]_0/K_{\text{P·CaM}}$ comparable to $[\text{MLCK}]_0/K_{\text{MLCK·CaM}}$, the method is most effective with intermediate and high-affinity peptides or proteins.

C. Intrinsic Fluorescence of Tryptophan-Containing Peptides

Some of the most direct measurements of calmodulin binding are obtained with the tryptophan-containing peptides. The intrinsic fluorescence of these peptides is observable under conditions largely excluding the fluorescence of calmodulin, which contains tyrosine but no tryptophan. The tryptophan emission spectra and quantum yields of ACTH, glucagon (Malencik and Anderson, 1982), GIP (Malencik and Anderson, 1983a), Polistes mastoparan (Malencik and Anderson, 1983b), mastoparan X, dynorphin$_{1-17}$ (Malencik and Anderson, 1984), and melittin (Maulet and Cox, 1983) are sensitive to calmodulin binding. The spectra of these complexes may shed light on the environment of the peptide binding site (see Section VIII). In addition, the fluorescence changes can be followed in titrations to determine equilibrium constants and stoichiometries (Fig. 3). The fraction of peptide bound by calmodulin, f_{b}, is calculated from

---

[2] $K_{\text{app}} = [([9\text{AC}]/K_1 + 1)/([9\text{AC}]/K_2+1)]$ K, where K_1 is the dissociation constant for the enzyme–9AC complex (20 μM), K_2 is the constant for the calmodulin–enzyme–9AC complex (6.5 μM), and K is the dissociation constant for the enzyme–calmodulin complex in the absence of 9-anthroylcholine.

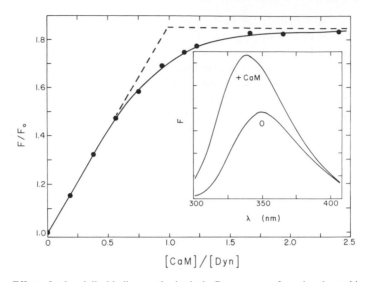

Fig. 3. Effect of calmodulin binding on the intrinsic fluorescence of porcine dynorphin$_{1-17}$. The inset shows the fluorescence emission spectra of 2.7 μM dynorphin$_{1-17}$ recorded before and after the addition of 8.0 μM porcine brain calmodulin. The stoichiometric fluorescence titration of 4.8 μM dynorphin$_{1-17}$ with calmodulin was monitored at 340 nm. A small correction was made for the background fluorescence of calmodulin in both experiments. Conditions: 1.0 mM CaCl$_2$, 0.05 M KCl, and 10 mM 3-(N-morpholino)propanesulfonic acid, pH 7.3, 25.0°. Excitation: 290 nm. Reprinted with permission from Malencik, D. A. and Anderson, S. R., *Biochemistry* **23**, 2420–2428. Copyright (1984) American Chemical Society.

$$f_b = (F_{obs}/F_0 - 1)/(F_\infty/F_0 - 1)$$

where F_∞ is the fluorescence of the complex and F_o, that of the unbound peptide (cf. review by Anderson, 1974). The fluorescence polarization and anisotropy, which relate to molecular weight, are also responsive to complex formation (Malencik and Anderson, 1984). Since the excitation wavelength used to exclude tyrosine fluorescence results in some loss of sensitivity, we do not use peptide concentrations less than 1 μM for these experiments. Thus the titrations are suitable for the accurate determination of dissociation constants greater than ~0.2 μM.

D. Dansyl Calmodulin

Dansyl calmodulin is an exceptionally responsive covalent conjugate prepared by labeling calmodulin with 5-(dimethylamino)-l-naphthalenesulfonyl chloride (Malencik and Anderson, 1982). The fluorescence emission maximum and quantum yield of dansyl calmodulin change dramatically on the binding of both calcium and proteins or peptides (Fig. 4). These effects have been used in rapid kinetic studies of calcium binding (Malencik *et al.*, 1981) and in titrations of calmodulin with smooth muscle myosin light chain kinase (Malencik *et al.*, 1982a), cyclic nucleotide phosphodiesterase, and calcineurin (Kincaid *et al.*, 1982). They are also the basis of a *highly sensitive* method for the determination of both calcium-dependent and calcium-independent peptide binding (Malencik and Anderson, 1982, 1983a,b, 1984; Malencik *et al.*, 1982a,b). Varying amounts of peptide are added to a cuvette solution which contains a fixed concentration of dansyl calmodulin ranging upward from 0.1 μM (Fig. 5). The data are fitted to the simple equilibrium

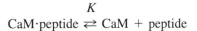

$$\text{CaM·peptide} \overset{K}{\rightleftarrows} \text{CaM + peptide}$$

for which the average degree of saturation of dansyl calmodulin with peptide (ϕ) is related to the fluorescence enhancement.

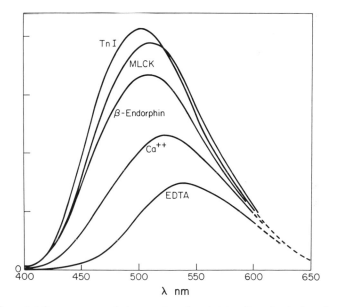

Fig. 4. Corrected fluorescence emission spectra of conjugate of porcine brain calmodulin with 5-(dimethylamino)naphthalenesulfonyl chloride. The spectra of 1.0 μM solutions of dansyl calmodulin were recorded in the presence of 1.2 mM EDTA, 0.85 mM $CaCl_2$, and 1.2 μM troponin I, 18 μM β-endorphin, or 2.0 μM myosin light chain kinase *plus* 0.85 mM $CaCl_2$. Conditions: 0.20 N KCl and 50 mM 3-(N-morpholino)propanesulfonic acid, pH 7.3, 25.0°. Excitation: 340 nm. Reprinted with permission from Malencik, D. A. and Anderson, S. R., *Biochemistry* **21**, 3480–3486. Copyright (1982) American Chemical Society.

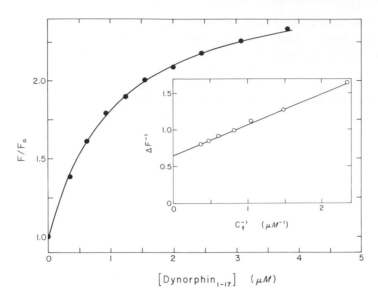

Fig. 5.   Fluorescence titration of dansyl calmodulin with dynorphin$_{1-17}$. The relative fluorescence was measured at 460 nm. The double reciprocal plot of the changes in fluorescence versus the concentration of unbound dynorphin$_{1-17}$ corresponds to 1 : 1 binding with $K_d = 0.65 \pm 0.05\ \mu M$ and a fluorescence enhancement of 2.5 (inset). The solution contained 0.50 μM dansyl calmodulin, 1.0 mM CaCl$_2$, 0.20 N KCl, and 5.0 mM 3-(N-morpholino)propanesulfonic acid, pH 7.3, 25.0°. Excitation: 340 nm. Negligible enhancement occurs in parallel experiments with samples containing 1.0 mM EDTA and no added calcium.

$$\phi = (F_{obs}/F_0 - 1)/(F_\infty/F_0 - 1)$$

F_∞ is the fluorescence of the complex, and F_0, that of dansyl calmodulin in the absence of peptide. This method is especially useful for peptides which have no tryptophan residues. In the case of the tryptophan-containing peptides, the dissociation constants obtained with dansyl calmodulin compare to those found when the peptides are titrated with unlabeled calmodulin, as described in Section II,C.

Dansyl calmodulin is also a valuable tool for the rapid detection of calmodulin binding proteins and peptides in crude extracts and partially purified fractions (see Section III,D). It has a wide range of possible application in cases where biological activity or radioimmune assays are unavailable. Fluorescent conjugates which rely on polarization measurements for the detection of binding are not very useful for small peptides since the magnitude of the observed changes is limited by the marginal difference between the molecular weights of the complex and free calmodulin.

The unusual sensitivity notwithstanding, recent fragmentation experiments show that dansyl calmodulin is heterogeneously labeled. Work is under way in our laboratory to determine the sequence locations of the responsive residues.

III. OVERVIEW OF CALMODULIN-BINDING PEPTIDES

A. The Search for a Recognition Sequence

The peptides may be ideal models for binding—presenting at once a reflection of the protein binding site in calmodulin and structural features characteristic of the still unknown calmodulin binding sites of calmodulin-dependent enzymes. In 1982 we made all possible forward and reverse sequence comparisons of the peptides then known to undergo efficient, calcium-dependent binding by calmodulin: ACTH, β-endorphin (Weiss *et al.,* 1980; Malencik and Anderson, 1982), glucagon, and substance P (Malencik and Anderson, 1982). Table I shows the sequence alignments giving the highest degree of apparent homology. We proposed a rudimentary recognition sequence for calmodulin consisting of a strongly basic tripeptide sequence, with at least two residues which are either Arg or Lys, three positions away from a pair of hydrophobic residues (Malencik and Anderson, 1982). Observations on related peptides lacking the specified features generally supported this interpretation. Application of the model resulted in the selection of peptides with increasingly higher affinities for calmodulin (Tables II and III). In 1983 we sought a peptide conforming to the paradigm, but lacking aromatic residues, for use in proton nuclear magnetic resonance experiments. Commercial peptide catalogs contained a likely candidate—*mas-*

TABLE I

Sequence Comparisons of Calmodulin-Binding Peptides[a]

			Arg	Pro	Lys	Pro	Gln	Gln	Phe	Phe	Gly	Leu	Met-NH$_2$
Substance P (complete)			Arg	Pro	Lys	Pro	Gln	Gln	Phe	Phe	Gly	Leu	Met-NH$_2$
Glucagon (14-27)	-Leu	Asp	Ser	Arg	Arg	Ala	Gln	Asp	Phe	Val	Gln	Trp	Leu	Met-
ACTH (14-27)	-Gly	Lys	Lys	Arg	Arg	Pro	Val	Lys	Val	Tyr	Pro	Asn	Gly	Ala-
β-Endorphin (reversed, 31-18)	Glu	Gly	Lys	Lys	Tyr	Ala	Asn	Lys	Ile	Ile	Ala	Asn	Lys	Phe-
Mastoparan (reversed, complete)	H$_2$N-Leu	Ile	Lys	Lys	Ala	Leu	Ala	Ala	Leu	Ala	Lys	Leu	Asn	Ile

[a]Alignments are in accord with the model suggested by Malencik and Anderson (1982).

TABLE II

Peptides Showing Slight or No Interaction with Dansyl calmodulin[a,b]

Peptide	Maximum concentration tested	Extent of binding
Angiotensin I	110 μM	Slight
Angiotensin III	52 μM	None
Neurotensin	54 μM	Slight
Physaelemin	68 μM	Slight
Insulin	49 μM	Slight
Xenopsin	60 μM	Slight
Insulin-like growth factor II 33-40	100 μM	Slight
Methionyl-lysyl-bradykinin	—	Slight (K_d = 0.5 mM)
Lysyl-bradykinin	—	Slight (K_d = 0.7 mM)
Bradykinin	100 μM	None
Bradykinin potentiator B	150 μM	None
Kytorphin (Tyr-Arg)	1.0 mM	Slight
Arg-Arg-Arg	1.7 mM	Slight
Kemptide	2.0 mM	Slight
Ribonuclease S-peptide	100 μM	None
Leu-enkephalin	180 μM	None
Met-enkephalin	180 μM	None

[a]Conditions: 1–10 μM dansyl calmodulin, 1.0 mM CaCl$_2$, 0.2 N KCl, 50 mM MOPS, pH 7.3 (25°).

[b]Data from Malencik and Anderson, 1982, 1983a; Malencik et al., 1982b; unpublished results included.

toparan. This tetradecapeptide from the venom of the Vespid wasp proved to have an affinity for calmodulin possibly surpassing that of smooth muscle myosin light chain kinase, which it effectively displaces in competition experiments (Malencik and Anderson, 1983b). Coincidentally, Barnette et al. (1983) discovered that mastoparan is a potent inhibitor of the calmodulin-dependent activity of cyclic nucleotide phosphodiesterase. Table I shows one of the two ways in which the sequence of mastoparan follows the model. A larger proportion of hydrophobic residues, a marked potential for α-helix formation, and a palindromic sequence spanning residues 4–11 distinguish mastoparan from the other four peptides.

The recognition sequence we envisioned resembles the general model presented by Prozialeck and Weiss (1982) and Weiss et al. (1982) for the binding of phenothiazines and other pharmacological agents. These authors noted that the most potent calmodulin inhibitors have two hydrophobic moieties separated from a positively charged amino nitrogen by at least three carbon atoms.

B. Deletion Peptides of ACTH, β-Endorphin, and Dynorphin$_{1-17}$

Systematic sequence deletion is the basis of a subtractive approach to the determination of minimum structure compatible with function. In the case of ACTH (Fig. 6), deletion peptides containing residues 1–24, 7–38, and 1–17 are efficient inhibitors of calmodulin-dependent cyclic nucleotide phosphodiesterase activity while peptides containing residues 1–10, 4–10, 1–11, or 18–39 are not. The effective analogs contain most of the basic tetrapeptide sequence (-Lys-Lys-Arg-Arg-) of ACTH. Yet the overall orientation of the basic cluster is also important; the peptide consisting of residues 11–24 is a weak inhibitor (Barnette and Weiss, 1983). The moderate binding of the α-melanocyte–stimulating hormone by calmodulin is consistent with the stabilizing effect of the N-terminal sequence (Table III; Malencik and Anderson, 1982). Perhaps the tryptophan residue at position 9 of ACTH interacts with calmodulin. The C-terminal region of β-endorphin (Fig. 6) is vital to its inhibitory properties while the N-terminal region, including the opioid pentapaptide sequence, is not. Deletion peptides of β-endorphin containing residues 6–31 (Barnette and Weiss, 1983), 2–31, 8–31, 12–31, and 14–31 (Giedroc *et al.,* 1983b; Puett *et al.,* 1983) inhibit cyclic nucleotide phosphodiesterase nearly as well as β-endorphin, with 50% inhibition occurring at peptide concentrations of 10 μ*M* or less. Peptides lacking the -Lys-Lys- sequence inhibit cyclic nucleotide phosphodiesterase either poorly (100 μ*M* concentrations required for 50% inhibition with 1–25 and 14–25) or not at all (10–19 and 14–23). Preliminary experiments suggested that the derivative representing residues 20–31 is only weakly inhibitory, even though it resembles several more effective peptides (Giedroc *et al.,* 1983b).

The dynorphins are strongly basic opioid peptides of unknown physiological function (Goldstein *et al.,* 1979). They undergo calcium-dependent binding by both calmodulin and troponin C (see Section VI). The effects of stepwise deletion from the C-terminal end of dynorphin$_{1-17}$ were independently determined in inhibition experiments with cyclic nucleotide phosphodiesterase (Barnette and Weiss, 1983; Sellinger-Barnette and Weiss, 1984) and in fluorescence titrations of dansyl calmodulin (Malencik and Anderson, 1984). Table IV summarizes the dissociation constants (K_d) obtained with dansyl calmodulin together with the peptide concentrations (IC$_{50}$) required to inhibit the calmodulin-sensitive activity of phosphodiesterase by 50%. The two sets of results generally follow each other. Both show that the N-terminal opioid pentapeptide has a major stabilizing effect and that binding is inefficient for C-terminal deletion peptides having fewer than nine amino acid residues. The titrations of dansyl calmodulin also indicate that the associations are highly ionic-strength dependent and that the removal of a basic residue generally has a larger effect than the removal of a neutral residue. As expected (see Section II,A), the values for IC$_{50}$ are characteristically larger than the dissociation constants.

TABLE III

Peptides Binding Calmodulin: Dissociation Constants and IC$_{50}$ Values

Peptide	K_d[a]	Method[b]	Reference	IC$_{50}$[c]	Reference
Melittin	3 nM	A	Comte et al., 1983	100 nM	Barnette et al., 1983
	≤0.1 nM	B	Malencik and Anderson, 1984		
Mastoparan	~0.3 nM	B	Malencik and Anderson, 1983b	20 nM	Barnette et al., 1983
Mastoparan X	~0.9 nM	B	Malencik and Anderson, 1983b	—	—
Polistes mastoparan	~3.5 nM	B	Malencik and Anderson, 1983b	90 nM	Barnette et al., 1983
Crabrolin	~5.5 nM	B	This work	—	—
Protein kinase inhibitor analogue	70 nM	B,D	Malencik et al., 1986	—	—
VIP	~50 nM	B,D	Malencik and Anderson, 1983a	—	—
GIP	~90 nM	B,C	Malencik and Anderson, 1983a	—	—
Secretin	0.14 μM	B,D	Malencik and Anderson, 1983a	—	—
Dynorphin$_{1-17}$	0.65 μM	C,D	Malencik and Anderson, 1984	2.4 μM	Barnette and Weiss, 1983
ACTH	1.5 μM	D	Malencik and Anderson, 1982	Inhibition	Weiss et al., 1980
	2.5 μM	C			
β-Endorphin	2.0 μM	D	Malencik and Anderson, 1982	3.2 μM	Sellinger-Barnette and Weiss, 1982
	4.6 μM	E	Sellinger-Barnette and Weiss, 1982		

	K_d[a]	Method[b]		[b]	
Substance P	1.9 μM	D	Malencik and Anderson, 1982	150 μM	Sellinger-Barnette and Weiss, 1982
Porcine glucagon	3.4 μM	C,D	Malencik and Anderson, 1982	No inhibition	Weiss et al., 1985
Human pancreatic polypeptide	3.7 μM	D	This work	—	—
Gramicidin S	4.5 μM	D	This work	—	—
TnI analog	4.8 μM	D	Malencik and Anderson, 1984	—	—
Granuliberin-R	5.1 μM	D	This work	30 μM	Barnette et al., 1983
Apamin	31 μM	D		10 μM	Barnette et al., 1983
Catfish glucagon	18 μM	C	This work	—	
α-MSH	59 μM	D	Malencik and Anderson, 1982	>200 μM	Sellinger-Barnette and Weiss, 1984
Somatostatin	65 μM	D	Malencik and Anderson, 1982	30 μM	Sellinger-Barnette and Weiss, 1984
Bombesin	91 μM	D	Malencik and Anderson, 1982	—	—
Catfish glucagon-like peptide	70 μM	D	This work	—	—

[a] Dissociation constants determined by methods B, C, and D generally apply to solutions containing 0.20 N KCl, 50 mM MOPS, 1.0 mM CaCl$_2$, pH 7.3 (25°). The [KCl] was 0.15 M for VIP, GIP, and secretin. The concentration of MOPS was reduced to 5 mM in the cases of dynorphin$_{1-17}$ and the TnI analogue.

[b] A, phosphodiesterase inhibition; B, competitive displacement of smooth muscle myosin light chain kinase; C, intrinsic peptide fluorescence; D, dansyl calmodulin; E, equilibrium dialysis. Refer to Section III for a description of the methods.

[c] Peptide concentration required for 50% inhibition of calmodulin-dependent cyclic nucleotide phosphodiesterase activity in solutions containing 50 mM glycylglycine, 25 mM ammonium acetate, 3.0 mM MgCl$_2$, and 0.1 mM CaCl$_2$, pH 8.0 (37°). Values of IC$_{50}$ for inhibition of basal activity are usually an order of magnitude or more larger.

```
1               10              20              30              40
S Y S M E H F R W G K P V G K K R R P V K V Y P N G A E D E S A E A F P L E F
```

Human ACTH

```
Y G G F M T S E K S Q T P L V T L F K N A I I K N A Y K K G E
```

Human β-endorphin

Fig. 6. ACTH and β-endorphin. The sequences are from Bloom (1981). See Table VI for abbreviations.

The binding of all the peptides described in this article is strongly calcium dependent. Even the highly charged dynorphin$_{1-17}$ molecule exhibits only slight association with calmodulin in the absence of calcium. Measurements of the intrinsic fluorescence anisotropy of dynorphin$_{1-17}$ indicate that $K_d \geqslant 20$ μM in solutions containing 0.20 N KCl and 0.1 mM EDTA (Malencik and Anderson, 1984).

TABLE IV

Deletion Peptides of Porcine Dynorphin$_{1-17}$: Interactions with Calmodulin

Sequence[a,b]				Dissociation constant (μM) of the complex with dansyl calmodulin determined at the indicated concentration of KCl (N)[c]				
1	5	10	15	0.20 N	0.10 N	0.05 N	0.025 N	IC$_{50}$ (μM)[b]
Y G G F L R R I R P K L K W D N Q				0.65	—	—	—	2.4
R R I R P K L K W D N Q				>40	4.3	1.8	—	120
Y G G F L R R I R P K L K				2.5	0.82	—	—	8.0
Y G G F L R R I R P K L				7.1	2.4	0.7	—	20
Y G G F L R R I R P K				14	3.6	1.6	—	50
Y G G F L R R I R P				—	18	11	—	100
Y G G F L R R I R				—	45	20	—	200
Y G G F L R R I				—	—	—	>200	>500

[a]Goldstein *et al.* (1979).

[b]Abbreviations are given in Table VI.

[c]Malencik and Anderson (1984). Determined in 5.0 mM MOPS, 1.0 mM CaCl$_2$, pH 7.8 (25.0°). The concentration of calmodulin was 0.2–2 μM, with the lower concentrations for the peptides giving the smaller dissociation constants.

[d]Barnette and Weiss (1983). Inhibition of calmodulin-sensitive phosphodiesterase was measured in 50 mM glycylglycine containing 25 mM ammonium acetate, 3 mM MgCl$_2$, and 0.1 mM CaCl$_2$, pH 8.0 (37°). The values of IC$_{50}$ for the inhibition of basal activity were usually an order of magnitude or more larger.

C. The Glucagon–Secretin Family

Glucagon, secretin, vasoactive intestinal peptide (VIP), and the gastric inhibitory peptide (GIP) belong to a family of structurally related peptides (Table V) first associated with the gastroenteropancreatic–neuroendocrine system and, later, the brain. VIP, possibly representing the ancestral molecule, occurs in neurons of both the central and peripheral nervous systems (cf. review by Barrington, 1982). Members of this family may have a role in the regulation of calmodulin-dependent enzymes since cAMP is apparently a second messenger of both VIP (Quik *et al.*, 1978) and glucagon (Exton *et al.*, 1971). Glucagon is also one of two calmodulin-binding peptides for which X-ray diffraction data are available. The structural homologies within this group facilitate a systematic study of the peptide-binding requirements of calmodulin.

Measurements using either dansyl calmodulin or the intrinsic tryptophan fluorescence of porcine glucagon revealed a moderate interaction (K_d = 3.4 μM at 25°) which is strongly calcium dependent and temperature dependent ($\triangle H_{assoc}$ = −17 kcal/mol) but insensitive to modest variations in ionic strength (Malencik and Anderson, 1982, 1983a, 1984). Although VIP, GIP, and secretin are comparable in size to glucagon, they bind calmodulin 10–70 times more strongly. Competition experiments with smooth muscle myosin light chain kinase show that VIP, with just 28 amino acid residues, binds calmodulin nearly as well as

TABLE V

Sequences of Peptides Belonging to the Glucagon Family[a]

	10	20	30	40

H S D A V F T D N Y T R L R K Q M A V K K Y L N S I L N-NH₂
Porcine VIP[b] (50 nM)

Y A D G T F I S D Y S I A M D K I R Q Q D F V N W L L A Q K G K K S D W K H N I T Q
Porcine GIP[c] (90 nM)

H S D G T F T S E L S R L R D S A R L Q R L L Q G K V-NH₂
Porcine Secretin[b] (0.14 μM)

H S Q G T F T S D Y S K Y L D S R R A Q D F V Q W L M N T
Porcine Glucagon[b] (3.4 μM)

H S E G T F T S N D Y S K Y L E T R R A Q D F V Q W L M (N,S)
Catfish Glucagon[d] (18 μM)

H A D G T Y T S D V S S Y L Q D Q A A K A F I T W L K S G Q P K P E
Catfish Glucagon-like Peptide[d] (70 μM)

[a] Abbreviations are given in Table VI.
[b] From Bloom (1981). The dissociation constant for the complex with calmodulin is given in parentheses.
[c] Includes sequence correction by Jornvall *et al.* (1981).
[d] Andrews and Ronner (1985).

184-residue rabbit skeletal muscle troponin I (Fig. 2A). The ratios of dissociation constants ($K_{MLCK \cdot CaM}/K_{P \cdot CaM}$) calculated from the experiments are 0.048 for TnI, 0.026 for VIP, 0.027 for a deletion peptide containing VIP residues 10–28, ~0.014 for GIP, and 0.010 for secretin. The titration of a 10 μM solution of calmodulin containing 4.3 μM glucagon with either VIP or secretin shows a nearly linear decrease in glucagon fluorescence with an endpoint of 1 mol peptide/mol calmodulin (Fig. 7). Virtually all of the VIP or secretin is bound up to saturation, showing that the dissociation constant for the glucagon–calmodulin complex is much larger than the dissociation constants for the other two peptides. The results are consistent with competitive displacement, i.e.,

$$\text{glucagon} \cdot \text{CaM} \rightleftarrows \text{glucagon} + \text{CaM}$$
$$\text{CaM} + \text{VIP} \rightleftarrows \text{VIP} \cdot \text{CaM}$$

The affinity of VIP for calmodulin approaches that for the cell surface VIP receptors within a factor of 10. Comparison of our results with those of Ottesen *et al.* (1982), who studied the binding of several peptides by the VIP receptors on

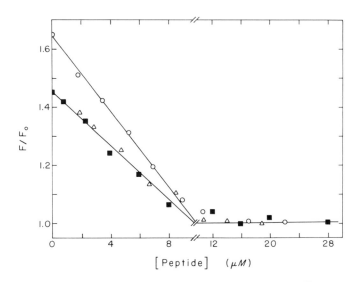

Fig. 7. Dissociation of porcine glucagon·calmodulin complex by VIP (■), secretin (△), and mastoparan (○). The solutions contained 10.0 μM calmodulin and either 4.3 μM glucagon for titrations with VIP and secretin or 2.9 μM glucagon for the titration with mastoparan. The abscissa gives the concentration of added peptide (exclusive of glucagon). F_0 is the fluorescence of glucagon in the absence of calmodulin. Conditions: 1.0 mM $CaCl_2$, 0.20 N KCl, and 50 mM 3-(N-morpholino)propanesulfonic acid, pH 7.3, 24.0°. Excitation, 295 nm; emission, 340 nm. (Taken in part from Malencik, D. A., and Anderson, S. R. and reprinted with permission from *Biochemistry* **22,** 1995–2001. Copyright (1983) American Chemical Society.

crude smooth muscle membranes from porcine uterus, indicates that VIP, secretin, and glucagon have the same relative orders of binding in the two systems. However, the differences between the peptides are larger with the receptors, indicating a higher degree of specificity than that found with calmodulin (Malencik and Anderson, 1983a).

The occurrence of a second cluster of basic amino acid residues is a salient feature of VIP, GIP, and secretin. In VIP and GIP, the arrangement of the basic and hydrophobic centers follows the pattern depicted in Table I, while in secretin an Arg-Leu sequence falls where a pair of hydrophobic residues might be expected. The effective binding of VIP_{10-28} suggests that the N-terminal region of VIP does not interact with calmodulin. Porcine secretin shows 50% sequence homology with porcine glucagon (Table V). Significant amino acid differences include those at positions 9 (Asp → Glu), 14 (Leu → Arg), and 21 (Asp → Arg). While working with this group, we calculated the average frequencies of occurrence of the component amino acids for all calmodulin-binding peptides known to give complexes with dissociation constants less than 10 μM (Malencik and Anderson, 1983a). The results showed that glutamic acid appears with strikingly low frequency and that glutamine (probably a replacement), arginine, and lysine occur with above average frequencies. Our most recent compilation (Table VI) emphasizes these characteristics and also demonstrates an increased rate of occurrence for several hydrophobic residues. Because of its similarity to secretin, glucagon was removed from Table VI. The amino acid compositions of effective deletion peptides were used in the case of ACTH, β-endorphin, and VIP.

Experiments with catfish glucagon and catfish glucagon-like peptide strengthened the ideas that clusters of basic amino acid residues are strongly stabilizing and that glutamic acid is destabilizing. Catfish glucagon exhibits ~83% sequence homology with porcine glucagon, having the same basic residues and nearly identical hydrophobic C-terminal sequence. Yet the dissociation constant for the catfish glucagon–calmodulin complex is only 18 μM. The decreased binding may relate to key amino acid differences at sequence positions 3 (Gln → Glu) and 16 (Asp → Glu, considering an apparent deletion in porcine glucagon). The 34-residue glucagon-like peptide, with three fairly well separated lysyl residues, has an order of affinity for calmodulin ($K_d = 70$ μM) comparable to those of smaller peptides containing a single basic residue (e.g., bombesin, <Glu-Gln-Arg-Leu-Gly-Asn-Gln-Trp-Ala-Val-Gly-His-Leu-Met-NH_2).

D. Venom Peptides: Melittin and the Mastoparans

The toxicity of insect venoms reflects the combined actions of several biologically active components: amines, polypeptides, and enzymes. Phospholipase A_2 and melittin, a 26-residue polypeptide (Fig. 8), account for 12 and 50%, respectively, of the dry weight of honey bee venom. Melittin is a surfactant which

TABLE VI

Amino Acid Frequencies in 13 Calmodulin-Binding Peptides[a]

Residue	No. of occurrences	Observed frequency[a]	Average frequency[b]
Asp(D)	10.5	4.4	5.5
Asn(N)	6.5	2.7	4.4
Glu(E)	3.0	1.2	6.2
Gln(Q)	15.0	6.4	3.9
Lys(K)	29.0	12.2	7.0
Arg(R)	21.0	8.9	4.7
His(H)	3.0	1.3	2.1
Pro(P)	11.0	4.6	4.6
Ser(S)	13.0	5.5	7.1
Thr(T)	11.0	4.6	6.0
Ala(A)	13.5	5.7	9.0
Gly(G)	17.5	7.4	7.5
Trp(W)	5.5	2.3	1.1
Phe(F)	11.0	4.6	3.5
Tyr(Y)	9.0	3.8	3.5
Val(V)	11.0	4.6	6.9
Leu(L)	27	11.4	7.5
Ile(I)	15	6.3	4.6
Met(M)	4.5	1.9	1.7
Cys(C)	0	0	2.8

[a] VIP_{10-28}, GIP, secretin, $ACTH_{1-24}$, β-endorphin$_{14-31}$, substance P, dynorphin$_{1-17}$, TnI analog, granuliberin-R, melittin, *Polistes* mastoparan, and average of mastoparan and mastoparan X.
[b] Tabulated by Klapper (1977) for 207 unrelated proteins.

interacts with membrane phospholipids, specifically making them better substrates for honey bee phospholipase (cf. review by O'Connor and Peck, 1978). The venoms of wasps and hornets contain the mastoparans (Hirai *et al.,* 1979a,b), cytoactive tetradecapeptides (Fig. 8) which stimulate phospholipase A_2 from diverse sources including bee venom, rattlesnake venom, and porcine pancreas (Argiolas and Pisano, 1983). The physiological consequences of these interactions may include hemolysis and, in the case of the mastoparans, degranulation of tissue mast cells.

Melittin also inhibits purified cardiac muscle myosin light chain kinase (Katoh *et al.,* 1982) and calmodulin-dependent cyclic nucleotide phosphodiesterase (Comte *et al.,* 1983; Barnette *et al.,* 1983). The mastoparans resemble melittin in their potent inhibition of both the phosphodiesterase (Barnette *et al.,* 1983) and smooth muscle myosin light chain kinase (Malencik and Anderson, 1983b). Whether any of these effects occur *in vivo* is unknown. However, the tenacity of binding and strong calcium dependence suggest the feasibility of physiologically significant peptide–calmodulin complexes.

All accounts place melittin and the mastoparans at the top of a hierarchy of calmodulin-binding peptides (Table III). Competition experiments with smooth muscle myosin light chain kinase unveil affinities for calmodulin which rival that of the enzyme. The results in Fig. 2B correspond to the following values of $K_{MLCK \cdot CaM}/K_{P \cdot CaM}$: TnI, 0.035 ± 0.01; *Polistes* mastoparan, 0.4 ± 0.02; mastoparan X, 1.5 ± 0.2; and mastoparan, 4.9 ± 0.3. Since the value of $K_{MLCK \cdot CaM}$ in the presence of 5 μM 9-anthroylcholine is ~ 1.4 nM, the calculated dissociation constants for these peptide–calmodulin complexes range from ~ 0.3 nM for mastoparan to ~ 3.5 nM for *Polistes* mastoparan. The addition of melittin to the myosin light chain kinase–calmodulin complex results in a stoichiometric decrease in the fluorescence of 9-anthroylcholine, with an endpoint of 1 mol mellitin/mol calmodulin. The linearity of this titration indicates that $K_{P \cdot CaM} \ll K_{MLCK \cdot CaM}$. In fact, our estimations consistently give values for $K_{P \cdot CaM}$ less than 0.1 nM. Observations supporting the extraordinary binding of melittin by calmodulin include its nearly irreversible adsorption to a matrix of calmodulin–Sepharose 4B; the correspondence of the fluorescence spectra and anisotropy of 1 : 1 : 1 mixtures of the enzyme, melittin, and dansyl calmodulin with those of the melittin–dansyl calmodulin complex; the distribution of $\sim 85\%$ melittin–dansyl calmodulin and $\sim 15\%$ enzyme–dansyl calmodulin obtained during gel filtration of solutions initially containing 20 μM of each protein component; and the comparatively high concentrations of mastoparan required to displace bound melittin (Malencik and Anderson, 1984). The complexes of calmodulin with melittin and the mastoparans respond relatively quickly to addition of either EDTA or competing peptides. The simultaneous occurrence of such high affinity and responsiveness has been demonstrated previously in the association–dis-

```
1                 10                  20                  30

I  N  L  K̲  A  L  A  A  L  A  K̲  K̲  I  L-NH₂

        Mastoparan

I  N  W  K̲  C  I  A  A  M  A  K̲  K̲  L  L-NH₂

        Mastoparan X

V  D  W  K̲  K̲  I  G  Q  H  I  L  S  V  L-NH₂

        Polistes Mastoparan

G  I  G  A  V  L  K̲  V  L  T  T  G  L  P  A  L  I  S  W  I  K̲  R̲  K̲  R̲  Q  Q-NH₂

                    Melittin
```

Fig. 8. Melittin and the mastoparans. The sequences are from Hirai *et al.* (1979a,b) and from Terwilliger and Eisenberg (1982). See Table VI for abbreviations.

sociation kinetics of the *Escherichia coli* lac repressor–operator complex (Berg *et al.*, 1981). The reported values of K_d, K_I, and IC_{50} for melittin vary (Table III). This probably results from the individual measurement conditions, the inherent difference between K_d and IC_{50} (see Section II,A), and possible adsorption losses in inhibition experiments conducted at low-peptide concentrations. In the absence of calcium, melittin forms one or more complexes with calmodulin having $K_d \geqslant 10$ μM (Maulet and Cox, 1983), while mastoparan X shows minimal interaction in the micromolar range (Fig. 9).

A marked propensity for α-helix formation sets the mastoparans and melittin apart from the other calmodulin-binding peptides (see Section VIII). Nonetheless, the addition of mastoparan to the complexes of calmodulin with porcine glucagon and dynorphin$_{1-17}$ results in stoichiometric competitive displacement (Fig. 7). Acetylation, performed under conditions specific for lysine, demonstrates that the positive charge of the venom peptides is important in binding. The value of IC_{50}, measured by inhibition of calmodulin-dependent phosphodiesterase, increases from $20 nM$ to 0.8 μM when mastoparan is acetylated. The effect of modification on melittin binding is less, with IC_{50} increasing from

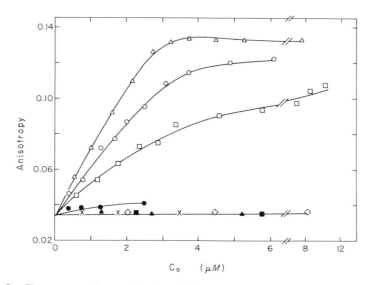

Fig. 9. Fluorescence anisotropy titrations of 3.0 μM mastoparan X with porcine brain calmodulin ($\triangle$), half-calmodulin fragment 72–148 ($\bigcirc$), thrombic fragments 1–106 ($\square$) and 107–138 ($\Diamond$), and dogfish parvalbumin ($\times$). Conditions: 1.0 mM CaCl$_2$, 0.20 N KCl, and 5.0 mM 3-(N-morpholino)propanesulfonic acid, pH 7.3, 25.0°. The solid symbols ($\bullet$, $\blacktriangle$, $\blacksquare$) represent the control experiments performed with 0.1 mM EDTA and no added calcium. Excitation: 294 nm with 2-nm band-pass; emission filters, Corning glass CSO-54. Reprinted with permission from Malencik, D. A., and Anderson, S. R., *Biochemistry* **23**, 2420–2428. Copyright (1984) American Chemical Society.

TABLE VII

Interaction of Lyophilized Animal Venoms with Dansyl Calmodulin[a]

Source	Concentration	$F/F_0{}^c$ (450 nm)	ΔA^d
Troponin I (184 residues)	2.7 μM	4.2	0.054
Melittin (26 residues)	3.0 μM	5.7	0.016
Crotalus horridus horridus	0.21 mg/ml[b]	3.3	0.034
Micrurus fulvius	0.83 mg/ml[b]	4.0	0.040
Naja mocambique mocambique	0.23 mg/ml[b]	7.7	0.065
Heloderma horridum	0.24 mg/ml[b]	4.0	0.019
Androctonus australis	0.41 mg/ml[b]	5.3	0.026
Palamnaeus gravimanus	0.32 mg/ml[b]	4.2	0.049
Bufo marinus	>0.76 mg/ml[b]	1.2	—
Diadema antillarum	>0.79 mg/ml[b]	1.1	—

[a]Conditions: 2.0 μM dansyl calmodulin in 0.20 N KCl–45 mM MOPS–1.0 mM CaCl$_2$, pH 7.3 (25°).

[b]Concentration of lyophilized venom needed to attain near maximal fluorescence enhancement.

[c]F is the fluorescence intensity of dansyl calmodulin at the indicated venom concentration, and F_0, the intensity in the absence of venom.

[d]ΔA is the increase in the fluorescence anisotropy of dansyl calmodulin at the indicated venom concentration.

0.1 to 0.4 μM, probably reflecting the presence of arginine (Barnette *et al.*, 1983). Melittin also undergoes a concentration-dependent self-association to form tetramers (Quay and Condie, 1983). Most of the experiments with calmodulin were performed under conditions where melittin is largely monomeric. The fluorescence polarization values are consistent with a complex of molecular weight 20,000.

The results in this section led us to speculate on the general occurrence of calmodulin-binding peptides in animal venoms and on the value of calmodulin as an antidote. We conducted a brief survey showing that the timber rattlesnake, eastern coral snake, cobra, Gila monster, and two scorpions produce venoms containing calmodulin-binding substances, while the poisonous marine toad and sea urchin do not. The fluorescence anisotropies obtained with dansyl calmodulin suggest that the molecular weights of these materials—presumably proteins and peptides—vary widely (Table VII). (The anisotropy changes measured with troponin I and melittin are included for comparison.) The effect of melittin on the erythrocyte can be demonstrated with washed human red blood cells ($5 \times 10^6/$cm^3) suspended in a solution containing 155 mM NaCl, 1.0 mM CaCl$_2$, and 10 mM MOPS, pH 7.3 (37°). Under these conditions, a 2 μM concentration of melittin is sufficient to produce 90% hemolysis (Argiolas and Pisano, 1983). We find that the addition of calmodulin to the suspended cells completely inhibits the

lytic activity of melittin when certain provisions are made (Malencik and Anderson, 1985): (1) the concentration of calmodulin equals or exceeds that of melittin, (2) calmodulin is added either prior to or simultaneously with melittin, and
(3) calcium is present. Rabbit muscle troponin C exerts protective effects both in
the presence and absence of calcium, while parvalbumin shows none.

Recent experiments show that crabrolin (FLPLIL*RK*IVTAL-NH$_2$), a venom
peptide from the European hornet (Argiolas and Pisano, 1984), also possesses
high affinity for calmodulin.

IV. COMPETITIVE RELATIONSHIPS AND STOICHIOMETRY

The association of calmodulin with its target enzymes usually occurs in the
ratio 1 : 1, as found with smooth muscle myosin light chain kinase (Fig. 1).
Whether the various enzymes recognize a common binding site on calmodulin is
unknown. However, a reasonable model for enzyme binding is one in which the
sites overlap considerably with some variation in local interactions. Different
calmodulin-dependent enzymes compete in binding (cf. review by Klee and
Vanaman, 1982). They are similarly affected by calmodulin antagonists, such as
the phenothiazines. On the other hand, chemical modification of calmodulin
does not affect the activation of all the enzymes in the same way (Thiry *et al.*,
1980). For example, dansyl calmodulin activates cyclic nucleotide phosphodiesterase (Kincaid *et al.*, 1982), but not smooth muscle myosin light chain
kinase, even though it binds the kinase efficiently (Malencik *et al.*, 1982).

The experiments in the preceding sections showed that the high-affinity calmodulin-binding peptides—VIP, mastoparan, etc.—displace both myosin light
chain kinase and several of the tryptophan-containing peptides in a competitive
manner. The order of effectiveness in inhibition of calmodulin-dependent enzyme activity usually follows the relative affinities determined in direct binding
measurements (Table III). Major exceptions occur with glucagon and substance
P. No inhibition of cyclic nucleotide phosphodiesterase was detected in solutions
containing up to 100 μ*M* porcine glucagon (Weiss *et al.*, 1985), while only 30%
inhibition of myosin light chain kinase was found in the presence of 39 μ*M*
glucagon (Malencik *et al.*, 1982a). The self-association of glucagon in concentrated solutions may contribute to these effects.

As with the enzymes, demonstrable saturation occurs when the high-affinity
peptides and calmodulin are combined in 1 : 1 proportion (Figs. 3, 7, 9, 10).
Equilibrium dialysis of calmodulin against varying concentrations of β-endorphin also gives a stoichiometry of 1 mol/mol (Sellinger-Barnette and Weiss,
1982). These authors also observed that the photochemical coupling of chlorpromazine to calmodulin (1 mol/mol) largely suppresses the binding of β-en-

dorphin, suggesting that the interaction of the peptide with calmodulin is similar to that of antipsychotic drugs. Calmodulin has two to three calcium-dependent phenothiazine binding sites per molecule (cf. review by Weiss *et al.*, 1980). Evidence that it also has more than one peptide binding site was obtained by Giedroc *et al.* (1983a), who used the cross-linking agent bis(sulfosuccinimidyl) suberate to couple β-endorphin to calmodulin. At high concentrations of both the reagent and the reactants, a limiting 2:1 β-endorphin–calmodulin complex is formed. Chlorpromazine-affinity columns fail to bind the 2:1 complex but retain the 1:1 complex to some extent (Giedroc *et al.*, 1983c). The different stoichiometries determined by equilibrium methods and covalent cross-linking are not necessarily inconsistent. The tenfold or more increase in the protein concentrations used for cross-linking plus the strongly stabilizing effect of a covalent reaction would explain the detection of a species which is a minor component in the equilibrium binding experiments. Demonstration of 1:1 stoichiometry by an equilibrium method indicates the existence of a single binding site of substantially higher affinity than any others.

9-Anthroylcholine resembles trifluoperazine in its ability to compete alike

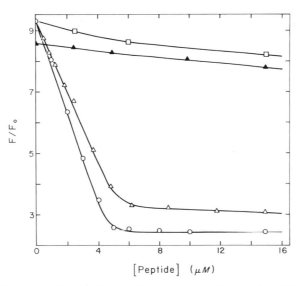

Fig. 10.   Fluorescence titration of calmodulin–9-anthroylcholine complex with mastoparan (□), melittin (○), troponin I (△), and troponin I in the presence of a fixed concentration (6.0 μM) of mastoparan (▲). F/F_0 is the ratio of the observed fluorescence intensity to the intensity of 9AC in the absence of protein or peptides. The varied concentration of added peptide is given on x-axis. Conditions: 5.0 μM porcine brain calmodulin, 10.0 μM 9-antyroylcholine, 1.0 mM $CaCl_2$, 0.20 N KCl, 50 mM 3-(N-morpholino)propanesulfonic acid, pH 7.3, 25.0°. Excitation, 360 nm; emission, 460 nm.

with cyclic nucleotide phosphodiesterase and troponin I in calmodulin binding. The calcium–calmodulin complex adsorbs four to six molecules of 9AC with an average dissociation constant of 440 μM and a 24-fold enhancement of ligand fluorescence. The stepwise addition of troponin I to calmodulin results in dissociation of 9AC and a linear decrease in fluorescence up to the endpoint of 1 mol TnI/mol calmodulin (LaPorte *et al.*, 1980). We discovered the comparatively high-affinity binding of 9AC by smooth muscle myosin light chain kinase while titrating calmodulin with the enzyme. Because of the greater affinities and fluorescence enhancement obtained, the binding of 9AC by calmodulin contributed little to our earlier experiments (see Section II,B). However, none of the calmodulin-binding peptides interacts with 9AC, probably because both are cationic. Titration of the calmodulin–9AC complex with melittin gives a linear decrease in fluorescence, comparable to that found with troponin I, and a stoichiometry of 1 mol melittin/mol calmodulin. Surprisingly, titration with mastoparan causes only a slight change in 9AC fluorescence. Yet equivalent concentrations of mastoparan almost completely block the effect of troponin I (Fig. 10). This suggests that mastoparan displaces the less effectively bound troponin I molecule but, because of its smaller size, does not overlap or otherwise affect as many of the 9AC binding sites. High concentrations of mastoparan also inhibit the binding of melittin by the calmodulin–9AC complex.

V. PEPTIDE BINDING BY PROTEOLYTIC FRAGMENTS OF CALMODULIN

The exceptional stabilities of the calmodulin–mastoparan complexes suggested that the mastoparans could be useful for the detection of peptide binding by fragments of calmodulin. We selected three proteolytically derived fragments of calmodulin for study: an N-terminal fragment containing residues 1–106, obtained by thrombin-catalyzed cleavage of calmodulin in the absence of calcium (Wall *et al.*, 1981); the corresponding C-terminal fragment, comprising residues 107–148; and a C-terminal half-calmodulin fragment, corresponding to residues 72–148, obtained as a result of endogenous proteolysis by Schreiber *et al.* (1981). Fragment 1–106 contains postulated calcium binding sites I, II, and III, while fragment 72–148 contains sites III and IV, probably the two high-affinity calcium binding sites of calmodulin (Andersson *et al.*, 1983).

We carried out fluorescence anisotropy titrations of mastoparan X, which contains tryptophan, with the three fragments as well as with calmodulin and dogfish parvalbumin. Only fragment 107–148 and parvalbumin failed to interact with mastoparan X in this concentration range. Figure 9 shows that the addition of either calmodulin or fragment 72–148 gives nearly stoichiometric (complete) binding at the peptide concentration used (3.0 μM). The titration with fragment

72–148 indicates that $K_d \leq 0.15 \ \mu M$ while that with fragment 1–106, the least strongly bound, gives a value of K_d of 0.9 μM. All the equilibria are strongly calcium dependent, with negligible interaction occurring in solutions containing EDTA. The fragments do not associate significantly with dynorphin$_{1-17}$ or porcine glucagon.

The binding of mastoparan X may follow that reported for the protein troponin I, which associates with calmodulin fragments 78–148 (Drabikowski *et al.*, 1977) and 1–106 but not with 1–90 and 107–148 (Walsh *et al.*, 1977). None of the known calmodulin fragments appears to affect the activity of calcineurin. Fragment 78–148 behaves as a calmodulin antagonist with cyclic nucleotide phosphodiesterase (Newton *et al.*, 1984) and as an activator with phosphorylase kinase (Kuznicki *et al.*, 1981). The latter result may relate to the ability of diverse proteins—histones, troponin C, calmodulin, etc.—to activate phosphorylase kinase by stimulating autophosphorylation (Malencik and Fischer, 1982). We also find that fragment 72–148 is a weak inhibitor of smooth muscle myosin light chain kinase. Highly purified fragment 1–106 does not affect the activity of any of these enzymes (Newton *et al.*, 1984).

The phenothiazines interact with calmodulin fragment 77–124 (Head *et al.*, 1982) and troponin C fragment 90–123, which is homologous to calmodulin sequence 80–113 (Reid *et al.*, 1983). They also recognize fragments 1–77 and 78–148, showing that phenothiazine binding sites occur in both the N-terminal and C-terminal halves of calmodulin (Newton *et al.*, 1984; Thulin *et al.*, 1984). This result may apply to peptides as well since Giedroc *et al.* (1983a) covalently cross-linked up to 2 mol β-endorphin/mol calmodulin. We have not yet examined the peptide-binding ability of fragment 1–77. However, the already diminished affinity of 1–106 suggests that, if there are two sites, the stronger occurs within the C-terminal half of calmodulin. ^{113}Cd NMR spectra indicated that the higher-affinity phenothiazine binding site probably lies between residues 78 and 148 (Thulin *et al.*, 1984).

VI. PEPTIDE ASSOCIATIONS WITH PROTEINS RELATED TO CALMODULIN

X-Ray crystallography and amino acid sequencing studies identified a family of calcium-binding proteins derived from a common ancestor similar to calmodulin (cf. review by Kretsinger, 1980). Observations on other proteins in this group—rabbit muscle troponin C, beef cardiac troponin C, the 16,000- and 21,000-dalton smooth muscle myosin light chains, and dogfish parvalbumin—indicate the extent to which the peptide associations are specific for calmodulin. Table VIII summarizes the results of binding experiments performed with dynorphin$_{1-17}$ and mastoparan X. Neither of these peptides, or even melittin,

TABLE VIII

Peptide Binding by Calmodulin Fragments and by Proteins Ancestrally Related to Calmodulin[a]

Protein or fragment	K_d for dynorphin$_{1-17}$	K_d for mastoparan X
CaM	0.65 μM	~0.9 nM
Ca 1-106	No interaction	0.9 μM
Ca 72-148	Slight interaction	≤0.15 μM
Ca 107-148	No interaction	No interaction
Skeletal muscle TnC	2.4 μM	~0.2 μM[b]
Cardiac TnC	9.0 μM	—
Smooth muscle myosin light chain (21K)	—	21 μM
Smooth muscle myosin light chain (16K)	—	Slight interaction
Parvalbumin	No interaction	No interaction

[a]Data from Malencik and Anderson, 1984; also unpublished results. Conditions: 0.20 M KCl, 10 mM MOPS, 1.0 mM CaCl$_2$, pH 7.3 (25°).
[b]Mastoparan was used in place of mastoparan X.

interacts with parvalbumin, which contains two calcium binding sites but is believed to have lost specific protein recognition sites (Blum *et al.*, 1977). Very strongly basic proteins and peptides such as histone H2A, salmine, and poly-L-arginine associate with parvalbumin, both in the presence and absence of calcium. Even though the calcium binding sites are absent, the 21,000-dalton smooth muscle myosin light chain retains vestigial peptide binding ability. The complex with mastoparan X corresponds to a K_d of 21 μM. Visible aggregation occurs when the light chain is added to melittin solutions. Difficulties encountered in inhibition studies of myosin light chain kinase may relate to these effects. For example, Perry (1980) found that inhibition of the enzyme by troponin I is less than expected, while Katoh *et al.* (1982) reported that the inhibition by melittin is reversed by increasing concentrations of the substrate. Since excess concentrations of the myosin light chain (≥10 μM) are used in these assays, the fraction of peptide or protein bound by it can be substantial.

The peptide-binding properties of troponin C are consistent with the belief that it is the protein most like calmodulin. Dynorphin$_{1-17}$ has similar affinity, within a factor of 4, for calmodulin and rabbit fast skeletal muscle troponin C. The dynorphins remind us of the inhibitory region of troponin I. Both contain strongly basic sequences with hydrophobic residues interspersed. Dynorphin$_{1-17}$ actually exerts an inhibitory effect in the rabbit skeletal muscle actomyosin adenosine triphosphatase assay (Malencik and Anderson, 1984) similar to that found for troponin I and its inhibitory analogs (Talbot and Hodges, 1981). Mastoparan X also interacts with rabbit muscle troponin C. However, the corresponding dissociation constant is at least two orders of magnitude larger than that estimated for the calmodulin complex. Neither mastoparan or melittin inhibits rabbit muscle actomyosin adenosine triphosphatase. Sellinger-Barnette and Weiss (1982) noted association of β-endorphin with both troponin C and S-100, but did not determine equilibrium constants. Troponin C does not bind glucagon.

VII. DO CALMODULIN AND THE cAMP-DEPENDENT PROTEIN KINASE ACT ON SIMILAR RECOGNITION SEQUENCES?

Phosphorylation of turkey gizzard myosin light chain kinase by the cAMP-dependent protein kinase markedly reduces its affinity for calmodulin, with 20- to 100-fold increases in K_m (Conti and Adelstein, 1981) and K_d (Malencik et al., 1982a). Excess concentrations of calmodulin in vitro block phosphorylation of specific serine or threonine residues in both myosin light chain kinase (Conti and Adelstein, 1981; Malencik et al., 1982a) and phosphorylase kinase (Cox and Edstrom, 1982). Sequenced phosphorylation sites in physiologically significant protein kinase substrates follow patterns, such as Arg-Arg-X-Ser and Lys-Arg-X-X-Ser (Huang et al., 1979; Carlson et al., 1979), similar to those occurring in calmodulin-binding peptides. We suggested that calmodulin and the cAMP-dependent protein kinase interact with common sequences in some proteins (Malencik and Anderson, 1982). Rabbit fast skeletal muscle troponin I, for example, contains phosphorylatable serine residues at positions 20 and 117 (cf. review by Perry, 1980). Phosphorylation of troponin I, as well as of histone H2A or the myelin basic protein, reduces its ability to compete with smooth muscle myosin light chain kinase for calmodulin (Malencik et al., 1982a). A synthetic analog (Ac-Gly-Lys-Phe-Lys-Arg-(Pro)₂-Leu-(Arg)₂-Val-Arg-NH₂) corresponding to sequence positions 104–115 of troponin I undergoes calcium-dependent binding by both troponin C (Cachia et al., 1983) and dansyl calmodulin (Malencik and Anderson, 1984; Table III).

We selected a peptide derived from the β-subunit of rabbit muscle phosphorylase kinase (Arg-Thr-Lys-Arg-Ser-Gly-Ser-Val-Tyr-Glu-Pro-Leu-Lys-Ile) for further study. Although the location of the calmodulin binding site in any enzyme is still unknown, this particular peptide originally seemed to resemble those which calmodulin binds well. It also corresponds to the phosphorylation site responsible for the activation of phosphorylase kinase by β-adrenergic stimulation (cf. review by Malencik and Fischer, 1982) and to the site which Cox and Edstrom (1982) found to be blocked in the presence of excess calmodulin. The peptide proved to be one of the most effective synthetic substrates for the cAMP-dependent protein kinase, with a K_m of 1 to 2 μM (Malencik and Anderson, 1983c). Phosphorylation occurs primarily at a single residue, probably serine at position 7 by analogy to phosphorylase kinase. It is accompanied by a 36% decrease in the intrinsic tyrosine fluorescence of the peptide, a property which we have used in continuous fluorometric assays for cAMP-dependent protein kinase. However, the affinity of the peptide for calmodulin is only fair—with a dissociation constant of 94 μM, determined in 50 mM MOPS with no added KCl, which increases to ~500 μM after phosphorylation (Malencik and Anderson, 1983a). The presence of glutamic acid and the apparent lack of repetitive secondary structure separate it from most of the effective calmodulin binding peptides.

Inhibition experiments also show that the optimum small peptide substrates for calmodulin and the cAMP-dependent protein kinase may differ. Neither mastoparan nor VIP inhibits protein kinase when present at the same concentration (10 μM) as the tetradecapeptide. Earlier experiments by Katoh et al. (1982) indicated slight inhibition by melittin. However, since phosphorylation is thermodynamically irreversible, any overlap in specificity with protein kinases that does occur can be significant.

Microsequencing of the 8 000 Da heat-stable cAMP-dependent protein kinase inhibitor revealed that its inhibitory region contains the following amino acid sequence (Scott et al., 1985):

Y A F I A S G R T G R R N A I H D I L V S S A

Observations on synthetic deletion peptides derived from it provided the first clear evidence for overlap in the peptide binding specificities of calmodulin and protein kinase. All of the resulting peptides which are known to interact effectively with either of the proteins contain the -Arg-Thr-Gly-Arg-Arg- sequence. However, the amino acid sequences adjoining the N- and C-terminal ends of this basic center apparently have different roles in the binding of calmodulin and protein kinase. For example, the duodecapeptide containing residues 4–23 of the preceding sequence forms a calcium-dependent complex with calmodulin corresponding to $K_d = 0.07$ μM and a complex with the catalytic subunit of cAMP-dependent protein kinase giving $K_I = 0.8$ μM. The derivative representing residues 1–19, in contrast, gives corresponding values of $K_d = 5.8$ μM and $K_I = 0.18$ μM. Since the duodecapeptide binds calmodulin at least one order of magnitude *more* effectively than does the native protein kinase inhibitor, key amino acid residues involved in calmodulin binding may be inaccessible in the latter.

A synthetic substrate educed from the ''pseudo-substrate'' site of the heat-stable protein kinase inhibitor—G R T G R R N S I H D I L—is readily phosphorylated at both its seryl and threonyl residues, with average $K_m = 220$ μM. The effect of peptide phosphorylation on calmodulin binding, with K_d increasing from 7.0 μM to 63 μM, is larger than that found for two other phosphorylatable peptides, but still smaller than that characteristic of smooth muscle myosin light chain kinase (*cf.* Malencik et al., 1986).

VIII. PEPTIDES AS STRUCTURAL AND SPECTROSCOPIC PROBES OF CALMODULIN

The most efficient calmodulin-binding peptides exhibit several common features at the primary level of structure. These include the occurrence of clusters of basic amino acid residues in conjunction with hydrophobic sequences (see Section III). The environment of the protein binding site in calmodulin may stabilize specific secondary structure in the peptides, which are largely unfolded in dilute aqueous solutions (cf. review by Blundell and Wood, 1982). In some cases, at the expense of binding energy, the conformation of the bound peptide may differ from the most stable conformation. The prediction rules of Chou and Fasman (cf. 1978 review) suggest several distinctly different secondary structures for calmodulin binding peptides, as illustrated for substance P, mastoparan, and dynorphin$_{1-13}$ (Fig. 11). The predicted structure of substance P contains a rigid N-terminal β-turn connected to a β-strand or α-helix. The average β-sheet and α-helix potentials for the C-terminal heptapeptide are 1.15 and 1.10, respectively. The Phe-Phe sequence that we treated as a part of the recognition sequence appears in the core of the latter structure (Malencik and Anderson,

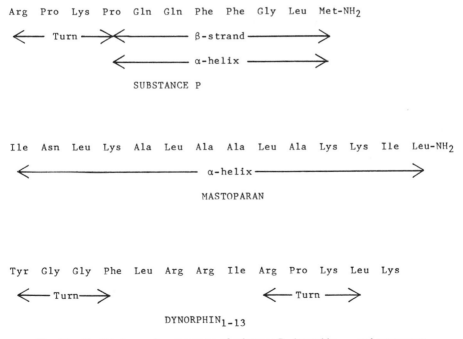

Fig. 11. Predicted secondary structures of substance P, dynorphin$_{1-13}$, and mastoparan.

1983a). Sequences having similar probabilities of occurrence in the α-helix and β-strand conformation also appear in the glucagon–secretin group. The ability of glucagon to exist in more than one conformation, predicted by Chou and Fasman in 1975, is well established (cf. review by Blundell, 1983). X-Ray crystallographic analysis shows a trimeric arrangement of glucagon molecules in which each protomer, although not rigidly defined, is approximately α-helical from positions 6–28. Glucagon is 90% α-helical in 2-chloroethanol and 50% α-helical in 25 mM sodium dodecylsulfate below pH 4, but forms fibrils of antiparallel β-sheet in acidic aqueous solutions. We suggested that, in some cases, a β-strand conformation is involved in calmodulin binding (Malencik and Anderson, 1983a). This was based on both the predicted structures and the low frequency of occurrence of glutamic acid in the peptides. Glutamic acid is strongly stabilizing in α-helices, but destabilizing in β-strands and β-turns (see review by Chou and Fasman, 1978). Recent work in our laboratory by Y. Shalitin demonstrates calcium-dependent binding of gramicidin S by calmodulin. The association is competitive with glucagon binding. Gramicidin S is known to form an internal β-pleated sheet structure (Krauss and Chan, 1982).

However, the avidity of calmodulin for α-helical structures became apparent when the binding of melittin and the mastoparans was reported (Comte et al., 1983; Barnette et al., 1983; Malencik and Anderson, 1983b). Mastoparan has the highest potential for α-helix formation of all the peptides and, unlike several, contains no nuclei for predicted β-strands. The α-helical nucleus at positions 5–10 corresponds to $\langle P_\alpha \rangle = 1.35$, while the sequence as a whole gives $\langle P_\alpha \rangle = 1.2$. X-Ray studies of melittin reveal a tetrameric assembly in which each monomer has the overall shape of a bent rod with two α-helical segments (Terwilliger and Eisenberg, 1982). The circular-dichroism spectra of the melittin–calmodulin (Fig. 12; Maulet and Cox, 1983) and β-endorphin–calmodulin (Giedroc et al., 1983b) complexes are consistent with α-helix formation, presumably within the peptides. Smaller changes in circular dichroism, such as those which we find with VIP, are more difficult to interpret. The secondary folding of melittin and glucagon gives amphipathic helices. The hydrophobic and polar amino acid side chains of crystalline melittin are almost perfectly aligned on opposite faces of the helical segments. This pattern of organization is considered important in the binding of a number of peptides by cell surface receptors and membranes (cf. review by Epand, 1983; Hammonds et al., 1982) and also by calmodulin (Giedroc et al., 1983b; Maulet and Cox, 1983).

That an α-helical propensity is not an absolute requirement for calmodulin binding is evident with the dynorphins. The predicted secondary structure of dynorphin$_{1-13}$ is irregular, with two β-turns (Malencik and Anderson, 1983a). However, secondary folding of polypeptides is only one of several ways to generate amphipathic structure. For example, hydrophobic photolabeling demonstrated that association of dynorphin$_{1-13}$ with lipid vesicles is due to its segmental organization into a hydrophobic message (Tyr-Gly-Gly-Phe-) and hydro-

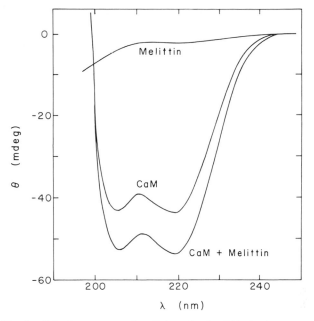

Fig. 12. Circular dichroism spectra of melittin and calmodulin in the presence of calcium. Ellipticities are expressed in millidegrees. Conditions: 10.0 μM melittin and/or 10.0 μM bovine brain calmodulin, 1.0 mM $CaCl_2$, 0.10 N NaCl, 20 mM 2-{[tris(hydroxymethyl)-methyl]amino}cthanesulfonic acid, pH 7.0. Light path, 1 cm. Reprinted with permission from Maulet, Y., and Cox, J. A., *Biochemistry* **22**, 5680–5686. Copyright (1983) American Chemical Society.

philic address (-Leu-Arg-Arg-Ile-Arg-Pro-Lys-Leu-Lys) (Gysin and Schwyzer, 1983). Experiments with $ACTH_{1-24}$ indicated that the N-terminal decapeptide unit enters lipid bilayers, where it assumes an α-helical conformation, while the hydrophilic C-terminal tetradecapeptide remains on the polar interface as an irregular structure, with peptide bonds oriented perpendicularly to the bilayer (Gremlich et al., 1983, 1984; Gysin and Schwyzer, 1984). Similar phenomena may be involved in the binding of specific peptides by calmodulin.

The peptides seem to have minimal effect on calmodulin structure. Its intrinsic tyrosine fluorescence is unaffected by peptide binding (Malencik and Anderson, 1982, 1984). Electron paramagnetic resonance spectra of a spin-labeled calmodulin–β-endorphin complex (Giedroc et al., 1983b) and proton nuclear magnetic resonance spectra on the calmodulin–mastoparan complex (Muchmore et al., 1985) indicate only subtle changes in the protein. The unusual responsiveness of dansyl calmodulin is consistent with shielding of the dansyl groups from the solvent, either directly by the bound peptides or indirectly through conformational rearrangement.

The fluorescence emission maxima of the tryptophan-containing peptides shift

TABLE IX

Fluorescence Emission Maxima of Tryptophan-Containing Peptides[a]

Peptide	λ_{max} (0 CaM)	λ_{max} (sat. CaM, Ca^{2+})
Porcine glucagon	350 nM	340 nM
Dynorphin$_{1-17}$	349 nM	338 nM
Polistes mastoparan	349 nM	325 nM
Mastoparan X	348 nM	329 nM
Melittin	348 nM	336 nM

[a]Malencik and Anderson (1982, 1983a,b, 1984; also unpublished results).

toward shorter wavelengths on complex formation (Table IX). This phenomenon also occurs in some calmodulin-binding proteins, including rabbit skeletal muscle myosin light chain kinase (Johnson *et al.*, 1981). The magnitude of the change varies among the peptides, suggesting that the individual tryptophan residues may occupy different subsites which have varying degrees of exposure to the water molecules of the solvent. In general, fluorescence emission at lower wavelengths is favored by nonpolar media and by polar media of limited mobility. Both conditions are known to occur within the binding sites of proteins. In fact, X-ray crystallography of the fluorescent complex of chymotrypsin with 8-anilino-1-naphthalenesulfonate (ANS), a popular probe of "hydrophobic" binding sites, disclosed a rigid hydrophilic interaction (Weber *et al.*, 1979). The proposal of a hydrophobic protein binding domain in calmodulin is based in part on the calcium-dependent binding of fluorescent probes such as ANS and 9-anthroylcholine (LaPorte *et al.*, 1980; Tanaka and Hidaka, 1980). The fluorescence spectra of these complexes, as well as our spectra of the complexes of calmodulin with tryptophan-containing peptides, are consistent with either a hydrophobic or rigid hydrophilic environment. The sequences of the model peptides suggest that the peptide binding site of calmodulin has both hydrophobic and anionic sequences close together in the three-dimensional structure.

X-Ray crystallography of the calmodulin–Ca_4^{2+} complex has revealed a structure consisting of two globular lobes connected by an exposed α-helix comprising amino acid residues 65–92. Each of the two lobes contains two bound calcium ions (Babu *et al.*, 1985). The exposed α-helix is a possible interaction site for drugs, peptides, and calmodulin-dependent enzymes. Methionine residues known to be essential for enzyme activation occur at positions 71, 72, and 76 (Walsh and Stevens, 1978). The anionic sequence at positions 78–84 (Asp-Thr-Asp-Ser-Glu-Glu-Glu), which happens to occur in both of the peptide-binding fragments discussed in Section V, stands out as a possible subsite for ion pairing with the basic sequences found in all the efficient calmodulin-binding peptides. Such an association could place the peptide hydrophobic domain in close proximity to nonpolar amino acid side chains within calmodulin sequence 88–92. NMR spectra suggested that both trifluoperazine (Gariepy and Hodges, 1983) and a synthetic peptide analog corresponding to the inhibitory region of rabbit skeletal muscle troponin I (Cachia *et al.*, 1983) interact with the N-terminal region of site III of troponin C.

IX. PEPTIDES AS USEFUL CALMODULIN ANTAGONISTS

Peptide calmodulin antagonists may find application to more complex biological systems. We have begun collaborative work with P. Hoar and G. Kerrick on the functionally skinned muscle fiber system. Tension measurements on these fibers, which are permeable to proteins and other agents, permit functional classification of the muscle from which they are obtained. Trifluoperazine (50 μM) and the catalytic subunit of cAMP-dependent protein kinase (1 μM) inhibit contraction of skinned smooth muscle fibers, but have little effect on similarly prepared skeletal muscle or scallop adductor muscle fibers (Kerrick *et al.*, 1981). This is consistent with the unique regulation of smooth muscle contraction by a myosin light chain kinase–phosphatase system (cf. reviews by Adelstein and Klee, 1980; Hartshorne and Siemankowski, 1981). Figure 13 shows the effects of melittin (10 μM) on the contraction of skinned chicken gizzard smooth muscle fibers. Control experiments (lower trace) demonstrate the gradual increase in

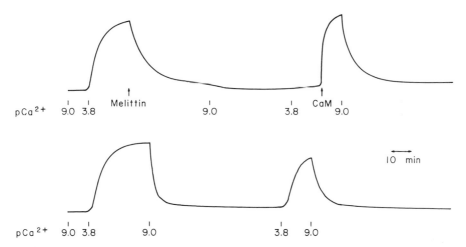

Fig. 13. Effect of melittin (10 μM) on maximal Ca^{2+}-activated tension in skinned chicken gizzard smooth muscle fibers. The solutions contained 70 mM K^+, 2 mM Mg-ATP, 1 mM Mg^{2+}, 7 mM EGTA, 1–0.14 mM free Ca^{2+}, and propionate as the major anion. Ionic strength was adjusted to 0.15 M and the pH maintained at 7.0 with imidazole propionate. The temperature was maintained at 19°.

tension which occurs when a mounted fiber, previously relaxed in a solution of 1% Triton X-100 containing low free calcium ($pCa^{2+} = 9.0$), is transferred to a solution containing 0.16 mM free calcium ($pCa^{2+} = 3.8$). The maximum tension is maintained over the time interval used later in the inhibition experiments. Gradual relaxation takes place when the contracted fiber is returned to a solution corresponding to $pCa^{2+} = 9.0$. The tension typically attains lower maximum values on repeated contraction of the original fiber. The addition of 10 μM melittin to a freshly contracted fiber (upper trace) results in gradual relaxation consistent with prevailing phosphatase action on the phosphorylated light chains. The tension remains low during transfers through solutions corresponding, in order, to $pCa^{2+} = 9.0$ and $pCa^{2+} = 3.8$. However, rapid recovery of tension occurs in the presence of 5.0 μM calmodulin and 0.16 mM free calcium. The enhancement of tension over that measured in the first contraction is at least partially due to an effect of added calmodulin also detected in the absence of melittin (not shown). Subsequent reduction of the free calcium concentration to $pCa^{2+} = 9.0$ produces normal relaxation. Mastoparan and VIP also inhibit contraction, but to a lesser extent consistent with their relative affinities for calmodulin. S-Peptide (Table II) has no effect.

Preliminary experiments indicate that the contraction of skinned skeletal muscle fibers is unaffected by 10 μM melittin. However, we plan to repeat the measurements using solutions containing a creatine phosphate–creatine phosphokinase ATP regenerating system. This system is not necessary with smooth muscle due to its low ATPase activity.

X. SUMMARY

A number of small peptides which range upward in size from 9 or 10 amino acid residues and contain certain common structural features—i.e., clusters of two or more basic amino acid residues, associated hydrophobic sequences, a low incidence of glutamyl residues, and a propensity for either an α-helical or a "random coil" conformation—undergo efficient calcium-dependent binding by calmodulin *in vitro*. Competition experiments indicate that the various peptides generally interact with identical or closely overlapping sites which are also related to the binding sites for phenothiazine drugs and calmodulin-dependent enzymes. We have hypothesized that sequences similar to those found in the peptides occur in accessible positions on the surfaces of enzymes and other proteins which recognize calmodulin and that they are directly involved in the interaction. The sequence of a high-affinity calmodulin-binding fragment recently prepared by cyanogen bromide cleavage of skeletal muscle myosin light chain kinase—*K R R* W *K K* N F I A V S A A N *R* F *K K* I S S S G A L M—is consistent with this prediction (Blumenthal *et al.*, 1985). Several of the peptides described may eventually prove useful in X-ray diffraction studies to locate and define the protein binding site of calmodulin.

The fact that many of the peptides are also biologically active could result either from a distant evolutionary relationship between calmodulin and some of the cell surface receptors or from a fortuitous convergence to similar types of highly efficient binding. It may relate to the finding that surprising structural similarities often occur among functionally diverse protein molecules (Doolittle, 1985). Both calmodulin and certain of the peptides are parts of ancient signaling systems. For example, ACTH and β-endorphin have been detected in extracts of a unicellular eukaryote, *Tetrahymena pyriformis* (LeRoith *et al.*, 1982). Since most of the peptides are known only to be extracellular in function, the peptide–calmodulin associations presented here do not necessarily occur *in vivo*. However, the high affinity and strong calcium dependence obtained with peptides, such as melittin and the mastoparans, point to the feasibility of physiologically relevant peptide–calmodulin interactions.

ACKNOWLEDGMENTS

We thank Philip C. Andrews, John C. Brown, Robert S. Hodges and John Pisano for providing peptide samples; Robert H. Kretsinger, John Penniston, David Puett, and Benjamin Weiss for manuscripts and reprints; Phyllis Hoar and W. G. L. Kerrick for tension measurements on smooth muscle fibers; and David Herpich and Stu Mott of the Oregon Turkey Growers for their continuing interest and help. The Muscular Dystrophy Association, the Oregon Affiliate of the American Heart Association, and the National Institutes of Health (AM 13912) have provided ongoing support.

REFERENCES

Adelstein, R. S., and Klee, C. B. (1980). Smooth muscle myosin light chain kinase. *In* "Calcium and Cell Function" (W. Y. Cheung, ed.), Vol. 1, pp. 167–194. Academic Press, New York.

Anderson, S. R. (1974). Use of fluorometry in studies of binding of small molecules by proteins. *In* "Experimental Techniques in Biochemistry" (J. M. Brewer, A. J. Pesce, and R. B. Ashworth, eds.), pp. 248–261. Prentice-Hall, Englewood Cliffs, New Jersey.

Andersson, A., Forsén, S., Thulin, E., and Vogel, J. J. (1983). Cadmium-113 nuclear magnetic resonance studies of proteolytic fragments of calmodulin: Assignment of strong and weak cation binding sites. *Biochemistry* **22**, 2309–2313.

Andrews, P. C., Ronner, P., and Dixon, J. E. (1985). *J. Biol. Chem.* **260**, 3910–3914.

Argiolas, A., and Pisano, J. J. (1983). Facilitation of phospholipase A$_2$ activity by mastoparans, a new class of mast cell degranulating peptides from wasp venom. *J. Biol. Chem.* **258**, 13697–13702.

Argiolas, A., and Pisano, J. J. (1984). Isolation and Characterization of Two New Peptides, Mastoparan C and Crabrolin, from the Venom of the European Hornet, *Vespa crabro*. *J. Biol. Chem.* **259**, 10106–10111.

Babu, Y. S., Sack, J. S., Greenhough, T. G., Bugg, C. E. Means, A. R., and Cook, W. J. (1985). Three-dimensional structure of calmodulin. *Nature* **315**, 37–40.

Barnette, M. S., and Weiss, B. (1983). Interaction of neuropeptides with calmodulin: A structure-activity study. *Psychopharmacol. Bull.* **19**, 387–392.

Barnette, M. S., Daly, R., and Weiss, B. (1983). Inhibition of calmodulin activity by insect venom peptides. *Biochem. Pharmacol.* **32**, 2929–2933.

Barrington, E. J. W. (1982). Evolutionary and comparative aspects of gut and brain peptides. *Br. Med. Bull.* **38**, 227–232.

Berg, O. G., Winter, R. B., and von Hippel, P. H. (1981). Diffusion-driven mechanisms of protein translocation on nucleic acids. 1. Models and theory. *Biochemistry* **20**, 6929–6947.

Bloom, F. E. (1981). Neuropeptides. *Sci. Am.* **245**, 148–168.

Blum, H. E., Lehky, P., Kohler, L., Stein, E. A., and Fischer, E. H. (1977). Comparative properties of vertebrate parvalbumins. *J. Biol. Chem.* **252**, 2834–2838.

Blumenthal, D. K., Takio, K., Edelman, A. M., Charbonneau, H., Titani, K., Walsh, K.-A., and Krebs, E. G. (1985). Identification of the calmodulin-binding domain of skeletal muscle myosin light chain kinase. *Proc. Natl. Acad. Sci. USA* **82**, 3187–3191.

Blundell, T. L. (1983). The conformation of glucagon. *In* "Handbuch der experimentellen Pharmakologie" (P. J. Lefèbvre, ed.), Vol. 66, Part I, pp. 37–55. Springer-Verlag, Berlin and New York.

Blundell, T. L., and Wood, S. (1982). The conformation, flexibility, and dynamics of polypeptide hormones. *Annu. Rev. Biochem.* **51**, 123-154.

Cachia, P. J., Sykes, B. D., and Hodges, R. S. (1983). Calcium-dependent inhibitory region of troponin: A proton nuclear magnetic resonance study on the interaction between troponin C and the synthetic peptide N^α-acetyl[FPhe^{106}]TnI(104–115)amide. *Biochemistry* **22**, 4145–4152.

Carlson, G. M., Bechtel, P. J., and Graves, D. J. (1979). Chemical and regulatory properties of phosphorylase kinase and cyclic AMP-dependent protein kinase. *Adv. Enzymol. Relat. Areas Mol. Biol.* **50**, 41–115.

Cheung, Y. (1980). Calmodulin plays a pivotal role in cellular regulation. *Science* **207**, 19–27.

Chou, P. Y., and Fasman, G. D. (1975). The conformation of glucagon: Prediction and consequences. *Biochemistry* **14**, 2536–2541.

Chou, P. Y., and Fasman, G. D. (1978). Empirical predictions of protein conformation. *Annu. Rev. Biochem.* **47**, 251–276.

Comte, M., Maulet, Y., and Cox, J. A. (1983). Ca^{++}-dependent high-affinity complex formation between calmodulin and melittin. *Biochem. J.* **209**, 269–272.

Conti, M. A., and Adelstein, R. S. (1981). The relationship between calmodulin and phosphorylation of smooth muscle myosin kinase by the catalytic subunit of 3′:5′ cAMP-dependent protein kinase. *J. Biol. Chem.* **256**, 3178–3181.

Cox, D. E., and Edstrom, R. D. (1982). Inhibition by calmodulin of the cAMP-dependent protein kinase activation of phosphorylase kinase. *J. Biol. Chem.* **257**, 12728–12733.

Doolittle, R. F. (1985). The genealogy of some recently evolved vertebrate proteins. *Trends Biochem. Sci.* **10**, 233–237.

Drabikowski, W., Kuznicki, J., and Grabarek, Z. (1977). Similarity in Ca^{++}-induced changes between troponin-C and protein activator of 3′:5′-cyclic nucleotide phosphodiesterase and their tryptic fragments. *Biochim. Biophys. Acta* **485**, 124–133.

Epand, R. M. (1983). The amphipathic helix: Its possible role in the interaction of glucagon and other peptide hormones with membrane receptor sites. *Trends Biochem. Sci.* **8**, 205–207.

Exton, J. H., Robison, J. A., Sutherland, E. W., and Park, C. R. (1971). Studies on the role of adenosine 3′,5′-monophosphate in the hepatic actions of glucagon and catecholamines. *J. Biol. Chem.* **246**, 6166–6177.

Finn, F. M., and Hofmann, K. (1973). The S-peptide-S-protein system: A model for hormone-receptor interaction. *Acc. Chem. Res.* **6**, 169–176.

Gariepy, J., and Hodges, R. S. (1983). Localization of a trifluoperazine binding site on troponin C. *Biochemistry* **22**, 1586–1594.

Giedroc, D. P., Puett, D., Ling, N., and Staros, J. V. (1983a). Demonstration by covalent cross-linking of a specific interaction between β-endorphin and calmodulin. *J. Biol. Chem.* **258**, 16–19.

Giedroc, D. P., Ling, N., and Puett, D. (1983b). Identification of β-endorphin residues 14–25 as a region involved in the inhibition of calmodulin-stimulated phosphodiesterase activity. *Biochemistry* **22**, 5584–5591.

Giedroc, D. P., Staros, J. V., Ling, N., and Puett, D. (1983c). Functional characterization of the β-endorphin-calmodulin association. *Fed. Proc., Fed. Am. Soc. Exp. Biol.* **42**, 2204.

Goldstein, A., Tachibana, S., Lowney, L. L., Hunkapillar, M., and Hood, L. (1979). Dynorphin$_{1-13}$, an extraordinarily potent opioid peptide. *Proc. Natl. Acad. Sci. U.S.A.* **76**, 6666–6670.

Grand, R. J. A., Perry, S. V., and Weeks, R. A. (1979). Troponin C-like proteins (calmodulins) from mammalian smooth muscle and other tissues. *Biochem. J.* **177**, 521–529.

Gremlich, H.-V., Fringeli, V.-P., and Schwyzer, R. (1983). Conformational changes of adrenocorticotropin peptides upon interaction with lipid membranes revealed by infrared attenuated total reflection spectroscopy. *Biochemistry* **22**, 4257–4264.

Gremlich, H.-V., Fringeli, V.-P., and Schwyzer, R. (1984). Interaction of adrenocorticotropin-(11–24)-tetradecapeptide with neutral lipid membranes revealed by infrared attenuated total reflection spectroscopy. *Biochemistry* **23**, 1808–1810.

Gysin, B., and Schwyzer, R. (1983). Head group and structure specific interactions of enkephalins and dynorphin with liposomes: Investigation by hydrophobic photolabeling. *Arch. Biochem. Biophys.* **225**, 467–474.

Gysin, B., and Schwyzer, R. (1984). Hydrophobic and electrostatic interactions between adrenocorticotropin-(1-24)-tetracosapeptide and lipid vesicles. Amphiphilic primary structures. *Biochemistry* **23**, 1811–1818.

Hammonds, R. G., Hammonds, A. S., Ling, N., and Puett, D. (1982). β-Endorphin and deletion peptides. A correlation of opiate receptor affinity with helix potential. *J. Biol. Chem.* **257**, 2990–2995.

Hartshorne, D. J., and Siemankowski, R. F. (1981). Regulation of smooth muscle actomyosin. *Annu. Rev. Physiol.* **43**, 519–530.

Head, J. F., Masure, H. R., and Kaminer, B. (1982). Identification and purification of a phenothiazine binding fragment from bovine brain calmodulin. *FEBS Lett.* **137**, 71–74.

Hirai, Y., Yasuhara, T., Yoshida, H., Nakajima, T., Fujino, M., and Kitada, C. (1979a). A new mast cell degranulating peptide "mastoparan" in the venom of *Vespula lewisii*. *Chem. Pharm. Bull.* **27**, 1942–1944.

Hirai, Y., Kuwada, M., Yasuhara, T., Yoshida, H., and Nakajima, T. (1979b). A new mast cell degranulating peptide homologous to mastoparan in the venom of Japanese hornet (*Vespa xanthoptera*). *Chem. Pharm. Bull.* **27**, 1945–1946.

Hokfelt, T., Johansson, O., Ljungdahl, A., Lundberg, J. M., and Schultzberg, M. (1980). Peptidergic neurons. *Nature (London)* **284**, 515–521.

Huang, T.-S., Feramisco, J. R., Glass, D. B., and Krebs, E. G. (1979). Specificity considerations relevant to protein kinase activation and function. *In* "From Gene to Protein: Information Transfer in Normal and Abnormal Cells" (T. R. Russell, K. Brew, H. Faber, and J. Schultz, eds.), pp. 449–461. Academic Press, New York.

Itano, T., Itano, R., and Penniston, J. (1980). Interactions of basic polypeptides and proteins with calmodulin. *Biochem. J.* **189**, 455–459.

Johnson, J. D., Holroyd, M. J., Crouch, T. H., Solaro, R. J., and Potter, J. D. (1981). Fluorescence studies of the interaction of calmodulin with myosin light chain kinase. *J. Biol. Chem.* **256**, 12194–12198.

Jornvall, H., Carlquist, M., Kwauk, S., Otte, S. C., McIntosh, C. H. S., Brown, J. C., and Mutt, V. (1981). Amino acid sequence and heterogeneity of gastric inhibitory polypeptide (GIP). *FEBS Lett.* **123**, 205–210.

Katoh, N., Raynor, R. L., Wise, B. C., Schatzman, R. C., Turner, R. C., Helfman, D. M., Fain, J. N., and Kuo, J.-F. (1982). Inhibition by melittin of phospholipid-sensitive and calmodulin-sensitive Ca^{++}-dependent protein kinases. *Biochem. J.* **202**, 217–224.

Keller, C. H., Olwin, B. B., Heideman, W., and Storm, D. R. (1982). The energetics and chemistry for interactions between calmodulin and calmodulin-binding proteins. *In* "Calcium and Cell Function" (W. Y. Cheung, ed.), Vol. 3, pp. 103–127. Academic Press, New York.

Kerrick, W. G. L., Hoar, P. E., Cassidy, P. S., Bolles, L., and Malencik, D. A. (1981). Calcium-regulatory mechanisms. Functional classification using skinned fibers. *J. Gen. Physiol.* **77**, 177–190.

Kincaid, R. L., Vaughan, M., Osborne, J. C., and Tkachuk, V. A. (1982). Ca^{++}-dependent interaction of 5-dimethylamino-naphthalene-1-sulfonyl-calmodulin with cyclic nucleotide phosphodiesterase, calcineurin, and troponin I. *J. Biol. Chem.* **257**, 10638–10643.

Klapper, M. H. (1977). The independent distribution of amino acid near neighbor pairs into polypeptides. *Biochem. Biophys. Res. Commun.* **78**, 1018–1024.

Klee, C. B., and Vanaman, T. C. (1982). Calmodulin. *Adv. Protein Chem.* **35**, 213–321.

Krauss, E. M., and Chan, S. L. (1982). *J. Am. Chem. Soc.* **104**, 6953–6961.

Krebs, E. G., and Beavo, J. A. (1979). Phosphorylation-dephosphorylation of enzymes. *Annu. Rev. Biochem.* **48**, 923–960.

Kretsinger, R. H. (1980). Structure and evolution of calcium-modulated proteins. *CRC Crit. Rev. Biochem.* **8**, 119–174.

Krieger, D. T. (1983). Brain peptides: What, where and why? *Science* **222**, 975–985.

Kuznicki, J., Grabarek, Z., Brzeska, H., and Drabikowski, W. (1981). Stimulation of enzyme activities by fragments of calmodulin. *FEBS Lett.* **130**, 141–145.

LaPorte, D. C., Wierman, B. M., and Storm, D. R. (1980). Calcium-induced exposure of a hydrophobic surface on calmodulin. *Biochemistry* **19**, 3814–3819.

LeRoith, D., Liotta, A. S., Roth, J., Shiloach, J., Lewis, M. E., Pert, C. B., and Krieger, D. T. (1982). Corticotropin and β-endorphin-like materials are native to unicellular organisms. *Proc. Natl. Acad. Sci. U.S.A.* **79**, 2086–2090.

Malencik, D. A., and Anderson, S. R. (1982). Binding of simple peptides, hormones, and neurotransmitters by calmodulin. *Biochemistry* **21**, 3480–3486.

Malencik, D. A., and Anderson, S. R. (1983a). Binding of hormones and neuropeptides by calmodulin. *Biochemistry* **22**, 1995–2001.

Malencik, D. A., and Anderson, S. R. (1983b). High affinity binding of the mastoparans by calmodulin. *Biochem. Biophys. Res. Commun.* **114**, 50–56.

Malencik, D. A., and Anderson, S. R. (1983c). Characterization of a fluorescent substrate for the adenosine 3',5'-cyclic monophosphate-dependent protein kinase. *Anal. Biochem.* **132**, 34–40.

Malencik, D. A., and Anderson, S. R. (1984). Peptide binding by calmodulin and its proteolytic fragments and by troponin C. *Biochemistry* **23**, 2420–2428.

Malencik, D. A., and Anderson, S. R. (1985). Effects of calmodulin and related proteins on the hemolytic activity of melittin. *Biochem. Biophys. Res. Commun.* (in press).

Malencik, D. A., and Fischer, E. H. (1982). Structure, function, and regulation of phosphorylase kinase. *In* "Calcium and Cell Function" (W. Y. Cheung, ed.), Vol. 3, pp. 161–188. Academic Press, New York.

Malencik, D. A., Anderson, S. R., Shalitin, Y., and Schimerlik, M. I. (1981). Rapid kinetic studies on calcium interactions with native and fluorescently labeled calmodulin. *Biochem. Biophys. Res. Commun.* **101,** 390–395.

Malencik, D. A., Anderson, S. R., Bohnert, J. L., and Shalitin, Y. (1982a). Functional interactions between smooth muscle myosin light chain kinase and calmodulin. *Biochemistry* **21,** 4031–4039.

Malencik, D. A., Huang, T.-S., and Anderson, S. R. (1982b). Binding of protein kinase substrates by fluorescently labeled calmodulin. *Biochem. Biophys. Res. Commun.* **108,** 266–272.

Malencik, D. A., Scott, J. D., Fischer, E. H., Krebs, E. G., and Anderson, S. R. (1986). Association of peptide analogues of the heat-stable inhibitor of cAMP-dependent protein kinase with calmodulin. *Biochemistry* **25** (submitted).

Maulet, Y., and Cox, J. A. (1983). Structural changes in melittin and calmodulin upon complex formation and their modulation by calcium. *Biochemistry* **22,** 5680–5686.

Means, A. R. (1981). Calmodulin: Properties, intracellular localization, and multiple roles in cell regulation. *Recent Prog. Horm. Res.* **37,** 333–367.

Muchmore, D. P., Malencik, D. A., and Anderson, S. R. (1985). In preparation.

Newton, D. L., Oldewurtel, M. D., Krinks, M. H., Shiloach, J., and Klee, C. B. (1984). Agonist and antagonist properties of calmodulin fragments. *J. Biol. Chem.* **259,** 4419–4426.

Norman, J. A., Drummond, A. H., and Moser, P. (1979). Inhibition of calcium-dependent regulator-stimulated phosphodiesterase activity by neuroleptic drugs is unrelated to their clinical efficacy. *Mol. Pharmacol.* **16,** 1089–1094.

O'Connor, R., and Peck, M. L. (1978). Venoms of Apidae. *In* "Handbuch der experimentellen Pharmakologie" (S. Bettini, ed.), Vol. 48, pp. 613–653. Springer-Verlag, Berlin and New York.

Ottesen, B., Staun-Olsen, P., Gammeltoft, S., and Fahrenkrug, J. (1982). Receptors for vasoactive intestinal polypeptide on crude smooth muscle membranes from porcine uterus. *Endocrinology* **110,** 2037–2043.

Perry, S. V. (1980). Structure and function relationships in troponin and calmodulin regulatory systems. *In* "Muscle Contraction: Its Regulatory Mechanisms" (S. Ebashi *et al.,* eds.), pp. 207–220. Jpn. Sci. Soc. Press, Tokyo/Springer-Verlag, Berlin and New York.

Prozialeck, W. C., and Weiss, B. (1982). Inhibition of calmodulin by phenothiazines and related drugs: Structure-activity relationships. *J. Pharmacol. Exp. Ther.* **222,** 509–516.

Puett, D., Giedroc, D. P., and Tollefson, S. (1983). Des-(1-13) human β-endorphin interacts with calmodulin. *Peptides (N.Y.)* **4,** 191–194.

Quay, S. C., and Condie, C. C. (1983). Conformational studies of aqueous melittin: Thermodynamic parameters of the monomer-tetramer self-association reaction. *Biochemistry* **22,** 707–716.

Quik, M., Iverson, L. L., and Bloom, S. R. (1978). Effect of vasoactive intestinal peptide (VIP) and other peptides on cAMP accumulation in rat brain. *Biochem. Pharmacol.* **27,** 2209–2213.

Reid, R. E., Gariepy, J., and Hodges, R. S. (1983). Interaction of neuroleptic drugs with a synthetic calcium-binding peptide analog of site III of rabbit skeletal troponin C. Phenothiazine selective binding. *FEBS Lett.* **154,** 60–64.

Richards, F. M., and Wyckoff, H. W. (1971). Bovine pancreatic ribonuclease. *Enzymes* **4,** 647–806.

Schreiber, W. E., Sasagawa, T., Titani, K., Wade, R. D., Malencik, D., and Fischer, E. H. (1981). Human brain calmodulin: Isolation, characterization, and sequence of a half-molecule fragment. *Biochemistry* **20,** 5239–5245.

Scott, J. D., Fischer, E. H., Demaille, J. G., and Krebs, E. G. (1985). Identification of an inhibitory region of the heat-stable protein inhibitor of the cAMP-dependent protein kinase. *Proc. Natl. Acad. Sci. USA* **82,** 4379–4383.

Sellinger-Barnette, M., and Weiss, B. (1982). Interaction of β-endorphin and other opioid peptides with calmodulin. *Mol. Pharmacol.* **21,** 86–91.

Sellinger-Barnette, M., and Weiss, B. (1984). Interaction of various peptides with calmodulin. *Adv. Cyclic Nucleotide Protein Phosphorylation Res.* **16,** 261–276.

Snyder, S. H., and Innis, R. B. (1979). Peptide neurotransmitters. *Annu. Rev. Biochem.* **48,** 755–782.

Talbot, J. A., and Hodges, R. S. (1981). Synthetic studies on the inhibitory region of rabbit skeletal troponin I. Relationship of amino acid sequence to biological activity. *J. Biol. Chem.* **256,** 2798–2802.

Tanaka, T., and Hidaka, H. (1980). Hydrophobic regions function in calmodulin-enzyme(s) interactions. *J. Biol. Chem.* **255,** 11,078–11,080.

Terwilliger, T. C., and Eisenberg, D. (1982). The structure of melittin. II. Interpretation of the structure. *J. Biol. Chem.* **257,** 6016–6022.

Thiry, P., Vandermeers, A., Vandermeers-Pinct, M.-C., Rathe, J., and Christophe, J. (1980). The activation of brain adenylate cyclase and brain cyclic nucleotide phosphodiesterase by seven calmodulin derivatives. *Eur. J. Biochem.* **103,** 409–414.

Thulin, E., Andersson, A., Drakenberg, T., Forsén, S., and Vogel, H. J. (1984). Metal ion and drug binding to proteolytic fragments of calmodulin: Proteolytic, cadmium-113, and proton nuclear magnetic resonance studies. *Biochemistry* **23,** 1862–1870.

Wall, C. M., Grand, R. J. A., and Perry, S. V. (1981). Biological activities of the peptides obtained by digestion of troponin C and calmodulin with thrombin. *Biochem. J.* **193,** 307–316.

Walsh, M., and Stevens, F. C. (1978). Chemical modification studies on the Ca^{++}-dependent protein modulator: The role of methionine residues in the activation of cyclic nucleotide phosphodiesterase. *Biochemistry* **17,** 3924–3930.

Walsh, M., Stevens, F. C., Kuznicki, J., and Drabikowski, W. (1977). Characterization of tryptic fragments obtained from bovine brain protein modulator of cyclic nucleotide phosphodiesterase. *J. Biol. Chem.* **252,** 7440–7443.

Weber, L. D., Tulinsky, A., Johnson, J. D., and El-Bayoumi, M. A. (1979). Expression of functionality of α-chymotrypsin. The structure of a fluorescent probe-α-chymotrypsin complex and the nature of its pH dependence. *Biochemistry* **18,** 1297–1303.

Weiss, B., Prozialeck, W., Cimino, M., Sellinger-Barnette, M., and Wallace, T. L. (1980). Pharmacological regulation of calmodulin. *Ann. N.Y. Acad. Sci.* **356,** 319–345.

Weiss, B., Prozialeck, W. C., and Wallace, T. L. (1982). Interaction of drugs with calmodulin: Biochemical, pharmacological, and clinical implications. *Biochem. Pharmacol.* **31,** 2217–2226.

Weiss, B., Prozialeck, W. C., Sellinger-Barnette, M., Winkler, J. D., and Schechter, L. (1985). Calmodulin antagonists: Structure activity relationships. *In* "Calmodulin Antagonists and Cellular Physiology" (H. Hidaka and D. J. Hartshorne, eds.). Academic Press, New York (in press).

Chapter 2

Molecular and Regulatory Properties of Calmodulin-Dependent Phosphodiesterase from Brain

RANDALL L. KINCAID
MARTHA VAUGHAN

Laboratory of Cellular Metabolism
National Heart, Lung and Blood Institute
National Institutes of Health
Bethesda, Maryland

43

I. INTRODUCTION

All cells, and particularly nerve cells, must maintain a delicate balance between stability and flexibility, between relatively long-term preservation of functional integrity and the necessity to respond rapidly and effectively to signals from their environment. The biochemical mechanisms that underlie this important duality require not only temporal but spatial coordination via "second messenger" systems. In the late 1950s, Sutherland and Rall (1958) discovered cAMP and pioneered investigation of its role in mediating cellular responses to hormones such as epinephrine and glucagon (Sutherland and Rall, 1960). It is now known that a great variety of cellular processes are influenced by cAMP (and/or cGMP), either directly or indirectly. The critical role of Ca^{2+} in stimulus–transduction pathways was recognized initially with demonstration of its importance in contractility (Sandow, 1970) and secretory phenomena (Rubin, 1970). More recently, the establishment of the Ca^{2+}-binding protein calmodulin as a mediator of Ca^{2+}-dependent enzyme activation has provided a molecular mechanism for many of the actions of Ca^{2+} on cellular function (see Cheung, 1980; Means and Dedman, 1980; Klee and Vanaman, 1982). Not surprisingly, the cyclic nucleotide and Ca^{2+}–calmodulin systems have points of interaction wherein the action of one system can modify operation of the other (Berridge, 1975). In brain tissue, for example, enzymes that synthesize and degrade cAMP (adenylate cyclase and a cyclic nucleotide phosphodiesterase) are both activated by calmodulin in the presence of Ca^{2+} (Teo and Wang, 1973; Brostrom *et al.*, 1975). Furthermore, the mediators of cAMP action, the cyclic AMP–dependent protein kinases, can alter the function of some calmodulin-dependent enzymes (e.g., myosin light chain kinase, phosphorylase kinase) at least *in vitro* (Conti and Adelstein, 1981; Cohen, 1982). Thus, these two systems can influence and coordinate cellular activities as apparently diverse as stimulus–response coupling, regulation of intermediary metabolism, and control of contractility (Fig. 1).

The first indication that these two types of regulatory systems might be directly related can be found in the initial reports of a heat-stable activator protein that greatly stimulated the activity of a soluble phosphodiesterase (Cheung, 1970) and of a Ca^{2+}-activated phosphodiesterase (Kakiuchi and Yamazaki, 1970). It was later demonstrated that enzyme stimulation required the simultaneous presence of the activator protein and Ca^{2+} (Teo and Wang, 1973; Wolff and Brostrom, 1974). With purification and characterization of the activator as a Ca^{2+}-binding protein in a number of laboratories, growing interest in calmodulin, as it was named, led to its identification as a modulator of the function of a wide variety of enzymes and structural proteins. The physical properties of calmodulin and its effects on enzyme activities are well reviewed in this treatise and elsewhere (Klee and Vanaman, 1982). This chapter, however, deals pri-

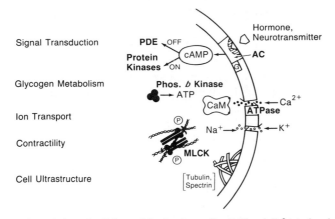

Signal Transduction

Glycogen Metabolism

Ion Transport

Contractility

Cell Ultrastructure

Fig. 1. Dual regulation of cellular activity by the cyclic AMP and Ca^{2+}/calmodulin "second messenger" systems. Abbreviations: PDE, cyclic nucleotide phosphodiesterase; AC, adenylate cyclase; Phos. *b* Kinase, phosphorylase *b* kinase; CaM, calmodulin; ATPase, Mg^{2+}-activated, Ca^{2+}-dependent ATPase; MLCK, myosin light chain kinase.

marily with the molecular and regulatory properties of the phosphodiesterase and is concerned with calmodulin only in that context. Section I provides an overview of the calmodulin-activated phosphodiesterase from brain, including its purification and its physical and enzymatic properties. Section II deals with its interaction with calmodulin, with emphasis on the experimental approaches that have been used in this laboratory, and also discusses current models for the reversible activation of this enzyme by Ca^{2+} and calmodulin. The final section briefly outlines some potentially important immunocytochemical approaches and offers some thoughts regarding the role of phosphodiesterase in the nervous system.

II. ENZYMATIC AND PHYSICAL PROPERTIES
OF PHOSPHODIESTERASE

The reaction catalyzed by phosphodiesterases—hydrolysis of the ribose 3',5'-phosphodiester bond (Fig. 2)—is highly exothermic, releasing $\sim$12 kcal/mol of nucleotide; this fact has led to speculation regarding its possible involvement in synthetic or other energy-consuming pathways (Rudolph *et al.*, 1971). Both cGMP and cAMP (also cIMP) will serve as substrates for the calmodulin-activated phosphodiesterase, whereas cCMP and cUMP will not. Catalysis requires a divalent cation, presumably Mg^{2+} under physiological conditions; Ca^{2+} alone has little or no effect on activity. Since the initial description of a cAMP phosphodiesterase that was stimulated by a heat-stable protein activator, a recurring

1. CATALYZES THE REACTION:

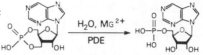

2. EFFECTORS OF ENZYME ACTIVITY:

ACTIVATORS- Ca^{2+}/ CALMODULIN, PHOSPHOLIPID, PROTEASES (IRREVERSIBLE)

INHIBITORS- K^+, POLYAMINES

3. SUBUNIT STRUCTURE AND Ca^{2+}/ CALMODULIN REGULATION:

HOMODIMER, SUBUNIT = 60 K

Fig. 2. Properties of the Ca^{2+}/calmodulin-dependent 3′,5′-cyclic nucleotide phosphodiesterase. The figure shows the hydrolytic reaction catalyzed by the enzyme, several effectors of possible physiologic relevance, and the subunit structure of the basal and activated enzymes.

question has been whether the enzyme has substantial intrinsic activity in the absence of such stimulation. If so, such "basal" activity might be important in regulation of cyclic nucleotide levels in the absence of Ca^{2+}/calmodulin stimulation. Over the years there have been widely discrepant values reported for the degree of calmodulin stimulation (and thus for the apparent basal activity) in both crude and purified enzyme preparations. Such differences could be accounted for by many factors, including the presence of other (i.e., noncalmodulin-stimulated) phosphodiesterases, the conditions of assay or purification, the action of effectors present in crude materials, and, perhaps most important, the contribution of proteolytically activated forms of this enzyme. Although this last factor will be discussed below, it is worth mentioning inasmuch as it has been assumed by many to account for much, if not all, of the variation in basal activities (Tucker *et al.*, 1981). Cheung (1971) first reported the activation of partially purified enzyme in assays when they contained trypsin or when crude snake venom was used as a source of 5′-nucleotidase in the assays. Most investigators have, therefore, taken special precautions during purification and assay to prevent proteolysis. Despite this, substantial differences in calmodulin activation are reported for different, highly purified enzyme preparations, even when their apparent molecular weights are the same (Klee *et al.*, 1979a; Morrill *et al.*, 1979; Sharma *et al.*, 1980; Kincaid *et al.*, 1981a, 1984a; Tucker *et al.*, 1981; Hansen and Beavo, 1982).

The differences in the extent of calmodulin activation prompted us to investigate possible molecular explanations for basal and calmodulin-dependent phosphodiesterase activities. Calmodulin-dependent activity is, by definition, the difference between total activity in the presence of calmodulin and Ca^{2+} and activity in their absence. In designing an assay for these studies, we chose buffers and assay components to maximize both basal and total activities, since additions which greatly increased calmodulin dependence might obscure important changes reflected in basal activity. For example, crystalline ovalbumin was used as a protein stabilizer in preference to other proteins (e.g., bovine serum albumin), which appeared to selectively reduce basal activity. The ability of high concentrations of bovine serum albumin to increase the degree of calmodulin activation has also been noted by others (Wolff and Brostrom, 1979). 2-[bis(2-hydroxyethyl)amino]ethanesulfonic acid (BES), pH 7.0, was selected over other buffers for similar reasons. Furthermore, since one might expect that relevant changes in basal activity would occur at physiologic concentrations of cyclic nucleotide, we chose to use 0.5 μM cGMP as substrate. During early attempts to purify this enzyme, we observed reproducible differences in calmodulin activation at different stages of purification, as well as with extensively purified (>75%) enzyme (Kincaid *et al.*, 1981a).

An unusual and potentially important observation was that the hydrodynamic properties of the phosphodiesterase changed within hours of tissue homogenization. When freshly prepared supernatants were rapidly subjected to gel filtration, the phosphodiesterase behaved like a protein of 50,000–60,000 Da. After 24 hr at 4°C, some activity appeared as a larger molecular form, and by 48 hr virtually all of the enzyme chromatographed with a nominal M_r of 120,000–140,000 (Fig. 3). Concomitant with the increase in apparent molecular size, there was a decrease in basal activity with no change in total enzyme activity, i.e., calmodulin-dependent activity was increased. Apparently the activity of the larger form of the enzyme was dependent on calmodulin (which stimulated it four- to eightfold), whereas that of the smaller form was relatively independent (stimulated two- to fourfold). It seemed unlikely that components present in the supernatant could be directly responsible for maintaining the smaller molecular form, because rapid dialysis of fresh supernatant prior to gel filtration did not give rise to the larger form, and because the isolated smaller form after 1 or 2 days invariably chromatographed as the larger species. The most plausible explanation appeared to be that, with spontaneous aggregation of the enzyme, a large fraction of activity was converted to a calmodulin-dependent form. For both the smaller and larger species of enzyme, activity was quantitatively bound to calmodulin-Sepharose, thus arguing against a calmodulin-independent enzyme species that was unable to interact with calmodulin.

It was shown that enzymes purified from bovine brain by two different methods differed markedly in their degree of calmodulin stimulation (four- to fivefold versus eight- to tenfold), although both showed a single peptide of 60,000 daltons on polyacrylamide gel electrophoresis under denaturing conditions (Kincaid *et al.*, 1981a). In both methods, the final purification step was calmodulin-Sepharose affinity chromatography, but the homogenization and first purifica-

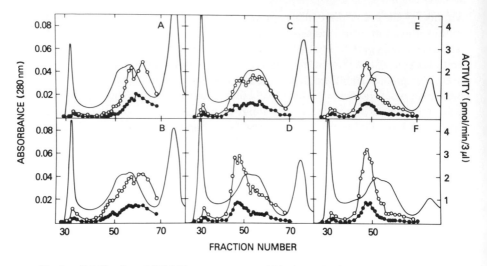

Fig. 3. Gel filtration of 100,000 xg supernatant at 12-hr intervals after homogenization. Supernatant (350 μl) was applied to a column (0.9 × 40 cm) of Sephacryl S-300 and eluted at a flow rate of 22 ml/hr. Gel filtration was begun at A, 2 hr; B, 14 hr; C, 26 hr; D, 38 hr; E, 50 hr; F, 62 hr after homogenization. Phosphodiesterase activity assayed in the presence of 1 mM EGTA (●) or Ca^{2+} plus calmodulin (○). From Kincaid *et al.* (1981a).

tion step were done at pH 5–6 in one method and pH 7–8 in the other. It would appear, therefore, that subtle differences in purification procedures can markedly affect the basal activity of the enzyme preparation. In addition, with storage of essentially homogeneous enzyme at low protein concentration (80 μg/ml) at pH 5.8 in the presence of (NH$_4$)$_2$SO$_4$ and 20% glycerol, there was a gradual decline in calmodulin activation to less than 50% without substantial loss in total activity or detectable degradation of the enzyme. On analysis by adsorption electrophoresis [a procedure involving adsorption to calmodulin-Sepharose combined with native gel electrophoresis (Kincaid and Vaughan, 1979)] at different acrylamide concentrations, the "calmodulin-insensitive" enzyme migrated with a much higher mobility than it did immediately after its purification. Ferguson plot analysis indicated that this was due to a shift from a primarily dimeric to a monomeric form of the enzyme. Since the method required that the enzyme was quantitatively adsorbed to immobilized calmodulin prior to entering these gels, this again indicated that monomeric forms were capable of interaction with calmodulin.

Although these data suggested a relationship between the physical form of the enzyme (e.g., monomeric, dimeric) and calmodulin sensitivity, direct estimates of enzyme size under the same conditions used for activity measurements were not possible. To approach this question experimentally, we used radiation inactivation. Samples of purified phosphodiesterase were frozen under conditions used for enzyme assay (enzyme concentration $\sim 10^{-9}$ M) and exposed to increasing doses of high-energy radiation. From the loss of activity observed at each dose, an approximation of the minimum size of the component responsible for a particular function can be calculated [for a review of this method, see Kempner and Schlegel (1979)]. For all soluble proteins previously studied, a first-order inactivation process had been observed. Indeed, the inactivation of calmodulin-dependent activity was apparently first-order with the slope expected for a protein of M_r 105,000, or roughly that of a dimer (Kincaid *et al.*, 1981b). The decay of basal activity was, however, *nonlinear*, and activity actually *increased* with lower doses of radiation. For enzyme preparations that were activated three- to fourfold by calmodulin, the maximal increase in basal activity was 20–25%. After the elevation of basal activity with increasing doses, an exponential decay expected for a component of $M_r \sim 60,000$ was seen (Fig. 4). In qualitative terms, these findings suggested the activation of previously inac-

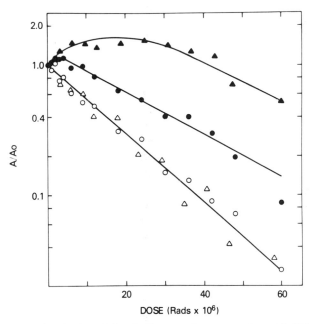

Fig. 4. Radiation inactivation of phosphodiesterase from preparations with different degrees of calmodulin activation. Purified phosphodiesterase was irradiated as indicated and assayed for basal (●, ▲) and calmodulin-dependent (○, △) activity. Data for a preparation stimulated three- to fourfold by calmodulin (circles) represent means of activities of three samples. Data from a preparation stimulated nine- to tenfold (triangles) are means of activities of two samples. Data are expressed relative to the activity of nonirradiated samples (A/A_0). From Kincaid *et al.* (1981b).

tive units by radiation; the minimal size of the unit expressing basal activity was that of a monomer. A physical hypothesis was proposed to explain these observations. Radiation of a dimer, which exhibited no activity without calmodulin, inactivated one subunit and generated an active monomer. At low doses of radiation, basal activity was increased; when all dimers had been irradiated, the decay of activity followed that expected for a monomer. This finding was consistent with the observation that the activities of noncovalently associated subunits are not mutually destroyed when a multimeric protein structure is irradiated (Kempner et al., 1980; Kempner and Miller, 1983). A quantitative mathematical model based on this hypothesis was developed and used to predict data from radiation inactivation of enzyme preparations with different proportions of basal and calmodulin-dependent activity. When a preparation that was activated approximately tenfold was irradiated (Fig. 4), both the maximal increase in basal activity (~100%) and the subsequent decline in activity as a function of dose closely approximated those predicted by the model; however, as expected, the decay of calmodulin-dependent activity was exactly that seen for the preparation with lower calmodulin stimulation. The difference between enzyme preparations that were activated to different degrees by calmodulin (both of which contained a single polypeptide of 60,000 daltons) apparently was the ratio of monomers to dimers under the conditions used for irradiation and assay. These results, in conjunction with those from gel filtration and electrophoresis, provided a consistent and plausible explanation for basal activity, at least in those particular enzyme preparations examined. It may be that certain purification or storage conditions (e.g., low pH, high ionic strength) affect the ability of the phosphodiesterase to self-associate at low protein concentrations and thereby alter its basal activity; the ability to interact with calmodulin is, however, maintained. It should be noted that this "monomer–dimer" model addressed exclusively the question of the nature of basal activity, not the mechanism by which calmodulin activates the enzyme. There was no evidence or suggestion that calmodulin activates by promoting dissociation of dimers, only that dimeric forms require calmodulin for activity and monomers do not.

Since brain has the highest specific activity of any mammalian tissue for calmodulin-activated phosphodiesterase and has relatively small amounts of other phosphodiesterases, it was chosen as the starting material for purification of the enzyme. However, phosphodiesterase represents a relatively minor fraction of the calmodulin-binding proteins in brain (1–3%) (Klee and Haiech, 1980; Kincaid et al., 1981a), and its separation from these other proteins presents a major difficulty. Although others have been able to purify phosphodiesterase using chromatography on agarose to which the dye Cibacron-Blue was immobilized (Morrill et al., 1979; Sharma et al., 1980), our attempts to use this matrix were not successful. The phosphoprotein phosphatase calcineurin (Stewart et al., 1982), which is the major calmodulin-binding protein in brain (Klee et al.,

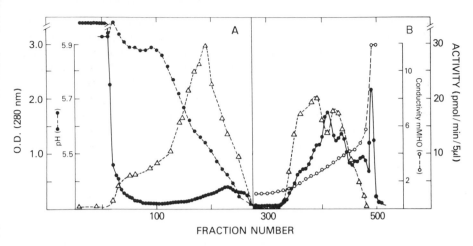

Fig. 5. DEAE–BioGel A chromatography of brain phosphodiesterase activity. Supernatant from bovine brain was adjusted to pH 6 and applied to a column of DEAE–BioGel A. After washing, elution was carried out with decreasing pH (A). When the pH of the effluent reached 5.3, elution was continued with a gradient of increasing ionic strength at pH 5.5 and finally with buffer containing $0.5 M$ NaCl (B). Fractions were analyzed for absorbance at 280 nm ($\bullet$), conductivity ($\bigcirc$), pH ($\bullet$---$\bullet$), and total phosphodiesterase activity ($\triangle$). From Kincaid *et al.* (1984a).

1979b), was only partially removed by repeated chromatography on the dye column. The hydrodynamic and chromatographic properties of calcineurin on several other matrices (DEAE–agarose, hydroxylapatite, phenyl–Sepharose) were sufficiently similar to those of the phosphodiesterase to preclude complete separation. We took advantage of the observation (Kincaid *et al.*, 1981a) that elution of anion-exchange media by a decrease in pH effectively eliminated >95% of other calmodulin-binding proteins, while a sizable fraction of phosphodiesterase (peak I) was recovered (Fig. 5). Interestingly, a second fraction of enzyme, comparable in total activity, was recovered only when we increased the ionic strength of the eluant. Since the enzymes from these two fractions, further purified on calmodulin-Sepharose, differed in isolectric point (Kincaid *et al.*, 1984a) in addition to exhibiting different anion-exchange properties, it was suggested that they may represent separate isozymes in brain tissue. This hypothesis was consistent with the observation that the proportion of the two forms was apparently invariant whether fresh or frozen tissue or bovine or ovine brain was used as the starting material. Recently, Sharma *et al.* (1984) used monoclonal antibodies to the brain phosphodiesterase to purify and identify the two isozymes. Whether these represent posttranslationally modified forms of the same peptide or different gene products possibly localized in discrete cell types is unclear.

A method was developed for large scale, rapid purification from mammalian brain of the peak I form of phosphodiesterase, the activity of which is virtually dependent on calmodulin. This procedure, which is outlined schematically in Fig. 6, uses only batchwise anion-exchange chromatography (as previously described) and sequential chromatography on two columns of calmodulin–Sepharose. These latter steps involve "overloading" the first affinity column (which quantitatively removes residual calcineurin) followed by chromatography of the unbound phosphodiesterase (>80% of that originally applied) on a second column of immobilized calmodulin (Kincaid *et al.*, 1982). The success of this sequential affinity column procedure (the efficacy of which can be demonstrated with mixtures of purified phosphodiesterase and calcineurin) apparently reflects a higher operational affinity or capacity of calmodulin–Sepharose for calcineurin than for phosphodiesterase.

This procedure has been used for purification of enzyme from bovine, ovine, and porcine brain with ~12% yield. The enzyme has no detectable (<3%) phosphatase activity and displays a single peptide of ~59 kDa on electrophoresis under denaturing conditions. The native enzyme behaves as a single species (M_r ~ 115,000) on high-performance gel filtration with constant specific activity

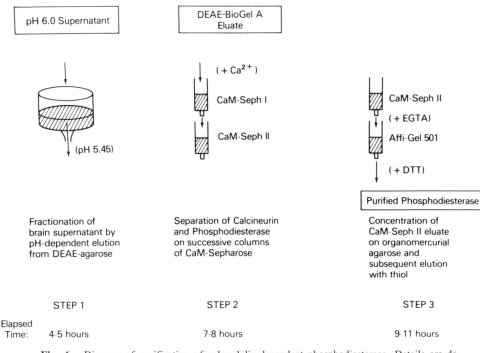

Fig. 6. Diagram of purification of calmodulin-dependent phosphodiesterase. Details are described in Kincaid *et al.* (1984a).

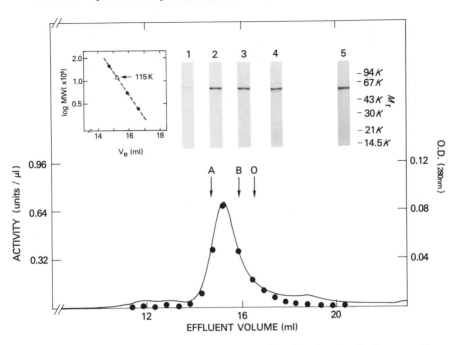

Fig. 7. High-performance gel filtration chromatography of bovine phosphodiesterase. Phosphodiesterase (0.1 mg in 250 μl) was injected onto a TSK-SW 3000 column (0.75 × 60 cm) and fractions (300 μl) assayed for enzyme activity. Arrows designate standards: aldolase (A), 161 kDa; bovine serum albumin (B), 67 kDa; and ovalbumin (O), 43 kDa. Inset: plot of M_r versus V_e. SDS gels: lanes 1–4, fractions from chromatogram; lane 5, enzyme applied to column. From Kincaid *et al.* (1984a).

across the peak of absorbance (Fig. 7). The hydrodynamic properties (Table I) (Stokes radius 4.3 nm; $s_{20,w} \sim 6.0$; $f/f_0 \sim 1.3$) suggest that this highly calmodulin-activated form is a slightly asymmetric dimer of identical subunits. Under denaturing conditions (i.e., 6 *M* guanidine·HCl) equilibrium sedimentation yields a polypeptide with $M_r \sim 56,000–57,000$. Interestingly, with increasing concentrations of guanidine·HCl, the circular dichroic spectrum of phosphodiesterase undergoes an unusually abrupt transition to that characteristic of random coil at low concentrations of this agent (half-effective concentration $\sim$ 0.8 *M*) suggesting that the tertiary and quaternary structure of this protein is very sensitive to denaturing conditions (R. Kincaid and J. Osborne, unpublished data).

As mentioned above and reported by us and others (Davis and Daly, 1978; Klee *et al.*, 1979a; Kincaid *et al.*, 1979), the magnitude of calmodulin stimulation greatly depends on the conditions of assay. When assayed with low concentrations of spermine (<10 μ*M*) or KCl (50–100 m*M*) the enzyme exhibits

TABLE I

Physical Properties of Purified Phosphodiesterases[a]

Property	Bovine	Ovine	Method
Molecular weight, subunit	59,000	59,000	SDS–gel electrophoresis
	57,000	56,200	Equilibrium sedimentation (6 M guanidine)
Molecular weight, native	124,300	112,200	Sedimentation equilibrium
	110,500	110,300	Stokes radius and $s_{20,w}$
	115,000	115,000	Gel filtration
Stokes radius (nm)	$4.35 \pm 0.03(3)$[b]	4.38	Gel filtration
$s_{20,w}$ (S)	$5.95 \pm 0.05(5)$	5.90	Sucrose density gradient
Partial specific volume	0.73	0.73	Amino acid composition
Frictional ratio (f/f_0)	1.31	1.30	Based on Stokes radius
	1.36	1.38	Based on $s_{20,w}$

[a]Adapted from Kincaid et al. (1984a).
[b]In parentheses, the number of determinations.

very little basal activity; stimulation by calmodulin is thirty- to fortyfold. Since these concentrations of K^+ and spermine are well within the physiologic range for brain tissue, the enzyme activity *in vivo* may be essentially completely dependent on calmodulin. The apparent affinity of purified phosphodiesterase for calmodulin in the presence of 300 μM Ca^{2+} was approximately 1×10^{-9} M (Fig. 8) under our assay conditions. As studied by others with partially purified phosphodiesterase (Levin and Weiss, 1977), the calmodulin-dependent activity was inhibited by trifluoperazine, which, with increasing concentrations, shifts the dose–response curve for calmodulin, consistent with competitive inhibition of the activator. Activation by Ca^{2+}/calmodulin occurred through an increase in maximal velocity of the enzyme with little change in substrate affinity. The preference for Mg^{2+} ($K_a \sim 0.1$ mM) as a required divalent cation appears to be consistent with a physiologic role for this ion, while Mn^{2+} was less effective, and Ca^{2+} by itself had no effect on activity.

During development of the purification procedure it was observed that approximately twice as much enzyme was recovered from ovine (~ 700 μg/kg) as from bovine brain although percentage yields were the same. Although the specific activity of the ovine enzyme was consistently ~ 30–40% lower than that from bovine brain, their apparent substrate affinities[1] (1–3 μM cGMP, 20–40 μM

[1] We (Kincaid, 1984a, Kincaid et al., 1984b) and others (Klee et al., 1979a) have observed anomalous substrate dependence for the basal activity of the purified enzyme. In the presence of calmodulin, the enzyme's activity followed Michaelis–Menten principles. The basal activity which exhibits an apparent "low-affinity" component (10–15 μM cGMP, 100–200 μM cAMP) has not been explained.

cAMP) and other kinetic or physical properties were quite similar. Based upon limited proteolysis, however, the enzymes from bovine and ovine brain clearly differ in primary structure. In the presence of EGTA, α-chymotrypsin degrades the bovine enzyme sequentially to peptides of 57 and 45 kDa, whereas peptides from the ovine enzyme are 55, 53, and 38 kDA (Kincaid *et al.*, 1984a, 1985). The activity of the final species, which are dimeric proteins of 80–90 kDa, is fully stimulated and not further enhanced by Ca^{2+} and calmodulin. Proteolysis in the presence of Ca^{2+} and calmodulin produces unique final peptides of 47 and 42 kDa from the bovine and ovine enzymes, respectively (Fig. 9) (Stith-Coleman *et al.*, 1982; Kincaid *et al.*, 1985). Thus, in addition to demonstrating species differences in primary sequence, these data provide direct evidence of structural changes in the enzyme upon interaction with calmodulin. This proposal is supported by recent studies using trypsin (Krinks *et al.*, 1984).

Since the initial rate of proteolysis was greater in the presence of Ca^{2+} and calmodulin than with EGTA, it appears that association with calmodulin alters the conformation of phosphodiesterase, exposing a new site with increased susceptibility to proteolytic attack. Upon addition of EGTA to such incubations, the peptides were cleaved to those observed when proteolysis was carried out without Ca^{2+} and calmodulin. These data indicate that calmodulin can remain associated with the proteolyzed phosphodiesterase, and that in this complex the proteolytic site cleaved in the absence of calmodulin is inaccessible. Whether calmodulin physically protects this site or induces a conformation in which the site is hidden is not known. In contrast, the phosphodiesterase proteolyzed in the

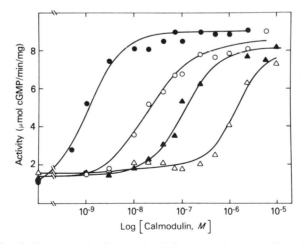

Fig. 8. Phosphodiesterase activation by calmodulin at several concentrations of trifluoperazine. Purified bovine phosphodiesterase (0.2 nM) was assayed at 30° with 0.5 μM cGMP, 5 mM MgCl$_2$, and the indicated concentrations of calmodulin without (●) or with 5 μM (○), 10 μM (▲) or 30 μM (△) trifluoperazine. Trifluoperazine was donated by Smith, Kline and French Laboratories.

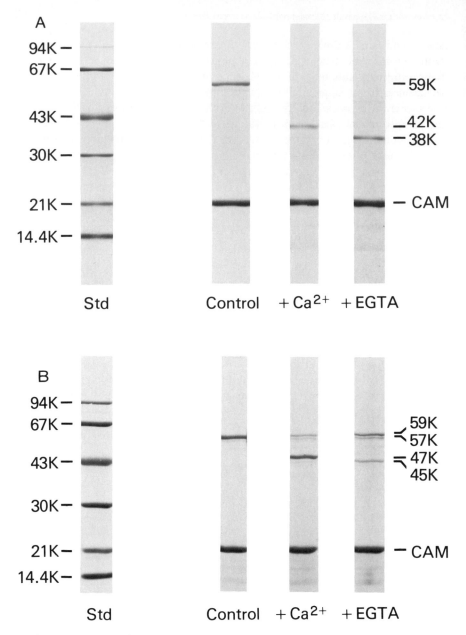

Fig. 9. Effect of Ca^{2+} and calmodulin on chymotryptic cleavage of bovine and ovine phosphodiesterases. Samples (4 μg) of ovine (A) or bovine (B) phosphodiesterase were incubated with chymotrypsin (0.15 μg) and calmodulin (6 μg) plus 1 mM Ca^{2+} or 1 mM EGTA as indicated. Samples were precipitated with trichloroacetic acid (10% w/v) and solubilized for sodium dodecyl sulfate electrophoresis. For control samples, trichloroacetic acid was added prior to incubation. From Kincaid *et al.* (1985).

absence of calmodulin did not bind to calmodulin–Sepharose, nor did it interact with dansyl calmodulin even at micromolar concentrations of the two proteins, which suggests abolition of its ability to form even a very weak association (Kincaid et al., 1985).

Despite differences in their ability to interact with calmodulin, the two proteolyzed forms of phosphodiesterase were fully activated, and no discernible difference in any enzymatic property was detected. Their affinities for substrate were essentially identical, as were their maximal velocities, and neither form was affected by modifiers of basal activity such as KCl or spermine (Table II). The somewhat higher activities observed for samples proteolyzed in the presence of Ca^{2+} and calmodulin may reflect stabilization of the enzyme during incubation.

The data on proteolytic activation may provide insights into the regulatory features of phosphodiesterase. It appears that a region of the protein exerts a self-inhibitory influence which is reversibly relieved upon activation by calcium and calmodulin (as well as other activators such as phospholipids). Proteolytic activation, in the presence or absence of calmodulin, destroys the function of this "inhibitory domain" and leads to irreversible activation of the enzyme. The calmodulin binding domain appears to be closely linked to this region, although

TABLE II

Effects of KCl, Spermine, and Trifluoperazine on Control and Proteolyzed Phosphodiesterase[a]

Assay conditions		Phosphodiesterase activity (μmol:min^{-1}·mg protein^{-1})		
Additions	CaM	Control	CHY/Ca^{2+}	CHY/EGTA
None	0	0.46	6.4	3.8
	+	4.1	6.6	4.5
KCl, 55 μM	0	0.07	6.7	4.2
	+	—	6.8	3.8
Spermine, 10 μM	0	0.15	5.3	3.7
	+	4.2	5.2	4.0
Trifluoperazine, 50 μM	0	0.96	5.4	2.7
	+	0.94	5.0	2.6
Trifluoperazine plus KCl	0	0.53	4.1	3.3
	+	0.56	5.1	2.8
Trifluoperazine plus spermine	0	0.44	5.0	3.2
	+	0.44	4.3	3.1

[a]Bovine phosphodiesterase was incubated with α-chymotrypsin and calmodulin (6 μg) in the presence of 1 mM $CaCl_2$ (CHY/Ca^{2+} or 3 mM EGTA (CHY/EGTA); incubation was terminated with addition of trypsin inhibitor. Control phosphodiesterase was incubated with trypsin inhibitor. Samples were assayed (0.5 μM cGMP) with and without calmodulin (CaM) as indicated. From Kincaid et al. (1985).

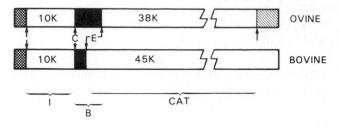

Fig. 10. Proposed arrangement of functional domains on calmodulin-dependent phosphodiesterase. Arrows indicate sites of cleavage in the presence of calmodulin and Ca^{2+} (C) or in the presence of EGTA (E). Brackets indicate hypothetical domains: I, an inhibitory domain, damage to which results in activation; B, a segment required for binding to calmodulin; CAT, a region containing the catalytic site. From Kincaid et al. (1985).

loss of calmodulin binding capacity does not necessarily accompany activation. Thus, proteolytic cleavage may directly damage such an "inhibitory domain" or may indirectly alter the conformation such that inhibitory constraint is lost.[2] A hypothetical arrangement of such functional domains is shown in Fig. 10. The fact that both calmodulin and proteolysis produce their activation through an increase in V_{max} argues for a catalytic region that responds primarily by increasing the efficiency with which substrate is hydrolyzed rather than by dramatic increases in the affinity for substrate.

III. USE OF CALMODULIN DERIVATIVES TO PROBE THE MECHANISM OF CALMODULIN ACTIVATION

Studies of the mechanism of phosphodiesterase activation by calmodulin have been largely of a mathematical and kinetic nature (Huang et al., 1981; Cox et al., 1981). It was shown that the Ca^{2+} requirement for enzyme stimulation decreased with increasing concentrations of calmodulin in the assay (Wolff and Broström, 1979) and proposed that interaction of the two proteins increased the affinity of calmodulin for Ca^{2+}, as has been shown for its interaction with troponin I (Keller et al., 1982). The number of bound Ca^{2+} ions required for phosphodiesterase activation is estimated to be three or four (Huang et al., 1981; Cox et al., 1981). It has been widely assumed that interaction of calmodulin with phosphodiesterase is synonomous with its activation, although interaction and activation could, of course, be sequential events with different requirements for Ca^{2+} binding to calmodulin. To investigate this possibility, we undertook to

[2] Since monomeric, nonproteolyzed forms can interact with calmodulin yet exhibit high basal activity, it might also be suggested that the inhibitory constraint existing in the native dimer is not present in the conformation of such intrinsically active monomers.

compare directly the Ca^{2+} requirements for physical interaction and for enzyme activation. We prepared two chemically modified derivatives of calmodulin which facilitated this comparison, as described below.

A. 5-Dimethylaminonaphthalene Sulfonyl (Dansyl) Calmodulin

Dansyl calmodulin, a fluorescent derivative, was first prepared by Tkachuk and his colleagues and used to study the effects of pH on Ca^{2+} interaction with calmodulin (Tkachuk et al., 1979). These studies were later extended to an examination of the Ca^{2+}-dependent interaction with several calmodulin-binding proteins including phosphodiesterase (Kincaid et al., 1982). Independently, the use of this derivative has been reported for studies of protein and peptide interactions with calmodulin (Malencik and Anderson, 1982). Although the physical properties of this derivative have not been extensively described, we have shown that >85% of the fluorescence is found in a single cyanogen bromide fragment and that the modified amino acid is lysine. Under some reaction conditions (e.g., in the presence of metal chelator), a substantial proportion of dansyl is incorporated into tyrosine with changes in the apparent Ca^{2+}-dependent fluorescent properties (R. Kincaid, unpublished data). Hence, characterization of the derivative is necessary for interpreting experimental data from heterogeneously-labeled populations, as has been emphasized in studies with another fluorescent derivative of calmodulin (Olwin et al., 1983).

In a fashion strikingly reminiscent of dansyl troponin C (van Eerd and Kawasaki, 1973), dansyl calmodulin exhibited a dramatic increase in fluorescence intensity with a shift in the maximum wavelength (λ_{max}) from ~525 to 495 nm, upon addition of saturating Ca^{2+} (Fig. 11). The effect of Ca^{2+} on fluorescence appeared to be specific since Na^+, K^+, and Mg^{2+} were ineffective and Mn^{2+} elicited a smaller maximal response (R. Kincaid, unpublished data). This Ca^{2+}-dependent "blue shift" in λ_{max} and increase in fluorescence intensity were further augmented by the inclusion of any of several calmodulin-binding proteins (Kincaid et al., 1982; Malencik and Anderson, 1982). For phosphodiesterase, the fluorescence was ~30–50% greater than that produced by Ca^{2+} alone. No change in fluorescence was observed without added Ca^{2+}, and the addition of excess EGTA fully reversed the Ca^{2+}-dependent increase in fluorescence. These data indicated that specific, reversible changes in the structure of dansyl calmodulin were produced by Ca^{2+} binding and interaction with phosphodiesterase. The increase in fluorescence intensity can probably be interpreted as the result of increased hydrophobicity in the environment of the dansyl group. The additional changes in the presence of phosphodiesterase may reflect further conformational events in calmodulin that shield the fluorescent group from aqueous quenching, direct interaction of the dansyl group with a hydrophobic domain on the enzyme, or both.

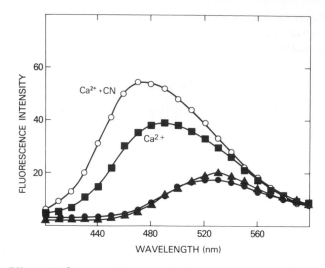

Fig. 11. Effects of Ca^{2+} and calcineurin on the fluorescence spectrum of dansyl calmodulin. The emission spectrum of 0.39 μM dansyl calmodulin was measured in the standard buffer containing 100 μM EGTA (●) and after successive additions of 215 μM $CaCl_2$ (~115 μM free Ca^{2+}) (■) and 215 μM EGTA (~0.4 μM free Ca^{2+} (▲). Spectrum of 0.39 μM dansyl calmodulin in the presence of 215 μM $CaCl_2$ (~115 μM free Ca^{2+}) and 0.44 μM calcineurin (○). Excitation wavelength was 335 nm. Adapted from Kincaid *et al.* (1982).

The effect of phosphodiesterase on fluorescence of dansyl calmodulin was maximal at a ratio of 1 mol of dansyl calmodulin per 60-kDa subunit, and it was Ca^{2+}-dependent (Kincaid *et al.*, 1982). However, since Ca^{2+}-independent interaction might take place without discernible change in fluorescence intensity, the polarization of fluorescence was monitored. This property, which reflects changes in the hydrodynamic properties of the fluorescent species (Weber, 1952), i.e., complex formation, was not changed by phosphodiesterase in the absence of Ca^{2+}, and in its presence, it was maximal with stoichiometric amounts of phosphodiesterase (Fig. 12). Since no Ca^{2+}-*independent* association was detected with 1–2 μM each of the two proteins, a lower limit for the dissociation constant of such a complex was estimated to be ~10^{-5} M, which is approximately four orders of magnitude above the nanomolar constant observed in the presence of Ca^{2+}. Also, the enzyme that had been proteolytically activated (in the absence of calmodulin), there was no evidence of association in the presence of calcium. Thus, the production of a high-affinity complex depends both on the availability of Ca^{2+} and on preservation of critical structural elements of phosphodiesterase required for calmodulin binding.

It has been proposed, as mentioned above, that the interaction of calmodulin and phosphodiesterase should greatly increase the affinity of calmodulin for Ca^{2+}; this increasing affinity may function to stabilize the complex when Ca^{2+}

concentration is decreasing (Haeich *et al.*, 1981). We sought to determine whether the Ca^{2+}-dependence of fluorescence changes in dansyl calmodulin would be altered by phosphodiesterase. No substantial difference was observed in the presence of phosphodiesterase; the half-maximally effective concentration of Ca^{2+} was ~3–5 μM both for changes in the fluorescence intensity of dansyl calmodulin alone (conformational event) and in polarization of fluorescence in the presence of phosphodiesterase (complex formation) (Fig. 13). These data would seem to suggest a direct correspondence between the initial alteration in calmodulin resulting from Ca^{2+}-binding and its ability to associate with enzyme. Since separate experiments indicated that the maximal increase in fluorescence of dansyl calmodulin occurred with the filling of two high-affinity Ca^{2+} sites, it appears that occupancy of these two sites is necessary and sufficient for complex formation and that phosphodiesterase does not produce a discernible increase in the affinity at these two sites. It is possible that in the presence of phosphodiesterase the affinity for Ca^{2+} is increased at a site that does not greatly influence changes in fluorescence. Since a small decrease in the fluorescence of dansyl calmodulin alone is observed with occupation of a third (and/or fourth) Ca^{2+} site (Kincaid *et al.*, 1984b), it is noteworthy that in the presence of phosphodiesterase this decrease still is observed. Regardless, it would appear that complex formation does not occur at free Ca^{2+} concentrations much lower

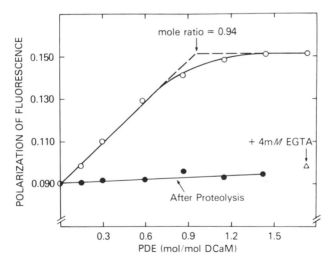

Fig. 12. Interaction of dansyl calmodulin with phosphodiesterase before and after proteolysis with α-chymotrypsin. Portions of control (○) and protease-treated (●) phosphodiesterase were added to 0.6 μM dansyl calmodulin plus 0.3 mM CaCl$_2$, and polarization of fluorescence was measured. Sample containing ~1.8 mol control phosphodiesterase (PDE) per mole dansyl calmodulin (D CaM) was adjusted to 4 mM EGTA (△). From Kincaid *et al.* (1985).

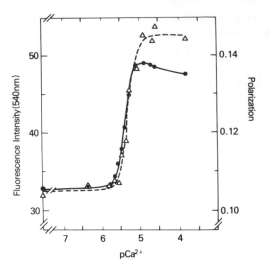

Fig. 13. Ca^{2+} dependence of changes in fluorescence intensity and polarization of fluorescence of dansyl calmodulin in the presence of PDE. Fluorescence intensity at 540 nm (●) and polarization of fluorescence (△) of 1.3 μM dansyl calmodulin were measured in the presence of 2.0 μM PDE. The calcium dependence of changes in fluorescence intensity of dansyl calmodulin alone was essentially identical to that shown by the closed circles. Values for free Ca^{2+} concentration have been recalculated using a Ca^{2+}-EGTA stability constant of 10^6 liters/M. Adapted from Kincaid *et al.* (1984b).

than those required for conformational change(s) in dansyl calmodulin itself; i.e., complexes would not be maintained when free Ca^{2+} is reduced substantially below that of the binding constants for the higher-affinity sites.

Phosphodiesterase activation, assayed under conditions essentially identical to those used in fluorescence studies, required somewhat higher concentrations of Ca^{2+} than were necessary for complex formation, suggesting that interaction could occur without activation (Kincaid *et al.*, 1984b). The concentration of phosphodiesterase used in the activation studies was, however, three orders of magnitude lower than that in the fluorescence experiments; hence an exact comparison is not possible. Recently, we examined the Ca^{2+} dependence of activation at the same phosphodiesterase concentration (0.5–1 μM) using a poorly hydrolyzed alternative substrate, 1-N^6-etheno cAMP. When assayed at these higher enzyme concentrations, the concentration of Ca^{2+} required for half-maximal activation ($\sim$15–20 μM) was three to four times that required for interaction; this finding supports the notion that these two phenomena can be distinguished in a functional sense. The finding is also consistent with a sequential mechanism, wherein filling of two Ca^{2+} sites permits formation of a calmodulin–enzyme complex which is inactive until the third and perhaps fourth

Ca^{2+} sites are occupied. Such a mechanism would imply the coexistence of active and inactive enzyme complexes at submaximal Ca^{2+} concentrations, with diffusion of Ca^{2+} the rate-limiting event in stimulation or attenuation of activity. While this model does not take into account other pertinent considerations (e.g., compartmentalization of enzyme and/or calmodulin), it appears to be an especially efficient means for regulation of activity in response to rapidly oscillating fluxes of Ca^{2+}.

B. 2-Pyridyldithiopropionyl (PDP) Calmodulin

PDP calmodulin, an activated sulfhydryl derivative, was first synthesized in this laboratory and used for preparation of novel disulfide-linked affinity matrices (Kincaid and Vaughan, 1983). Of course, this derivative potentially could react, as well, with free sulfhydryl groups on calmodulin-binding proteins to produce disulfide-linked complexes (Fig. 14). While comparing native and PDP calmodulin, a striking difference was noted; if incubated with 10–20 nM phosphodiesterase prior to a 50-fold dilution for enzyme assay, the sulfhydryl congener induced activation at much lower concentrations than did unmodified calmodulin ($K_a = 1$–2 nM). Indeed, the amount of PDP calmodulin required was

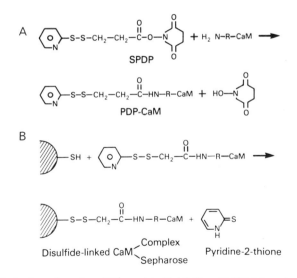

Fig. 14. Mechanism of reaction of N'-succinimidyl 3-(2-pyridyldithio)propionate (SPDP) with calmodulin and its use for preparation of disulfide-linked structures. (A) Nucleophilic attack of amino groups on calmodulin (CaM) on SPDP with production of pyridyldithiopropionyl (PDP) CaM and release of N-hydroxysuccinimide. (B) Interaction of PDP CaM with free thiol groups (e.g., thiol–Sepharose or cysteine groups on protein) to form disulfide-linked structures.

virtually stoichiometric with the enzyme (0.3 nM) (Kincaid, 1984). Formation of these active, presumably cross-linked complexes required the presence of Ca^{2+}. The activity of the complexes, however, appeared to be Ca^{2+} independent. That is, the enzyme was virtually fully activated after dilution with an excess of EGTA. To eliminate the possibility of residual Ca^{2+} being present under assay conditons, the cross-linked enzyme–calmodulin complex was subjected to gel-filtration chromatography in buffers containing metal chelators (Fig. 15). The apparent molecular weight of the complex was significantly greater than that of the native phosphodiesterase, consistent with formation of a tetrameric $\alpha_2\beta_2$

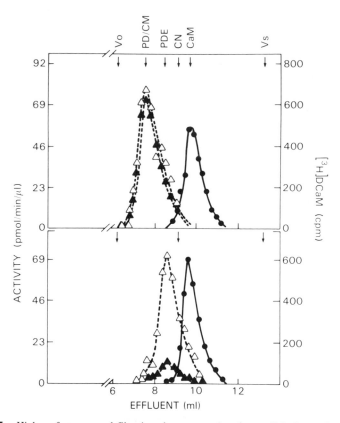

Fig. 15. High-performance gel filtration chromatography of cross-linked complex and control enzyme. Phosphodiesterase/PDP–calmodulin complex (upper panel) or phosphodiesterase which was incubated with reduced PDP calmodulin (lower panel) were chromatographed on a TSK-SW 3000 column in buffer containing 0.2 mM EDTA and 0.1 mM EGTA. Fractions were assayed for basal (▲) or total (△) enzyme activity. ³H-Dansyl calmodulin (●) was included as an internal standard. Further details given in Kincaid, 1984. Reprinted with permission from Kincaid, R. L., *Biochemistry* **23**, 1143–1147. Copyright (1984) American Chemical Society.

complex, first proposed by LaPorte *et al.* (1979), and the isolated fractions were still fully activated without added Ca^{2+}. Since activity eluted as a symmetrical peak and no Ca^{2+}-dependent activity was detected in any fractions, it was concluded that virtually quantitative cross-linking had occurred without production of heterogeneous species. When the PDP calmodulin was reduced with dithiothreitol prior to incubation with phosphodiesterase, there was no apparent complex formation and the activity of enzyme fractions was stimulated seven- to eightfold by Ca^{2+} plus calmodulin. Thus, formation of the phosphodiesterase/PDP–calmodulin complex required both Ca^{2+} and the activated (pyridyldithio)sulfhydryl moiety, as expected for a disulfide-linked complex.

The properties of the isolated complex were consistent with those of a covalently activated calmodulin–enzyme complex. The activity was not inhibited by several calmodulin antagonists such as trifluoperazine, and its behavior in the presence of EGTA was consistent with Michaelis–Menten kinetics, in contrast to the anomalous behavior observed in assays of basal activity (Kincaid, 1984a). In addition, the fact that Ca^{2+}-dependent activity was restored after incubation of the isolated complex with thiols indicated that maintenance of the activated state required preservation of intermolecular disulfide bonds.

Taken together, these results may provide insights into the mechanism of enzyme activation. After Ca^{2+}-dependent complex formation, there is a disulfide exchange reaction between PDP calmodulin and a cysteine in the calmodulin binding domain of phosphodiesterase. The "activated" conformation of the complex is stabilized by the disulfide bonds and results in high enzyme activity even after removal of Ca^{2+}. This finding suggests that, if the activated conformation of the enzyme (or calmodulin) can be maintained in some other way, i.e., by covalent linkage to calmodulin, Ca^{2+} is no longer required. Alternatively, it may be that Ca^{2+} has been sequestered in the complex in an unusual microenvironment, such that it is not accessible to chelation by EGTA, or that the cross-linked complex has an extraordinary affinity for Ca^{2+}. If the latter is true, it is estimated that this affinity would be nanomolar, an increase of approximately three orders of magnitude. Although these possibilities appear unlikely, direct studies using radiolabeled Ca^{2+} are needed to provide an unambiguous interpretation.

In addition to providing a probe for the mechanism of enzymatic activation, the pyridyldithio derivative may prove useful for studying the physical domains of interaction between calmodulin and phosphodiesterase (or other binding proteins) (Fig. 16). In such an approach, the disulfide-linked proteins would first be isolated as a complex and then selectively cleaved (e.g., with CNBr or proteases). Those unique disulfide-containing species could be purified, further cleaved if necessary, and treated with thiol to release the complementary binding domains. Only those regions in close proximity would be disulfide linked; hence this approach may enable us to identify the functional interaction sites.

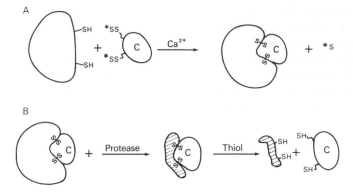

Fig. 16. A scheme for cross-linking of calmodulin-binding proteins to PDP–calmodulin and treatment with proteases and thiol to obtain interaction domains. In scheme A, PDP calmodulin (C), containing activated pyridyldithio groups (indicated as *SS), is cross-linked to binding protein (*S = pyridine-2-thione moiety). In scheme B, the complexes are treated with proteases and then reduced with thiol to give sulfhydryl-containing fragments.

IV. FUTURE DIRECTIONS

An understanding of the role of calmodulin and its activation of phosphodiesterase in normal and pathological cellular function is, of course, the ultimate goal of biochemical studies. Of necessity, the initial stages of such work involve the removal (and, perhaps, perturbation) of the proteins so that their purification and characterization can be carried out. However, some of the important questions regarding function and regulation can only be addressed in intact tissue. Relatively little is known about the regional distribution of the brain phosphodiesterase or of factors that influence its expression. In one early report, the amount of total phosphodiesterase activity was compared in dissected regions of adult rat brain, and higher amounts were found in cerebral cortex and hippocampus than in cerebellum (Weiss and Costa, 1968). Later ontogenetic studies showed dramatic increases in phosphodiesterase activity in fetal and newborn rat brain (Smoake *et al.*, 1974), which suggested some developmental regulation of this enzyme. While it has been thought that this enzyme may be primarily neuronal because of its high concentration in central nervous tissue, little information has appeared in the literature to directly confirm this assumption. Recent work in our laboratory using immunocytochemistry has, however, demonstrated a highly specific neuronal localization of this enzyme in rat brain. Although glial cells and many types of neurons do not apparently contain this enzyme, certain neurons (e.g., large pyramidal cells) show intensely stained dendritic projections, indicating, perhaps, that it has a specific functional role in particular areas of the brain. This suggests that the enzyme is not ubiquitous and is confined to or concentrated in certain types of differentiated cells. This conclusion is consistent with our findings that in several cultured cell lines (many of neuronal origin) the specific activity of this enzyme is only 1–3% of that in whole-brain tissues. Indeed, without the proper environmental signals this enzyme may not be expressed; perhaps only cells exhibiting a specific neuronal phenotype contain substantial levels of this enzyme. These possibilities can now be addressed, inasmuch as antibodies to the enzyme are available (Hansen and Beavo, 1982; Sharma *et al.*, 1984; R. L. Kincaid, unpublished data), and we may pose intriguing questions as to its physiologic role. Does the distribution of phosphodiesterase in specific neurons and/or regions thereof provide clues to its regulatory function *in vivo*? How does its localization/compartmentalization affect our hypotheses regarding the activation mechanism? This question seems especially pertinent since the enzyme appears to exist in much higher concentrations (10^{-6} M) in the cytoplasm of certain neurons than one might expect from its total content in whole-brain tissues. Are there clear developmental or environmentally related changes in its localization or expression? How does localization of this enzyme compare with that of other calmodulin-dependent or cyclic nucleotide-metabolizing enzymes? Our preliminary findings indicate that immunological approaches may, indeed, be fruitful in providing insights into regulation of the enzyme. Phosphodiesterase is found only in a particular subset of neurons in brain tissue, and in several of these (e.g., cerebellar Purkinje cells) both cyclic nucleotides and Ca^{2+} are thought to play a crucial role in their electrical activity. Furthermore, it appears that certain enzymes involved in cyclic nucleotide action (e.g., kinases) may share its pattern of distribution. It is hoped that careful examination of the functions subserved by the neuronal elements that contain phosphodiesterase may suggest its involvement in certain types of neuronal activity (e.g., integration of sensory inputs versus output of information). Through such an examination, we may also find clues to the related roles of Ca^{2+} and cyclic nucleotides in the regulation of nerve cell activity.

REFERENCES

Berridge, M. J. (1975). The interaction of cyclic nucleotides and calcium in the control of cellular activity. *Adv. Cyclic Nucleotide Res.* **6**, 1–98.
Brostrom, C. O., Huang, Y. C., Breckenridge, B. McL., and Wolff, D. J. (1975). Identification of a calcium-binding protein as a calcium-dependent regulator of brain adenylate cyclase. *Proc. Natl. Acad. Sci. U.S.A.* **72**, 64–68.
Cheung, W. Y. (1970). Cyclic 3',5'-nucleotide phosphodiesterase: Demonstration of an activator. *Biochem. Biophys. Res. Commun.* **38**, 533–538.
Cheung, W. Y. (1971). Cyclic 3'-5'-nucleotide phosphodiesterase: Evidence for and properties of a protein activator. *J. Biol. Chem.* **246**, 2859–2869.

Cheung, W. Y. (1980). Calmodulin plays a pivotal role in cellular regulation. *Science* **207**, 19–26.

Cohen, P. (1982). The role of protein phosphorylation in neural and hormonal control of cellular activity. *Nature (London)* **296**, 613–620.

Conti, M. A., and Adelstein, R. S. (1981). The relationship between calmodulin binding and phosphorylation of smooth muscle myosin kinase by the catalytic subunit of 3':5' cAMP-dependent protein kinase. *J. Biol. Chem.* **256**, 3178–3181.

Cox, J. A., Malnoe, A., and Stein, E. A. (1981). Regulation of brain cyclic nucleotide phosphodiesterase by calmodulin; a quantitative analysis. *J. Biol. Chem.* **256**, 3218–3222.

Davis, C. W., and Daly, J. W. (1978). Calcium-dependent 3':5'-cyclic nucleotide phosphodiesterase: Inhibition of basal activity at physiological levels of potassium ion. *J. Biol. Chem.* **253**, 8683–8686.

Haeich, J., Klee, C. B., and Demaille, J. G. (1981). Effects of cations on affinity of calmodulin for calcium: Ordered binding of calcium ions allows the specific activation of calmodulin-stimulated enzymes. *Biochemistry* **20**, 3890–3897.

Hansen, R. S., and Beavo, J. A. (1982). Purification of two calcium/calmodulin dependent forms of cyclic nucleotide phosphodiesterase by using conformation-specific monoclonal antibody chromatography. *Proc. Natl. Acad. Sci. U.S.A.* **79**, 2788–2792.

Huang, C. H., Chau, V., Chock, P. B., Wang, J. H., and Sharma, R. K. (1981). Mechanism of activation of cyclic nucleotide phosphodiesterase: Requirement for the binding of four Ca^{2+} to calmodulin for activation. *Proc. Natl. Acad. Sci. U.S.A.* **78**, 871–879.

Kakiuchi, S., and Yamazaki, R. (1970). Calcium dependent phosphodiesterase activity and its activating factor from brain. *Biochem. Biophys. Res. Commun.* **41**, 1104–1110.

Keller, C. H., Olwin, B. B., LaPorte, D. C., and Storm, D. R. (1982). Determination of the free-energy coupling for binding of calcium ions and troponin I to calmodulin. *Biochemistry* **21**, 156–162.

Kempner, E. S., and Miller, J. H. (1983). Radiation inactivation of glutamine dehydrogenase hexamer: Lack of energy transfer between subunits. *Science* **222**, 586–589.

Kempner, E. S., and Schlegel, W. (1979). Size determination of enzymes by radiation inactivation. *Anal. Biochem.* **92**, 2–10.

Kempner, E. S., Miller, J. H., Schlegel, W., and Hearon, J. Z. (1980). The functional unit of polyenzymes; determination by radiation inactivation. *J. Biol. Chem.* **255**, 6826–6831.

Kincaid, R. L. (1984). Preparation of an enzymatically active cross-linked complex between brain cyclic nucleotide phosphodiesterase and 3-(2-pyridyldithio) propionyl-substituted calmodulin. *Biochemistry* **23**, 1143–1147.

Kincaid, R. L., and Vaughan, M. (1979). Sequential adsorption-electrophoresis; combined procedure for purification of calcium-dependent cyclic nucleotide phosphodiesterase. *Proc. Natl. Acad. Sci. U.S.A.* **76**, 4903–4907.

Kincaid, R. L., and Vaughan, M. (1983). Affinity chromatography of brain cyclic nucleotide phosphodiesterase using 3-(2-pyridyldithio) propionyl-substituted calmodulin immobilized on Sepharose. *Biochemistry* **22**, 826–830.

Kincaid, R. L., Manganiello, V. C., and Vaughan, M. (1979). Effects of spermine on the activity and stability of calcium-dependent guanosine 3':5'-monophosphate phosphodiesterase. *J. Biol. Chem.* **254**, 4970–4973.

Kincaid, R. L., Manganiello, V. C., and Vaughan, M. (1981a). Calmodulin-activated cyclic nucleotide phosphodiesterase from brain: Changes in molecular size assessed by gel filtration and electrophoresis. *J. Biol. Chem.* **256**, 11345–11350.

Kincaid, R. L., Kempner, E. S., Manganiello, V. C., Osborne, J. C., Jr., and Vaughan, M. (1981b). Calmodulin-activated cyclic nucleotide phosphodiesterase from brain: relationship of subunit structure to activity assessed by radiation inactivation. *J. Biol. Chem.* **256**, 11351–11355.

Kincaid, R. L., Vaughan, M., Osborne, J. C., Jr., and Tkachuk, V. A. (1982). Ca^{2+}-dependent interaction of 5-dimethylaminonapthalene-1-sulfonyl calmodulin with cyclic nucleotide phosphodiesterase, calcineurin, and troponin I. *J. Biol. Chem.* **257**, 10638–10643.

Kincaid, R. L., Manganiello, V. C., Odya, C. W., Osborne, J. C., Jr., Stith-Coleman, I. E., Danello, M. A., and Vaughan, M. (1984a). Purification and properties of calmodulin-stimulated phosphodiesterase from mammalian brain. *J. Biol. Chem.* **259**, 5158–5166.

Kincaid, R. L., Danello, M. A. T., Osborne, J. C., Jr., Tkachuk, V. A., and Vaughan, M. (1984b). Calcium-dependent interaction of dansyl-calmodulin and phosphodiesterase: Relationship to Ca^{2+} requirement for enzyme activation. *Adv. Cyclic Nucleotide Protein Phosphorylation Res.* **16**, 77–87.

Kincaid, R. L., Stith-Coleman, I. E., and Vaughan, M. (1985). Proteolytic activation of calmodulin-dependent cyclic nucleotide phosphodiesterase. *J. Biol. Chem.* **260**, 9009–9015.

Klee, C. B., and Haeich, J. (1980). Concerted role of calmodulin and calcineurin in calcium regulation. *Ann. N.Y. Acad. Sci.* **356**, 43–53.

Klee, C. B., and Vanaman, T. C. (1982). Calmodulin. *Adv. Protein Chem.* **35**, 213–321.

Klee, C. B., Crouch, T. H., and Krinks, M. H. (1979a). Subunit structure and catalytic properties of bovine brain phosphodiesterase. *Biochemistry* **18**, 722–725.

Klee, C. B., Crouch, T. H., and Krinks, M. H. (1979b). Calcineurin: A calcium- and calmodulin-binding protein of the nervous system. *Proc. Natl. Acad. Sci. U.S.A.* **76**, 6270–6273.

Krinks, M. H., Haiech, J., Rhoads, A., and Klee, C. B. (1984). Reversible and irreversible activation of cyclic nucleotide phosphodiesterase: Separation of regulatory and catalytic domains by limited proteolysis. *Adv. Cyclic Nucleotide Protein Phosphorylation Res.* **16**, 31–47.

LaPorte, D. C., Toscano, W. A., Jr., and Storm, D. A. (1979). Cross-linking of Iodine-125-labelled, calcium-dependent regulatory protein to the Ca^{2+}-sensitive phosphodiesterase purified from bovine heart. *Biochemistry* **18**, 2820–2825.

Levin, R. M., and Weiss, B. (1977). Binding of trifluoperazine to the calcium dependent activator of cyclic nucleotide phosphodiesterase. *Mol. Pharmacol.* **13**, 690–697.

Malencik, D. A., and Anderson, S. R. (1982). Binding of simple peptides, hormones and neurotransmitters by calmodulin. *Biochemistry* **21**, 3480–3486.

Means, A. R., and Dedman, J. R. (1980). Calmodulin—an intracellular calcium receptor. *Nature (London)* **285**, 73–77.

Morrill, M. E., Thompson, S. T., and Stellwagen, E. (1979). Purification of a cyclic nucleotide phosphodiesterase from bovine brain using blue dextran-Sepharose chromatography. *J. Biol. Chem.* **254**, 4371–4374.

Olwin, B. B., Titani, K., Martins, T. J., and Storm, D. R. (1983). Immuno affinity purification and characterization of 5-[[[(iodoacetyl) amino] ethyl]amino]napthalene-1-sulfonic acid derivatized calmodulin. *Biochemistry* **22**, 5390–5395.

Rubin, R. P. (1970). The role of calcium in the release of neurotransmitter substances and hormones. *Pharmacol. Rev.* **22**, 389–428.

Rudolph, S. A., Johnson, E. M., and Greengard, P. (1971). The enthalpy of hydrolysis of various 3′,5′- and 2′,3′-cyclic nucleotides. *J. Biol. Chem.* **246**, 1271–1273.

Sandow, A. (1970). Skeletal muscle. *Annu. Rev. Physiol.* **32**, 87–138.

Sharma, R. K., Wang, T. H., Wirch, E., and Wang, J. H. (1980). Purification and properties of bovine brain calmodulin-dependent cyclic nucleotide phosphodiesterase. *J. Biol. Chem.* **255**, 5916–5923.

Sharma, R. K., Adachi, A.-M., Adachi, K., and Wang, J. H. (1984). Demonstration of bovine brain calmodulin-dependent cyclic nucleotide phosphodiesterase isozymes by monoclonal antibodies. *J. Biol. Chem.* **259**, 9248–9254.

Smoake, J. A., Song, S.-Y., and Cheung, W. Y. (1974). Cyclic 3′,5′-nucleotide phosphodiesterase.

Distribution and developmental changes of the enzyme and its protein activator in mammalian tissues and cells. *Biochim. Biophys. Acta* **341,** 402–411.

Stewart, A. A., Ingebritsen, T. S., Manalan, A. S., Klee, C. B., and Cohen, P. (1982). Discovery of a Ca^{2+} and calmodulin-dependent protein phosphatase: Probable identity with calcineurin (CaM-BP$_{80}$). *FEBS Lett.* **137,** 80–84.

Stith-Coleman, I. E., Danello, M. A., Kincaid, R. L., and Vaughan, M. (1982). Proteolysis of calmodulin-sensitive phosphodiesterase by α-chymotrypsin in the presence and absence of calmodulin. *Fed Proc., Fed. Am. Soc. Exp. Biol.* **41,** 760.

Sutherland, E. W., and Rall, T. R. (1958). Fractionization and characterization of a cyclic adenine ribonucleotide formed by tissue particles. *J. Biol. Chem.* **232,** 1077–1091.

Sutherland, E. W., and Rall, T. R. (1960). The relation of adenosine $3',5'$-phosphate and phosphorylase to the actions of catecholamines and other hormones. *Pharmacol. Rev.* **12,** 265–299.

Teo, T. S., and Wang, J. H. (1973). Mechanism of activation of a cyclic adenosine $3',5'$-monophosphate phosphodiesterase from bovine heart by calcium ions: Identification of the protein activator as a Ca^{2+}-binding protein. *J. Biol. Chem.* **248,** 5950–5955.

Tkachuk, V. A., Men'shikov, M. Yu., and Severin, S. E. (1979). Influence of pH on Ca^{2+}-binding properties of calmodulin and on its interaction with the Ca^{2+}-dependent form of cyclic nucleotide phosphodiesterase. *Proc. Acad. Sci. USSR* **246,** 240–244.

Tucker, M. M., Robinson, J. B., Jr., and Stellwagen, E. (1981). The effect of proteolysis on the calmodulin activation of cyclic nucleotide phosphodiesterase. *J. Biol. Chem.* **256,** 9051–9058.

Van Eerd, J.-P., and Kawasaki, Y. (1973). Effect of calcium (II) on the interaction between the subunits of troponin and tropomyosin. *Biochemistry* **12,** 4972–4980.

Weber, G. (1952). Polarization of the fluorescence of macromolecules. *Biochem. J.* **51,** 145–167.

Weiss, B., and Costa, E. (1968). Regional and subcellular distribution of adenyl cyclase and $3'5'$-cyclic nucleotide phosphodiesterase in brain and pineal gland. *Biochem. Pharmacol.* **17,** 2107–2116.

Wolff, D. J., and Brostrom, C. O. (1974). Calcium-binding phosphoprotein from pig brain: Identification as a calcium-dependent regulator of brain cyclic nucleotide phosphodieterase. *Arch. Biochem. Biophys.* **163,** 349–358.

Wolff, D. J., and Brostrom, C. O. (1979). Properties and functions of the calcium-dependent regulator protein. *Adv. Cyclic Nucleotide Res.* **11,** 27–88.

Chapter 3

Calmodulin-Dependent Protein Phosphatase

E. ANN TALLANT[1]
WAI YIU CHEUNG

Department of Biochemistry
St. Jude Children's Research Hospital
and University of Tennessee Center for the Health Sciences
Memphis, Tennessee

[1] Present address: Department of Pharmacology, University of Alabama at Birmingham, Birmingham, Alabama 35294.

CALCIUM AND CELL FUNCTION, VOL. VI

I. INTRODUCTION

Phosphorylation is an important cellular regulatory mechanism, controlling and coordinating various biological activities (Greengard, 1978; Krebs and Beavo, 1979; Cohen, 1982). Its relative importance in cellular metabolism is implicit in the number of processes which it mediates. Many of the enzymes involved in glycogen metabolism, glycolysis, gluconeogenesis, cholesterol and triglyceride metabolism, muscle contraction, neurotransmitter biosynthesis, and protein synthesis are regulated by phosphorylation of serine or threonine residues, or both (Krebs and Beavo, 1979; Lamy, 1982; Ingebritsen and Cohen, 1983b). A recent addition to this rather extensive list is phosphorylation of the β-adrenergic receptor and its involvement in desensitization of adenylate cyclase, suggesting that regulation of receptor-coupled adenylate cyclase systems is also mediated by phosphorylation (Sibley et al., 1984). Phosphorylation of tyrosine residues appears to be important in growth regulation, both in the induction of oncogenic transformation by retroviruses (Collett et al., 1980; Hunter and Sefton, 1980; Witte et al., 1980) and in the action of the receptors of epidermal growth factor (S. Cohen et al., 1981), insulin (Kasuga et al., 1982), and platelet-derived growth factor (Ek et al., 1982). More recently, the polyphosphorylation and hydrolysis of the phospholipid phosphatidylinositol have been implicated in the action of many hormones and neurotransmitters (Berridge, 1982, 1984). Thus, phosphorylation serves as a principal regulatory mechanism in the control of many cellular processes.

To function in a regulatory capacity, the relative concentrations of the phosphorylated and nonphosphorylated forms of the appropriate substrates must be properly maintained. The steady-state level of phosphorylation of a substrate is dependent upon a balance between the activities of kinases and those of phosphatases. Although more emphasis has been placed on the regulation of protein kinases, it is now evident that protein phosphatases are also subject to an intricate system of control.

The protein phosphatases involved in the regulation of glycogen metabolism, glycolysis, gluconeogenesis, fatty acid synthesis, cholesterol synthesis, protein synthesis, and muscle contraction have been most extensively studied. Cohen and his colleagues (Ingebritsen and Cohen, 1983a,b) have classified these phosphatases on the basis of their specificity for the α- or β-subunit of phosphorylase kinase and their susceptibility to inhibition by two heat-stable inhibitor proteins isolated from skeletal muscle (Huang and Glinsmann, 1975, 1976). They identified two types of enzymes that accounted for the majority of the protein phosphatase activity involved in various tissues. Type 1 protein phosphatases selectively dephosphorylate the β-subunit of phosphorylase kinase and are inhibited by nanomolar concentrations of inhibitor-1 and inhibitor-2. The catalytic subunit of this phosphatase ($M_r = 35,000$) has been extensively characterized by Lee and his co-workers (1976, 1980). An inactive form of this phosphatase, the Mg-ATP-dependent protein phosphatase, has also been isolated (Yang *et al.*, 1980; Vandenheede *et al.*, 1980; Resink *et al.*, 1983); it is composed of a 1 : 1 complex of the catalytic subunit of type 1 phosphatase and inhibitor-2, and can be reactivated by phosphorylation. Type 2 phosphatases selectively dephosphorylate the α-subunit of phosphorylase kinase and are insensitive to the inhibitor proteins. Three classes of type 2 enzymes have been characterized: type 2A enzymes, of which three species have been identified, all contain a catalytic subunit (M_r = 38,000) in complex with other subunits, probably regulatory in nature; type 2B phosphatase is Ca^{2+}/calmodulin-dependent; and type 2C is a Mg^{2+}-dependent enzyme with an apparent molecular weight of 43,000. Protein phosphatases which are distinct from type 1 and type 2 enzymes have also been identified. A Ca^{2+}-dependent phosphatase which dephosphorylates the mitochondrial enzyme pyruvate dehydrogenase has been isolated. This enzyme is different from type 2B phosphatase since it only binds a single molecule of Ca^{2+} (Reed and Pettit, 1981). In addition, the protein phosphatases involved in the dephosphorylation of phosphotyrosine appear to be distinct from the phosphoseryl and phosphothreonyl enzymes and probably comprise a unique class of enzymes (Foulkes *et al.*, 1983; Chernoff and Li, 1983; Nelson and Branton, 1984).

This chapter deals primarily with the general properties of the calmodulin-dependent phosphatase (a type 2B phosphatase) from bovine brain, comparing it wherever possible with the skeletal muscle enzyme. For a more detailed discussion of phosphatases in general, the reader is referred to three recent reviews (Lee *et al.*, 1980; Li, 1982; Ingebritsen and Cohen, 1983b).

II. HISTORICAL BACKGROUND

Calmodulin-dependent phosphatase was originally discovered as a major calmodulin-binding protein in bovine brain (Wang and Desai, 1976; Klee and Krinks, 1978). It was isolated as an "inhibitor" of calmodulin-dependent phosphodiesterase, the inhibition resulting from competition with phosphodiesterase for calmodulin (Wang and Desai, 1976; Klee and Krinks, 1978; Wallace *et al.*, 1978). Subsequently, the protein was purified to homogeneity and extensively characterized. Prior to identification of its intrinsic enzymic activity, it has been referred to as an inhibitor protein (Wang and Desai, 1976; Klee and Krinks, 1978; Wallace *et al.*, 1978), modulator-binding protein 1 (Sharma *et al.*, 1979), calmodulin-binding protein 1 (Sharma *et al.*, 1979), CaM-BP$_{80}$ (Wallace *et al.*, 1980; Wang *et al.*, 1980), and calcineurin (Klee *et al.*, 1979).

The role of phosphorylation and dephosphorylation in the control of glycogen metabolism in skeletal muscle has been extensively studied. Glycogenolysis is initiated by phosphorylase kinase, which catalyzes the phosphorylation of phosphorylase b and converts it to the activated phosphorylase a. Phosphorylase kinase is a tetramer, composed of $\alpha\beta\gamma\delta$, the δ-subunit being calmodulin; it is also activated by phosphorylation, on one serine residue of each of its α- and β-subunits, by a cyclic AMP-dependent protein kinase (Hayakawa *et al.*, 1973; Cohen, 1973; Cohen *et al.*, 1975, 1978; Nimmo and Cohen, 1978). Dephosphorylation of phosphorylase kinase in rabbit skeletal muscle is catalyzed by several different phosphatases (P. Cohen *et al.*, 1981; Cohen, 1982). One of them, which was specific for the α-subunit of phosphorylase kinase, was thought to require Mn^{2+} for activity; however, upon more extensive purification of the enzyme, the Mn^{2+} was replaced by micromolar Ca^{2+}, and the activity was further stimulated by calmodulin (Stewart *et al.*, 1982). In addition, the highly purified skeletal muscle phosphatase displayed a subunit structure strikingly similar to a calmodulin-binding protein from bovine brain (calcineurin), and a calcineurin preparation contained a protein phosphatase activity which was dependent upon Ca^{2+} and calmodulin. Stewart *et al.* (1982) therefore proposed that calcineurin was most likely a Ca^{2+}/calmodulin-dependent protein phosphatase. We had previously prepared a polyclonal antibody against calcineurin. On the basis of good correlation between calcineurin activity by radioimmunoassay and phosphatase activity in a sucrose gradient and on a nondenaturing polyacrylamide gel, we provided firm evidence that calcineurin is indeed a calmodulin-dependent protein phosphatase (Yang *et al.*, 1982). Further evidence substantiating the identification of calcineurin as a protein phosphatase, regulated by Ca^{2+}/calmodulin, was subsequently provided by Tonks and Cohen (1983), Manalan and Klee (1983), and Pallen and Wang (1983). Thus, a protein which was originally isolated on the basis of its ability to bind calmodulin was finally identified as a calmodulin-dependent protein phosphatase.

III. ASSAY

Calmodulin-dependent phosphatase was first isolated from bovine brain as an "inhibitor" of calmodulin-dependent phosphodiesterase and was thus assayed by means of its ability to suppress calmodulin-supported phosphodiesterase activity (Wang and Desai, 1976). The inhibition resulted from the Ca^{2+}-dependent binding of calmodulin to the phosphatase, making it unavailable for stimulating the phosphodiesterase (Klee and Krinks, 1978; Wallace *et al.*, 1978). Subsequently, the phosphatase was shown to be a specific "inhibitor" of other calmodulin-dependent enzymes, including brain adenylate cyclase (Wallace *et al.*, 1978), Ca^{2+},Mg^{2+}-ATPase (Larsen *et al.*, 1978; Lynch and Cheung, 1979), and phosphorylase kinase (Cohen *et al.*, 1979). As an alternative assay procedure, we have raised antibodies against the protein and developed a radioimmunoassay (Wallace *et al.*, 1980; Tallant *et al.*, 1983). The immunoassay is specific for the phosphatase and is accurate over a wide range of phosphatase levels. The inclusion of detergents also allows measurement of the protein in particulate fractions (Tallant *et al.*, 1983). However, the phosphatases in tissues other than brain may be different isoenzymes which cross-react poorly with the brain antibodies; thus, the radioimmunoassay may not correctly determine the phosphatase level in these tissues. A further procedure for monitoring the phosphatase is its identification on SDS–polyacrylamide gels; the 19,000-Da subunit (subunit B) can be detected by its shift in electrophoretic mobility in the presence of EGTA or Ca^{2+} (see Section VI), and the 60,000-Da subunit (subunit A), by its ability to bind [^{125}I]calmodulin (see Section VI) using a gel overlay technique (Klee *et al.*, 1983a).

The phosphatase is most conveniently measured by its ability to dephosphorylate phosphoproteins and phosphocompounds. One means of assay is the radioisotopic determination of ^{32}P released from ^{32}P-labeled proteins. A variety of protein substrates have been used, including phosphorylase kinase, inhibitor-1 (Stewart *et al.*, 1982, 1983), histone, casein (Yang *et al.*, 1982; Tallant and Cheung, 1984b), myosin light chain (Ingebritsen and Cohen, 1983a; Klee *et al.*, 1983a), and myelin basic protein (Gupta *et al.*, 1984). The phosphatase can also be monitored by its ability to release the phosphate group from *p*-nitrophenyl phosphate; *p*-nitrophenol is chromogenic and can be detected spectrophotometrically (Pallen and Wang, 1983). This method has the advantage of allowing continuous monitoring of the progress of the reaction.

In assessing calmodulin-dependent phosphatase activity in tissue extracts, one must distinguish it from the calmodulin-independent phosphatase activities which may predominate in many tissues (Ingebritsen *et al.*, 1983). Calmodulin-dependent activity can be differentiated by its dependence on Ca^{2+}/calmodulin; however, this requires the prior removal of any other Ca^{2+}-dependent phosphatase, such as pyruvate dehydrogenase phosphatase (Reed and Pettit, 1981), and of calmodulin. Calmodulin-dependent activity can also be measured in tissue extracts on the basis of its inhibition by trifluoperazine (Ingebritsen and Cohen, 1983a; Tallant and Cheung, 1983). Trifluoperazine is an antipsychotic phenothiazine which antagonizes the biological activity of calmodulin (Levin and Weiss, 1977); however, trifluoperazine, as well as other calmodulin antagonists, is known to affect calmodulin-independent activities, apparently because these compounds are hydrophobic (Roufogalis, 1982). Thus, some measure of reservation should be exercised when these compounds are used as calmodulin antagonists.

Alternatively, calmodulin-dependent phosphatase can be distinguished from type 1 enzymes by its resistance to inhibition by the type 1 phosphatase inhibitors, inhibitor-1 and inhibitor-2 (P. Cohen *et al.*, 1981; Ingebritsen and Cohen, 1983a). However, this trait does not differentiate calmodulin-dependent phosphatase from other type 2 enzymes.

A further distinguishing characteristic of calmodulin-dependent phosphatase is its relative specificity toward certain mammalian phosphoproteins (King et al., 1984). The enzyme in both liver and skeletal muscle accounted for approximately 60% of the total dephosphorylating activity toward inhibitor-1 and the α-subunit of phosphorylase kinase (Ingebritsen et al., 1983). The use of substrates which are specifically dephosphorylated by only this phosphatase may allow the development of a specific assay procedure.

Calmodulin-dependent phosphatase is susceptible to inactivation in tissue extracts (Stewart et al., 1982, 1983; Tallant and Cheung, 1984b; King and Huang, 1984; Pallen and Wang, 1984). Part of this inactivation is due to proteolysis, especially in the presence of Ca^{2+}. Proteolysis can be minimized by using the shortest possible assay times, by initiating the reaction with the enzyme rather than the substrate to avoid preincubation of the enzyme with Ca^{2+}, and by using various protease inhibitors, including EGTA (Stewart et al., 1982, 1983; Tallant and Cheung, 1984b). However, the phosphatase also undergoes a time-dependent deactivation in the presence of Ca^{2+} and calmodulin. This deactivation can be reversed or prevented by the inclusion of divalent metals such as Mn^{2+} or Ni^{2+} (see Section XI). In formulating an assay procedure for the phosphatase, especially in tissue extracts, one should therefore consider these various precautions.

IV. DISTRIBUTION

Calmodulin-dependent phosphatase activity has been found in extracts from bovine brain (Stewart et al., 1982; Yang et al., 1982), rabbit brain, skeletal muscle, heart muscle and adipose tissue, rat and rabbit liver (Ingebritsen et al., 1983), rabbit reticulocyte lysates (Foulkes et al., 1982), Xenopus oocytes (Foulkes and Maller, 1982), and human platelets (Tallant and Wallace, 1985). The highest level was detected in rat brain, which contains about 150 mg/kg (Tallant and Cheung, 1983). Erythrocyte membranes also contain a phosphatase activity which is dependent upon Ca^{2+} and is stimulated by calmodulin (Brissette et al., 1983). A calmodulin-binding protein with a molecular weight of 85,000 and a subunit composition similar to that of the brain and skeletal muscle phosphatases has been identified in bovine cardiac muscle (Wolf and Hofmann, 1980). Krinks et al. (1984) have also reported the isolation of calmodulin-dependent phosphatase from bovine cardiac muscle. The cardiac muscle enzyme was similar to the brain enzyme, although subunit B had a slightly smaller molecular weight.

The level of phosphatase in various bovine tissues has been determined by radioimmunoassay, with use of the antibody against the bovine brain protein (Wallace et al., 1980). The phosphatase was found in high levels in neural tissue, predominantly in the cerebrum, olfactory bulb, and cerebellum. Within the cerebrum, the level was especially high in the caudate nucleus and putamen (Table I). The levels of the enzyme in nonnervous tissues detected by radioimmunoassay were considerably lower than those found by enzyme assay (Ingebritsen *et al.*, 1983). In fact, Stewart *et al.* (1983) estimated that the concentration of the phosphatase in skeletal muscle was 12.5–25 mg/kg, which is comparable to the concentrations found in bovine brain by the immunoassay. One explanation for this discrepancy is that the antibody used in the radioimmunoassays was raised against the brain enzyme; the phosphatases in nonnervous tissues may be slightly different antigenically and may cross-react poorly with the brain antibody. Alternatively, the nonnervous tissues may contain high-

TABLE I

Level of Calmodulin-Dependent Phosphatase in Various Bovine Tissues[a]

Tissue	Phosphatase (mg protein per kg tissue)[b]
Adrenal	3.3 ± 0.2
Cerebellum	29.4 ± 12.0
Cerebrum	
Gray matter	36.0 ± 5.0
White matter	15.9 ± 2.5
Caudate nucleus	62.9 ± 10.9
Hippocampus	36.4 ± 7.3
Hypothalamus	31.3 ± 19.7
Putamen	84.8 ± 20.0
Thalamus	19.8 ± 7.0
Heart	0.7 ± 0.3
Kidney	
Cortex	2.3 ± 0.7
Medulla	3.4 ± 1.8
Liver	1.8 ± 0.5
Lung	3.2 ± 1.9
Medulla oblongata	2.5 ± 1.5
Olfactory bulb	17.2 ± 11.8
Pons	6.3 ± 3.9
Skeletal muscle	1.9 ± 0.6
Spleen	1.8 ± 0.6
Testis	3.1 ± 0.9
Tongue	2.8 ± 0.9
Thyroid	2.1 ± 1.5

[a]Adapted from Wallace *et al.* (1980).
[b]The level of the phosphatase was determined by radioimmunoassay in a 100,000 *g* supernatant fluid from each tissue.

er levels of proteases. The phosphatase is highly susceptible to proteolysis and may be degraded to various degrees in the nonnervous tissues, thus not being recognized by the antibody.

The subcellular distribution of the phosphatase has been determined in tissue extracts from brain, liver, and skeletal muscle. Using the radioimmunoassay, we found approximately 50% of the calmodulin-dependent phosphatase to be associated with a particulate fraction in both rat cerebrum (Tallant and Cheung, 1983) and bovine cerebrum (Tallant, 1983). Other workers have localized the protein in postsynaptic densities isolated from dog brain (Carlin *et al.*, 1981) and in synaptic plasma membranes and postsynaptic densities from chick retina (Cooper *et al.*, 1985). In contrast, in liver and skeletal muscle the enzyme was found exclusively in the cytosol (Ingebritsen *et al.*, 1983).

The subcellular localization of the phosphatase also has been determined immunocytochemically in various neural tissues by use of antisera to the brain enzyme. In the caudate–putamen of rodent brain, reaction product was found in association with neuronal elements and could not be detected in oligodendroglia or astrocytes (Wood *et al.*, 1980a,b). The label was primarily found in postsynaptic sites within neuronal somata and dendrites and, within the dendrites, in association with postsynaptic densities and dendritic microtubules (Fig. 1). The immunocytochemical localization pattern of calmodulin was also determined in the mouse' caudate-putamen and was found to be essentially identical to that of the phosphatase; hence, the two proteins probably share the same cellular compartment (Wood *et al.*, 1980a,b). In chick retina, the phosphatase was associated with photoreceptor synaptic terminals in the outer plexiform layer; in the inner plexiform layer, it was observed in dendrites of ganglion cells and in presynaptic terminals, some of which belonged to bipolar cells (Cooper *et al.*, 1982, 1985). In general, the label was primarily associated with presynaptic densities and vesicles in synaptic terminals (Fig. 2). In the dendrites, however, the label was found with postsynaptic densities and microtubules; these findings are reminiscent of those of Wood *et al.* (1980a,b) in rat caudate-putamen.

The ontogeny of the phosphatase was studied in developing nervous tissues with the radioimmunoassay (Tallant and Cheung, 1983) and immunocytochemistry (Cooper *et al.*, 1985). The levels of the protein increased significantly in rat cerebrum and cerebellum and in chick retina and brain during the periods corresponding to major synapse formation (Tallant and Cheung, 1983). Figure 3 shows the developmental profile of the phosphatase in chick retina; the phosphatase level increased markedly between day 10 and the time of hatching at day 21, the period corresponding to major synaptogenesis. As a comparison, the level of calmodulin was also measured in these samples; no appreciable change was noted. In a parallel study on developing chick retina, Cooper *et al.* (1985) showed a correlation between the appearance of synaptic vesicles and synaptic densities and the appearance of the phosphatase at these sites. Collectively, these

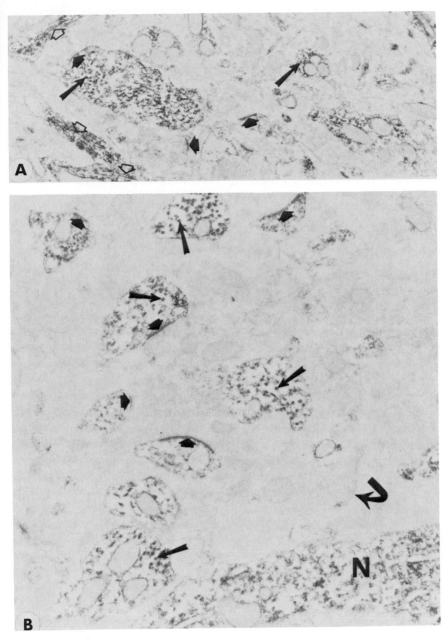

Fig. 1. Localization of calmodulin-dependent phosphatase in mouse caudate-putamen. Two examples (A and B) of fields showing antiphosphatase labeling of the PSD (closed short arrows) and microtubules cut in cross section (long straight arrows) or longitudinal section (short open arrows). Occasionally, an unlabeled PSD is observed (B, curved arrow) which serves as an internal control to compare with the appearance of labeled PSD. N, neuron which contains label deposited on various cellular organelles. (A) ×11,400; (B) ×22,800. From Wood *et al.* (1980a) and reproduced from the *Journal of Cell Biology*, 1980, vol. 84, pp. 66–76 by copyright permission of the Rockefeller University Press.

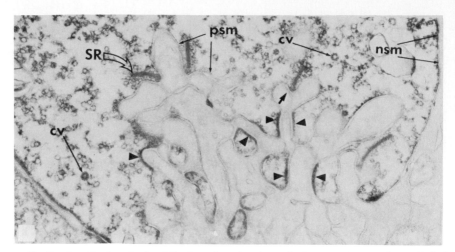

Fig. 2. Localization of calmodulin-dependent phosphatase in photoreceptor synaptic terminal of 2-week-old chick retina. Labeling is present at both ribbon and basal junctions. The synaptic ribbon (SR) and presynaptic density (arrowheads) of these two junctions are labeled, but postsynaptic densities (small arrow) are not. Most, but not all, synaptic vesicles together with coated vesicles (CV) are labeled. Compared to the nonsynaptic (nsm) regions of plasmalemma, little reaction product is detected along the presynaptic membrane (psm) at ribbon junctions. ×47,500. Experiment in collaboration with N. G. F. Cooper and B. J. McLaughlin; see also Cooper *et al.* (1985).

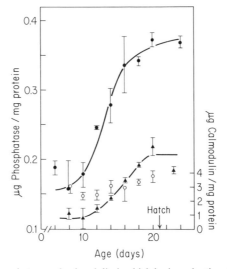

Fig. 3. Level of phosphatase and calmodulin in chick brain and retina during embryonic development. Phosphatase in chick brain (●) and retina (▲) as well as calmodulin in chick retina (○) were measured by radioimmunoassay in a detergent-solubilized extract of tissue from two animals. Reprinted with permission from Tallant and Cheung (1983). Calmodulin-dependent protein phosphatase: a developmental study. *Biochemistry* **22,** 3634. Copyright 1983 American Chemical Society.

observations suggest that the the phosphatase is developmentally linked to the formation of synaptic membranes and densities during the course of synaptogenesis.

V. PURIFICATION

Calmodulin-dependent phosphatase has been purified to apparent homogeneity from both bovine brain (Klee and Krinks, 1978; Wallace et al., 1979; Sharma et al., 1979) and rabbit skeletal muscle (Stewart et al., 1983). A common step in these procedures takes advantage of the Ca^{2+}-dependent, reversible association of the enzyme with calmodulin which has been immobilized on Sepharose 4B. Calmodulin-affinity chromatography was originally used to purify calmodulin-dependent phosphodiesterase from bovine brain (Watterson and Vanaman, 1976; Miyake et al., 1977). It was noted that the phosphodiesterase was only a minor component of the proteins that bound to the affinity column in a Ca^{2+}-dependent manner; the major Ca^{2+}-dependent calmodulin-binding protein in bovine brain extracts turned out to be the phosphatase. Calmodulin in the tissue extract must be removed prior to calmodulin-Sepharose chromatography in order to allow subsequent binding of the phosphatase to the immobilized calmodulin. In addition, the phosphatase must be separated from other calmodulin-binding proteins which copurify with the phosphatase on the calmodulin-affinity column.

We found that bovine brain phosphatase can be conveniently separated from both calmodulin and other calmodulin-dependent enzymes using an Affi-gel Blue column (Wallace et al., 1979). Calmodulin is eluted prior to the phosphatase. Calmodulin-dependent phosphodiesterase, which tends to copurify with the phosphatase by ion exchange chromatography, is tenaciously bound to the resin (Tallant et al., 1983). Other purification procedures utilize DEAE-cellulose chromatography prior to calmodulin-affinity chromatography (Klee et al., 1983a) or a combination of DEAE-cellulose and Affi-gel Blue chromatography (Sharma et al., 1983; King and Huang, 1984; Kincaid et al., 1984). Following use of the calmodulin-affinity column, a variety of chromatography steps have been employed to remove minor contaminants: DEAE-cellulose (Tallant et al., 1983), DEAE–Affi-gel Blue (King and Huang, 1984), gel filtration (Klee et al., 1983a; Sharma et al., 1983; Kincaid et al., 1984), and PBE-94, a chromatofocusing media (Klee et al., 1983a). The yield obtained by various procedures varies from 1.6 mg/kg of bovine brain (King and Huang, 1984) to 23 mg/kg of bovine brain (Sharma et al., 1983). That the enzyme is purified about 300 fold suggests that it constitutes 0.3% of the total soluble brain proteins (Wang et al., 1980; Tallant et al., 1983).

Bovine brain phosphatase in crude fractions is quite susceptible to attack by tissue proteases, resulting in an apparent higher basal activity and lower degree of calmodulin stimulation (see Section IX). Various inhibitors have been used to minimize the degree of proteolysis during purification, including phenylmethyl-sulfonyl fluoride, leupeptin, pepstatin, benzamidine, chymostatin, antipain, ap-rotinin, N-α-p-tosyl-L-lysine chloromethyl ketone, and the metal chelators EGTA and EDTA (Klee *et al.*, 1983a; Tallant *et al.*, 1983; King and Huang, 1984; Kincaid *et al.*, 1984). In addition, Klee *et al.* (1983a) have employed a hemoglobin-Sepharose column to remove contaminating proteases. Proteolytic degradation can also be minimized by shortening the time of the purification procedure.

Purification of the phosphatase from rabbit skeletal muscle (Stewart *et al.*, 1982, 1983) has many steps in common with the procedure for the brain enzyme. These include chromatography on DEAE-Sepharose, gel filtration on Sephadex G-200 and both Affi-gel Blue and calmodulin-Sepharose chromatography. Additional steps include an acid extraction at pH 6.1, to remove the protein–glycogen complex, and ammonium sulfate fractionation. The major calmodulin-binding proteins present in skeletal muscle are removed at various stages: myosin light chain kinase by chromatography on an Affi-gel Blue column and phosphorylase kinase and calmodulin-dependent glycogen synthase kinase by precipitation at pH 6.1 and ammonium sulfate fractionation. However, a much lower yield is obtained from skeletal muscle (0.15 mg/2.4 kg muscle) as compared to brain tissue.

The skeletal muscle phosphatase has also been purified by substrate-affinity chromatography on thiophosphorylated myosin light chain–Sepharose (Tonks and Cohen, 1983). This column had previously been used to purify other myosin light chain phosphatases (Pato and Adelstein, 1980), based upon their ability to interact with the thiophosphorylated substrate without dephosphorylating it at significant rates. Calmodulin-dependent phosphatase was retained by the thio-phosphorylated substrate column in the presence of Ca^{2+} and could be eluted with EGTA (Tonks and Cohen, 1983).

Calmodulin-dependent phosphatase from brain or skeletal muscle is stable for up to 6 months in the presence of 50–60% glycerol at $-20°$ (Tallant and Cheung, 1984b; Stewart *et al.*, 1983). Although activity is maintained upon storage at $-20°$ in the absence of glycerol, Sharma *et al.* (1983) noted a loss of activity upon repeated freezing and thawing.

VI. MOLECULAR PROPERTIES

Calmodulin-dependent phosphatase from bovine brain is a globular protein with a molecular weight of approximately 80,000 (Klee and Krinks, 1978; Wallace *et al.*, 1979; Sharma *et al.*, 1979; Klee *et al.*, 1983a). Various molecular properties of the phosphatase are summarized in Table II. Although small discre-

TABLE II

Physicochemical Properties of Bovine Brain Phosphatase

Property	Value	Reference[a]
Gel filtration (M_r)	80,000	1
Amino acid composition (M_r)	80,000	2
Sedimentation equilibrium (M_r)	84,000	3
Sedimentation and gel filtration (M_r)	86,000	3
Subunit A (M_r)		
SDS–gel electrophoresis	60,000	1
Amino acid composition	61,000	2
Subunit B (M_r)		
SDS–gel electrophoresis (Ca^{2+})	15,000	4
SDS–gel electrophoresis (EGTA)	16,500	5
Amino acid composition	19,000	2
Stokes radius (Å)	39–40.5	1,3
$s_{20,w}$ (S)	4.96	3
Isolectric point		
Holophosphatase	4.5–6.0	1,2
Subunit A	5.5	6
Subunit B	4.8	6
$E^{1\%}_{280\,nm}$	9.6–9.7	1,3,7

[a]References: (1) Wallace et al. (1979); (2) Klee et al. (1983a); (3) Sharma et al. (1979); (4) Klee et al. (1979); (5) Tallant et al. (1984b); (6) Klee and Haiech (1980); (7) Klee and Krinks (1978).

pancies in some of the reported values are evident, there is general agreement from many investigators. A possible exception is the reported value of the isoelectric point of the holoenzyme; a value of 6.0 was obtained by isoelectric focusing (Wallace et al., 1979) and a p*I* of 4.5 by chromatofocusing (Klee et al., 1983a). Both values indicate that the phosphatase is acidic, which is also evident from its high content of aspartic and glutamic acid residues, as shown in Table III (Klee et al., 1983a). The phosphatase displays a typical ultraviolet spectrum with maximum absorbance at 278 nm (Klee and Krinks, 1978; Wallace et al., 1979; Sharma et al., 1979), which is characteristic of a protein rich in tyrosine and poor in tryptophan (Klee et al., 1983a).

The phosphatase is a heterodimer, with subunit A of M_r 60,000 and subunit B of M_r 19,000 (Table II), present in a 1 : 1 ratio as determined by densitometric scanning of SDS–polyacrylamide gels (Wallace et al., 1979) and by cross-linking with dimethylsuberimidate (Klee and Krinks, 1978; Klee et al., 1979). The interaction of the two subunits is not dependent upon the presence of divalent cations (Klee et al., 1979). The subunits can be separated from one another by several techniques: gel filtration in the presence of 0.2% SDS (Shar-

TABLE III

Amino Acid Composition of Bovine Brain Calmodulin-Dependent Phosphatase[a]

Amino acid	Holophosphatase (M_r, 80,000)	Subunit B (M_r, 19,000)	Subunit A (M_r, 61,000)
Lys	48	15	33
His	17	2	15
Arg	34	6	28
Asp	92	28	64
Thr	35	3	32
Ser	49	11	38
Glu	77	22	55
Pro	32	3	29
Gly	53	14	39
Ala	47	5	42
Val	44	13	31
Met	21	6	15
Ile	32	11	21
Leu	65	14	51
Tyr	21	3	18
Phe	38	12	26
Cys	10	0	10
Trp	+	0	+
Tml	0	0	0

[a]Adapted from Klee *et al.* (1983a).

ma *et al.*, 1979) or 6 *M* urea (Sharma *et al.*, 1979; Winkler *et al.*, 1984); DEAE-cellulose chromatography in the presence of 6 *M* urea (Klee and Haiech, 1980); or antibody-affinity chromatography, using a monoclonal antibody to subunit A (Winkler *et al.*, 1984).

Phosphatase subunit A forms a Ca^{2+}-dependent complex with calmodulin, as has been demonstrated by cross-linking experiments using the bifunctional reagent dimethylsuberimidate and radiolabeled calmodulin (Richman and Klee, 1978; Klee *et al.*, 1979) and by the [^{125}I]calmodulin gel overlay technique (Carlin *et al.*, 1981; Klee *et al.*, 1983a). In addition, Sharma *et al.* (1979) showed that subunit A, separated from subunit B by gel filtration in the presence of 6 *M* urea, retains its ability to form a Ca^{2+}-dependent complex with calmodulin and thus to inhibit calmodulin-supported phosphodiesterase activity. The identification of subunit A as the catalytic subunit (see Section VII) and the ability of subunit A to be stimulated by calmodulin, independent of subunit B, also suggest that subunit A contains the calmodulin-binding domain (Winkler *et al.*, 1984).

Phosphatase subunit B is a Ca^{2+}-binding protein, capable of binding four calcium ions per molecule (Klee *et al.*, 1979). Although the exact binding constant has not been determined, a $K_d \leq 10^{-6} M$ was estimated in the presence of physiological concentrations of Mg^{2+} (Klee *et al.*, 1979). Subunit B undergoes shifts in its electrophoretic mobility in SDS–polyacrylamide gels, depending upon the presence of EGTA or Ca^{2+} (Klee *et al.*, 1979); its molecular weight corresponds to 15,000 in the presence of Ca^{2+} and 16,500 in the presence of EGTA (Table II). Several other Ca^{2+}-binding proteins, including calmodulin and troponin C, undergo similar changes in their electrophoretic mobility (Burgess *et al.*, 1980; Klee *et al.*, 1979; Wallace *et al.*, 1982). It is believed that the difference in the apparent molecular weight of these proteins is due to a Ca^{2+}-induced conformational change which makes the protein more compact. Upon binding Ca^{2+}, the phosphatase undergoes a shift in its UV absorption spectrum that indicates a change in the environment of both tyrosine and tryptophan residues (Klee *et al.*, 1979). In addition, the interaction of the enzyme with its substrate appears to be dependent upon Ca^{2+}. The phosphatase was retained by a substrate-affinity column (thiophosphorylated myosin light chain) in the presence of Ca^{2+}, but not in the presence of EGTA (Tonks and Cohen, 1983).

Aitken *et al.* (1984) have deduced the primary sequence of subunit B, which is composed of 168 amino acids and has a molecular weight of 19,200, as shown in Fig. 4. The sequence shows 35% identity with calmodulin and 29% with troponin C; the greatest homology is in the region of the four putative Ca^{2+}-binding loops. The overall sequence is compatible with that of a structure having four Ca^{2+}-binding "EF-hands", as described by Kretsinger (1980), suggesting that subunit B is a member of a homologous family of Ca^{2+}-binding proteins (Aitken *et al.*, 1984). As do other members of this class of proteins, subunit B contains a large number of acidic residues and no tryptophan or cysteine (Klee and Haiech, 1980; Aitken *et al.*, 1984). The amino acid composition of subunit A has also been determined (Klee and Haiech, 1980; Klee *et al.*, 1983a); it is rich in acidic amino acids and contains several cysteine residues (Table III).

The amino-terminus of phosphatase subunit B is blocked by acylation with myristic acid, a 14-carbon saturated fatty acid (Aitken *et al.*, 1982). The myristoyl moiety is linked to the N-terminal glycine through an amide bond. The presence of this fatty acid undoubtedly contributes to the hydrophobicity of subunit B, making it difficult to elute from an alkylphenyl column (Klee *et al.*, 1981). The catalytic subunit of bovine heart cAMP-dependent protein kinase and the p15 protein from Rauscher and Moloney's murine leukaemia viruses also contain an N-terminal myristic acid (Carr *et al.*, 1982; Henderson *et al.*, 1983). The N-terminus of the kinase and the phosphatase are identical (Myr-Gly-Asn) while the p15 protein begins with Myr-Gly-Gln (Aitken *et al.*, 1982; Carr *et al.*, 1982; Henderson *et al.*, 1983). Aitken *et al.* (1982) suggested that the myristic acid of the kinase and phosphatase may play a role in substrate recognition; alternatively, the fatty acid may aid in a reversible interaction with membranes or

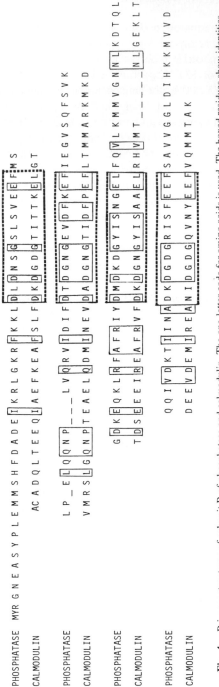

Fig. 4. Primary structures of subunit B of phosphatase and calmodulin. The one-letter code for amino acids is used. The boxed residues show identities, and the broken lines denote the regions of the four putative Ca^{2+}-binding domains. The sequence of calmodulin refers to the δ-subunit of phosphorylase kinase. Adapted from Aitken *et al.* (1984).

in maintaining subunit–subunit interactions. The p15 viral protein is an inner core structural protein; its myristyl group may function to anchor its precursor protein (Pr65gag) to the cell membrane or to stabilize the viral membrane, or both (Henderson *et al.*, 1983). The amino-terminus of phosphatase subunit A is also blocked; the identity of the blocking group has not been determined (Klee *et al.*, 1983a).

The brain phosphatase contains a hydrophobic region which binds tightly to a phenyl-Sepharose column (Gopalakrishna and Anderson, 1983) and a norchlorpromazine–Affi-Gel column (Klee *et al.*, 1983b). Both hydrophobic columns bind the enzyme in a Ca^{2+}-independent fashion, and denaturing conditions are required for its elution. Calmodulin also binds to phenyl-Sepharose (Gopalakrishna and Anderson, 1982) and norchlorpromazine–Affi-gel (Jamieson and Vanaman, 1979). However, the binding of calmodulin to both these resins is dependent upon Ca^{2+} and can be reversed with EGTA; these properties have been exploited for its purification. The binding of Ca^{2+} to calmodulin exposes a hydrophobic site (LaPorte *et al.*, 1980) that is believed to interact with a complementary hydrophobic site on a calmodulin-dependent enzyme. Perhaps the hydrophobic region on the phosphatase provides a complementary site for interaction with calmodulin.

Calmodulin-dependent phosphatase contains 0.2–0.4 mol of phosphate per mole of enzyme (King and Huang, 1984). However, the phosphatase was not phosphorylated by the catalytic subunit of bovine heart cAMP-dependent protein kinase (Tallant and Cheung, 1984b). It remains to be determined whether phosphorylation is involved in regulation of phosphatase activity.

The phosphatase is an excellent substrate *in vitro* for protein-carboxyl methylation (Gagnon, 1983). In fact, it is a better substrate than calmodulin (Gagnon *et al.*, 1981). Calmodulin contains an endogenous trimethyllysine (Watterson *et al.*, 1980); however, it is also rich in acidic amino acids, which can be carboxyl methylated. Amidation (Liu and Cheung, 1976) or carboxyl methylation (Gagnon *et al.*, 1981) of some of these carboxylic groups results in inactivation of calmodulin's biological activity. Calmodulin-dependent phosphodiesterase and Ca^{2+}/calmodulin–protein kinase are carboxyl methylated, and their activities appear to be affected by this modification (Billingsley *et al.*, 1984). It remains to be determined if carboxyl methylation of the phosphatase affects its activity.

A calmodulin-dependent phosphatase similar to the brain enzyme has been isolated from several different tissues (Table IV). The skeletal muscle enzyme has a molecular weight of 98,000 and is composed of three subunits: A ($M_r = 61,000$), A' ($M_r = 58,000$), and B ($M_r = 15,000$); the molar ratio of (A + A') to B is 1 : 1 (Stewart *et al.*, 1982, 1983). Stewart *et al.* (1982) postulated that A and A' could be different isozymes; A could be a modified form of A', or A' could be derived from A by limited proteolysis. Limited proteolysis of subunit A of brain

TABLE IV

Molecular Properties of Calmodulin-Dependent Phosphatase from Various Tissues

Property	Brain (bovine)	Skeletal muscle (rabbit)	Platelet (human)	Cardiac muscle (bovine)
Molecular weight				
Holoenzyme	80,000[a]	98,000[b]	75,000[c]	85,000[d]
Subunits				
A'		58,000[e]		
A	60,000[a]	61,000[e]	60,000[c]	61,000[f]
B	16,500[g]	15,000[e]	16,500[c]	15,000[f]
Subunit molar ratio	A:B, 1:1[a]	(A'+A):B, 1:1[e]	N.D.[h]	N.D.[h]
App K_D Ca^{2+} (μM)	0.35[g]	0.5[b]	N.D.[h]	N.D.[h]
App K_D calmodulin (nM)	30[g]	30[b]	N.D.[h]	N.D.[h]

[a]Wallace et al. (1979).
[b]Stewart et al. (1983).
[c]Tallant and Wallace (1985).
[d]Wolf and Hofmann (1980).
[e]Stewart et al. (1982).
[f]Krinks et al. (1984).
[g]Tallant and Cheung (1984b).
[h]Abbreviation: N.D., not determined.

phosphatase gave a fragment of 59,000 Da (Tallant and Cheung, 1984a), suggesting that muscle subunit A' may well represent a proteolyzed fragment of A.

A calmodulin-dependent phosphatase has been partially purified from human platelets (Tallant and Wallace, 1985); it has a molecular weight of 75,000 and consists of two subunits of 60,000 and 16,500 Da, respectively. The platelet enzyme shows antigenic identity to the brain phosphatase. A protein similar to the brain enzyme has also been isolated from bovine heart (Wolf and Hofmann, 1980; Krinks et al., 1984); the cardiac protein has a molecular weight of 85,000 and is composed of subunits with molecular weights of 60,000–61,000 and 15,000.

VII. IDENTIFICATION OF THE CATALYTIC SUBUNIT

Several techniques were used to isolate and identify the catalytic subunit of the phosphatase (Winkler et al., 1984). An immunoaffinity column was prepared with a monoclonal antibody to subunit A and was subsequently used to isolate the subunit (Fig. 5). Subunit A was also separated from subunit B by gel filtration chromatography in the presence of 6 M urea (Fig. 6). Subunit A was catalytically active; subunit B was not. The activity of subunit A was stimulated by calmodulin, by subunit B, or by both (Merat *et al.*, 1985). The effect of subunit B was greater than that of calmodulin; the effects of both were synergistic (Table V).

Additional evidence that subunit A is catalytically active was obtained by limited proteolysis of the phosphatase (Tallant and Cheung, 1984a). After extensive trypsinization, a proteolytic fragment of subunit A was present, whereas subunit B disappeared; moreover, the sample retained full phosphatase activity. Apparently, subunit A contains the catalytic site, and subunit B is not necessary for enzymic activity. The finding that subunit A and not subunit B contains the catalytic site is not unexpected; subunit B is structurally very similar to calmodulin, a regulatory protein with no known intrinsic enzyme activity.

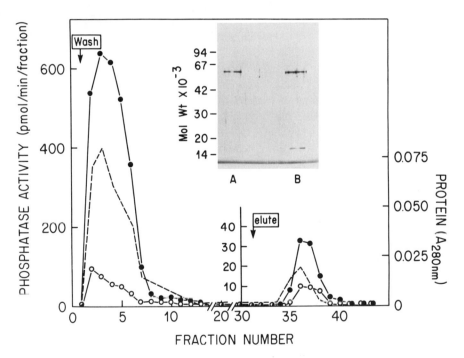

Fig. 5. Isolation of phosphatase subunit A on a monoclonal antibody-Sepharose 4B column. Phosphatase (0.4 mg) was loaded onto a column containing 5 ml of packed immunoglobulin-Sepharose, washed with 0.15 M NaCl and eluted with 50 mM diethylamine (pH 11). Phosphatase activity was measured using ^{32}P-labeled casein in the presence of 0.2 mM Ca^{2+} and 1 μg calmodulin (●) or 1 mM EGTA and 1 μg calmodulin (○). The inset shows the protein pattern on an SDS–polyacrylamide gel of the phosphatase eluted in fraction 37 (A) and the phosphatase applied to the column (B). From Winkler *et al.* (1984).

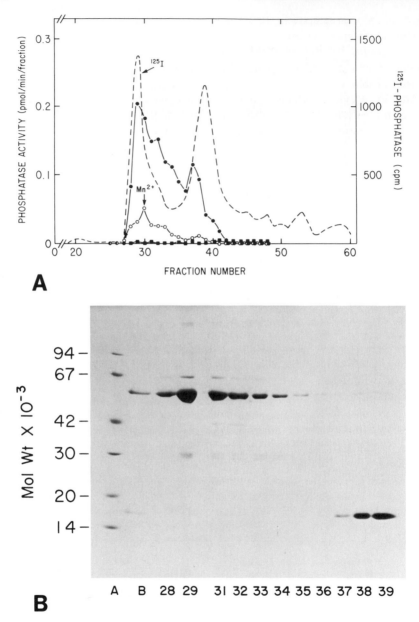

Fig. 6. Resolution of phosphatase subunits A and B by gel filtration in the presence of urea. Phosphatase (0.4 mg) supplemented with [^{125}I]phosphatase (28,000 CPM) was incubated for 3.5 hr at 4° in 100 mM 2-(N-morpholino)ethane sulfonic acid (pH 7.0) containing 15 mM EDTA, 5 mM dithiothreitol, 6 M urea, and 100 mM LiBr; it was then chromatographed on a Sephacryl S-200 column (1.0 × 116.5 cm). Aliquots were assayed for phosphatase activity in the presence of 0.1 mM

TABLE V

Stimulation of Subunit A Activity by Calmodulin or by Subunit B[a]

Additions	Phosphatase activity (nmol/mg/min)
B	0
B + CaM	0
A	0.3 (1)
A + CaM	0.9 (3)
A + B	3.7 (12)
A + B + CaM	9.0 (30)
A·B (native phosphatase)	4.3 (14)
A·B·CaM (holoenzyme)	22.4 (75)

[a]The subunits were isolated as described previously (Merat et al., 1985). The specific activities were calculated on the basis of subunit A protein concentration. The relative activities are given in parentheses. The specific activities of native phosphatase (A·B) and the holoenzyme (A·B·CaM) are included for comparison. From Merat et al. (1985).

Calmodulin-dependent phosphatase is unique in that the holoenzyme consists of a catalytic subunit A interacting with two structurally similar Ca^{2+}-binding subunits—calmodulin and subunit B. An analogous enzyme is phosphorylase kinase, which has a subunit structure of $\alpha\beta\gamma\delta$; the δ-subunit is calmodulin, which confers Ca^{2+} sensitivity to the enzyme. Kinase activity can be further stimulated by exogenous calmodulin or by another structurally similar Ca^{2+}-binding protein, troponin C (Cohen, 1980).

VIII. STIMULATION BY CALCIUM AND CALMODULIN

Calmodulin, in the presence of Ca^{2+}, stimulates the activity of both the brain and skeletal muscle phosphatases (Yang et al., 1982; Tallant and Cheung, 1984b; Stewart et al., 1982, 1983). Half-maximal activation of the brain enzyme was observed at 20–30 nM calmodulin, in the presence of a saturating concentra-

---

Ca^{2+}, 0.8 mM Mn^{2+}, and calmodulin (●), 0.1 mM Ca^{2+} and 0.8 mM Mn^{2+} (○), and 0.1 mM Ca^{2+} and calmodulin (■), as shown in (A). Recovery of calmodulin-stimulated activity was 40%. In (B), aliquots of each fraction were analyzed by SDS–polyacrylamide gel electrophoresis. Lane A, molecular weight markers; lane B, phosphatase not incubated with 6 M urea; lanes 28–39, corresponding to the column fractions. From Winkler et al. (1984).

tion of Ca^{2+} (Tallant and Cheung, 1984b; Li, 1984). The skeletal muscle enzyme was also half-maximally activated at 30 nM calmodulin, in the presence of 3 μM Ca^{2+} and 1 mM Mn^{2+} (Stewart et al., 1983). At a saturating level of calmodulin (290 nM), half-maximal activation of the brain enzyme was observed at 0.35 μM Ca^{2+} (Tallant and Cheung, 1984b). A Hill coefficient of 4.9 indicated a high degree of cooperativity. At a lower level of calmodulin (29 nM), half-maximal activation of the brain enzyme was obtained at 0.41 μM Ca^{2+} (Tallant and Cheung, 1984b). This is comparable to the value of 0.5 μM obtained for the skeletal muscle enzyme assayed with 30 nM calmodulin (Stewart et al., 1983) (see Table IV).

Calmodulin stimulation of phosphatase activity was inhibited by trifluoperazine, a calmodulin antagonist (Yang et al., 1982; Stewart et al., 1982, 1983; Tallant and Cheung, 1984b). Trifluoperazine (50 μM) abolished calmodulin stimulation of the brain phosphatase without affecting the basal activity (Yang et al., 1982; Tallant and Cheung, 1984b). In contrast, both the basal and calmodulin-stimulated activities of the muscle enzyme were abolished by 35–45 μM trifluoperazine (Stewart et al., 1982, 1983). These investigators postulated that inhibition of basal phosphatase activity might be due to interaction between subunit B and trifluoperazine. The brain phosphatase does interact with phenothiazines; it was retained by a norchlorpromazine–Affi-gel column in either the presence or absence of Ca^{2+} (Klee et al., 1983b). However, it is not known whether this interaction is through subunit A or subunit B.

The interaction between calmodulin and the phosphatase is reversible, depending upon the level of Ca^{2+}. This reversibility is implicit in the successful use of calmodulin-affinity chromatography for purification of the phosphatase; the enzyme binds to the immobilized calmodulin in the presence of Ca^{2+} and elutes when the Ca^{2+} is chelated by EGTA. Formation of a Ca^{2+}-dependent complex between calmodulin and phosphatase has also been demonstrated by a variety of techniques: glycerol gradient centrifugation (Richman and Klee, 1978; Manalan and Klee, 1983), sucrose density gradient centrifugation (Tallant and Cheung, 1984b), gel filtration chromatography (Wang and Desai, 1977; Tallant and Cheung, 1984b), and cross-linking with dimethylsuberimidate (Klee et al., 1979). The rate of formation and dissociation of the calmodulin–phosphatase complex is rapid, as the phosphatase activity can be readily reversed by changes in the Ca^{2+} concentration (Tallant and Cheung, 1984b).

The maximally activated calmodulin–phosphatase complex is comprised of 1 mol of calmodulin and 1 mol of phosphatase. This stoichiometry was determined by cross-linking with dimethylsuberimidate (Klee et al., 1979), by gel filtration (Tallant and Cheung, 1984b), by fluorescence studies using dansyl calmodulin (Kincaid et al., 1982), and by monitoring of the molar ratio required to fully stimulate phosphatase activity (King and Huang, 1984; Li, 1984).

The formation of a Ca^{2+}-dependent equimolar complex between calmodulin and the phosphatase is compatible with the proposed mechanism for calmodulin-stimulated enzymes. This mode of action was first established with the phosphodiesterase system. First, it involves the binding of Ca^{2+} to calmodulin; this binding forms an active Ca^{2+}–calmodulin complex, which then interacts with the apoenzyme to form the activated holoenzyme (Teo and Wang, 1973; Lin *et al.*, 1974; Cheung, 1980). In addition, the micromolar Ca^{2+} requirement and the nanomolar calmodulin requirement of the phosphatase are in the same range as those obtained for other calmodulin-dependent enzymes, such as phosphodiesterase (Huang *et al.*, 1981) and myosin light chain kinase (Blumenthal and Stull, 1980).

Calmodulin exerts its effect by increasing the maximal velocity of the phosphatase without altering its affinity for the substrate (Table VI). This has been demonstrated for the brain enzyme with casein (Tallant and Cheung, 1984b) and *p*-nitrophenyl phosphate (Pallen and Wang, 1984; Li, 1984). Similar effects were found with the muscle phosphatase, with either inhibitor-1, phosphorylase kinase, or myosin light chain as substrate (Stewart *et al.*, 1983). Although the kinetic parameters are altered by various divalent cations, as will be discussed in Section X, the changes induced by calmodulin are independent of the divalent cation (Pallen and Wang, 1984; Li, 1984).

TABLE VI

Kinetic Parameters of Phosphatase

Enzyme	Substrate	Calmodulin	K_m (μM)	V_{max} (nmol/mg/min)
Bovine	Casein[a]	−	2.7	1.7
brain		+	2.1	41
	Histone[b]	+	4.2	33.3
	p-Nitrophenyl phosphate[c]	−	12,500	1087
		+	13,000	3030
Rabbit	Inhibitor-1[d]	−	2.5	170
skeletal		+	2.5	2080
muscle	Phosphorylase kinase[d]	−	5.9	1040
	Myosin light chain[d]	+	3.7	790
Bovine	Regulatory subunit of type	+	5	1000
heart	II cAMP-dependent			
	protein kinase[e]			

[a]Assayed with Ca^{2+}; Tallant and Cheung (1984b).
[b]Assayed with Ca^{2+}; King and Huang (1983).
[c]Assayed with Mn^{2+}; Li (1984).
[d]Assayed with Mn^{2+}; Stewart *et al.* (1983).
[e]Assayed with Mn^{2+}; Blumenthal and Krebs (1983).

IX. STIMULATION BY LIMITED TRYPSINIZATION

Calmodulin-dependent phosphatase can be activated and rendered calmodulin-independent by limited trypsinization (Manalan and Klee, 1983; Tallant and Cheung, 1984a). Trypsinized phosphatase, when fully activated, had a molecular weight of 60,000 and was a heterodimer of subunit B and a 43,000-Da fragment of subunit A, as shown in Fig. 7 (see also Tallant and Cheung, 1984a; Manalan and Klee, 1983). The trypsinized phosphatase was insensitive to calmodulin, since it could no longer bind to calmodulin-Sepharose (Tallant and

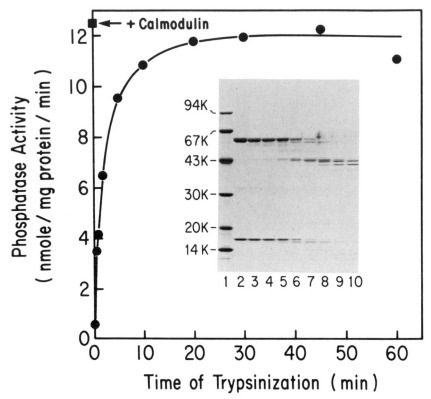

Fig. 7. Time course of trypsinization of phosphatase. Phosphatase (31 μg) was incubated at 30° with trypsin (0.3 μg) in the presence of 0.1 mM Ca^{2+}. At the times indicated, one aliquot was added to a 4-fold excess of soybean trypsin inhibitor to be assayed for phosphatase in the presence of 0.2 mM Ca^{2+}; phosphatase (23.3 nM) was assayed using ^{32}P-labeled casein as a substrate. Another aliquot was denatured in 1% SDS and 5% 2-mercaptoethanol and electrophoresed on an SDS–polyacrylamide gel containing 1 mM EGTA. Lane 1, molecular weight markers; lanes 2–10, phosphatase trypsinized for 0, 0.5, 1, 2, 5, 10, 20, 40, and 60 min, respectively. Reprinted with permission from Tallant and Cheung (1984a). Activation of bovine brain calmodulin-dependent protein phosphatase by limited trypinization. *Biochemistry* **23,** 975. Copyright 1984 American Chemical Society.

Cheung, 1984a), form a complex with calmodulin on glycerol gradients, or bind to [^{125}I]calmodulin in SDS–polyacrylamide gels (Manalan and Klee, 1983).

Trypsinization of the phosphatase proceeded in two phases: a rapid activation by limited proteolysis and a slow inactivation upon prolonged proteolysis (Tallant and Cheung, 1984a). Inactivation was retarded by Ca^{2+} implying that subunit B, the Ca^{2+}-binding subunit, might have protected the catalytic site. Inactivation was further retarded in the presence of Ca^{2+} and calmodulin, implying that the holoenzyme contains a more protected catalytic site. Manalan and Klee (1983) also found that calmodulin decreased the rate of tryptic digestion; trypsinization in the presence of calmodulin occurred more slowly, and the fragments which were generated were still able to interact with calmodulin. In an earlier study, Wallace et al. (1979) showed that the holoenzyme was more thermally stable.

The trypsin-treated enzyme, although insensitive to stimulation by calmodulin, was stimulated by Ca^{2+} with casein as a substrate (Tallant and Cheung, 1984a). On the other hand, the activity toward myosin light chain was inhibited by Ca^{2+} (Manalan and Klee, 1983). The mechanism by which Ca^{2+} affects the proteolyzed phosphatase is not understood, but may be due to the Ca^{2+}-binding properties of subunit B. Activity of the trypsinized phosphatase toward myosin light chains was further inhibited by Ca^{2+} and calmodulin (Manalan and Klee, 1983). In view of the structural homology between calmodulin and myosin light chain, it appears that calmodulin may be acting as a competitive inhibitor at the catalytic site.

Subunit A appears to have at least two regulatory domains. One domain, relatively resistant to trypsin, interacts with subunit B; the other domain, highly susceptible to tryptic attack, interacts with calmodulin. This calmodulin-binding domain inhibits the enzyme activity. The inhibition can be relieved either by proteolysis or by a conformation change induced by the binding of calmodulin (Manalan and Klee, 1983; Tallant and Cheung, 1984a).

Stimulation of a calmodulin-dependent enzyme by limited proteolysis was first noted with cyclic nucleotide phosphodiesterase (Cheung, 1967, 1971; Lin and Cheung, 1980). Subsequently, many calmodulin-dependent enzymes have been shown to be stimulated by limited proteolysis, which makes them insensitive to calmodulin. These include skeletal muscle phosphorylase kinase (Depaoli-Roach et al., 1979), erythrocyte Ca^{2+}-ATPase (Niggli et al., 1981), myosin light chain kinase (Walsh et al., 1982), NAD$^+$ kinase (Meijer and Guerrier, 1982), and adenylate cyclase (Keller et al., 1980).

X. REGULATION BY DIVALENT CATIONS

The activity of calmodulin-stimulated phosphatase is dependent upon divalent cations. The skeletal muscle enzyme was originally reported to require divalent cations for activity, and the assay mixture routinely contained 1 mM Mn^{2+} (Nimmo and Cohen, 1978). Upon further examination of the metal-ion requirement, Stewart et al. (1982) demonstrated that micromolar levels of Ca^{2+} replaced Mn^{2+} and that calmodulin further stimulated phosphatase activity. Although Mn^{2+} is capable of substituting for Ca^{2+} in the formation of an active metal–calmodulin complex (Lin et al., 1974; Wolff et al., 1977), it had a secondary effect on the skeletal muscle enzyme, independent of calmodulin. After chromatography on calmodulin-Sepharose, phosphatase activity was very low in the presence of Ca^{2+} and calmodulin; the activity was greatly increased by the inclusion of Mn^{2+} (Stewart et al., 1982, 1983).

The effects of divalent cations on phosphatase activity have been investigated in a number of other laboratories (King and Huang, 1983, 1984; Pallen and Wang, 1983, 1984; Tallant and Cheung, 1984a; Merat et al., 1984; Gupta et al., 1984; Li, 1984). At a neutral pH, the activity of the phosphatase was significantly increased by Ni^{2+} and Mn^{2+} as well as other divalent cations; the order of effectiveness in stimulating the phosphatase, in the presence or absence of calmodulin, was Ni^{2+} $\gg$ Mn^{2+} $\gg$ Co^{2+} > Ca^{2+}, Sr^{2+} or Ba^{2+} (King and Huang, 1983; Pallen and Wang, 1984; Li, 1984). Maximal activation was achieved at 100 μM Mn^{2+}, with a K_a of 40 μM; the concentration of Ni^{2+} required for maximum activity was 500 μM, with a K_a of 100 μM (Pallen and Wang, 1984; Li, 1984). Mg^{2+}, at a neutral pH, had no effect on phosphatase activity (Pallen and Wang, 1984; Li, 1984); however, at an alkaline pH (8.6), Mg^{2+} also stimulated phosphatase activity (Li, 1984). The order of effectiveness of the metal ions at pH 8.6 was Mg^{2+} $\gg$ Ni^{2+} > Mn^{2+} $\gg$ Co^{2+}, using p-nitrophenyl phosphate as a substrate. It should be noted that the stimulatory concentration of Mg^{2+} was high (20 mM), the K_a being 5 mM. However, since the cellular concentrations of Ni^{2+} and Mn^{2+} are extremely low and Mg^{2+} is present at millimolar levels, Li (1984) has suggested that Mg^{2+} may represent the physiological regulator.

The pH dependence of phosphatase activity, in the presence or absence of calmodulin, is thus dependent upon the metal ion. In the presence of Ca^{2+}, Mn^{2+}, or Ni^{2+}, the phosphatase showed optimal activity at a neutral pH (Tallant and Cheung, 1984b; Pallen and Wang, 1983; Li, 1984). Optimal activity in the presence of calmodulin occurred at pH 6.8–7.2; in the absence of calmodulin, the pH optimum was slightly broadened although the maximum was still between pH 6.8 and 7.2. With Mg^{2+} as a cofactor, phosphatase activity reached a maximum at pH 8.6, in the presence or absence of calmodulin (Li, 1984). The pH profile of bovine cardiac protein phosphatase-3C and its catalytic entity (M_r = 35,000) is similarly dependent upon Mg^{2+} (Chernoff and Li, 1983). This similarity suggests that there may be some common features in their catalytic sites (Li, 1984).

The kinetic parameters of the phosphatase have been determined in the presence of Mn^{2+} or Ni^{2+} at a neutral pH, and of Mg^{2+} at an alkaline pH (Pallen and Wang, 1984; Li, 1984). The K_m of the Mg^{2+}-stimulated activity was significantly increased over that found with Mn^{2+}, which was higher than the K_m with Ni^{2+}. The maximal velocity of the enzyme was higher with Mg^{2+} than with either Mn^{2+} or Ni^{2+}; the V_{max} of the Mn^{2+}- and Ni^{2+}-stimulated activities were almost identical. Calmodulin increased the V_{max} of the Ni^{2+}-, Mn^{2+}-, or Mg^{2+}-supported activities without affecting the K_m; thus, the effect of calmodulin is independent of the divalent cation.

Stimulation of phosphatase activity by Ni^{2+} is time-dependent and invariably preceded by a pronounced lag phase (King and Huang, 1983; Pallen and Wang, 1984). The addition of Ni^{2+} quenched the fluorescence of the phosphatase approximately 30%, and the quenching occurred at a faster rate than did the enzyme activation (King and Huang, 1983). In addition, subunit A, separated from subunit B by gel filtration in the presence of $8~M$ urea, displayed a similar Ni^{2+}-induced quenching of its tryptophan fluorescence; this finding suggests that the primary Ni^{2+} binding site is on subunit A (King et al., 1985). Pallen and Wang (1984) noted that the rate of initial activation of the phosphatase as a function of Ni^{2+} concentration was hyperbolic. Hence, the rate limiting step in the activation process may be a conformational change induced by the metal binding, rather than the binding of the metal itself.

EDTA did not reverse the Ni^{2+}-induced activation of the phosphatase, although it readily reversed the Mn^{2+}-stimulated activity (King and Huang, 1983; Pallen and Wang, 1984). The Ni^{2+}-induced quenching of phosphatase fluorescence was also not reversed by a 10-fold excess of the chelator (King and Huang, 1983). Thus, the activated enzyme appears to have a much higher affinity for Ni^{2+} than for Mn^{2+}, even though the nonactivated phosphatase has a higher affinity for Mn^{2+} than for Ni^{2+}. Neither the Mn^{2+}- nor the Ni^{2+}-stimulated activity of the phosphatase was reversed by dilution or by extensive dialysis against metal-free buffer; both metal ions may be very tightly bound to the activated phosphatase (Pallen and Wang, 1984).

Mn^{2+} and Ni^{2+} appear to compete in stimulating phosphatase activity (Pallen and Wang, 1984). Since Ni^{2+} activates the enzyme to a higher level than does Mn^{2+}, the two activities could be differentiated. When the two metal ions were added simultaneously, an intermediate activity was obtained, suggesting that the two ions compete for the same binding site. Moreover, the Ni^{2+}-activated enzyme did not bind Mn^{2+}, and the Mn^{2+}-activated enzyme did not bind Ni^{2+}. Apparently, once one metal is bound to the phosphatase, it cannot be easily displaced by the other.

Phosphatase subunit A, isolated by gel filtration in the presence of $6~M$ urea and $15~mM$ EDTA, required the addition of a divalent metal ion for activity (Winkler et al., 1984; Merat et al., 1984). Mn^{2+} was most effective; Ni^{2+} or Co^{2+} were less effective; and Ca^{2+}, Mg^{2+}, and Zn^{2+} were ineffective (Merat *et al.*, 1984). The Mn^{2+}-stimulated activity of subunit A was not easily reversed by EGTA (Merat *et al.*, 1984), in agreement with the inability of EDTA to reverse activity of the Mn^{2+}-stimulated holoenzyme (Pallen and Wang, 1984). Subunit A which was stimulated by limited proteolysis was also dependent upon Mn^{2+} for activity (Merat *et al.*, 1984). Isolation of subunit A in the presence of 6 *M* urea and EDTA may have removed a tightly bound metal ion, thus requiring the subsequent addition of an exogenous ion for activity. Subunit A isolated by immunoaffinity chromatography under milder conditions did not require Mn^{2+} for activity (Winkler *et al.*, 1984).

Trypsinized holophosphatase, which is calmodulin independent, was stimulated severalfold by Mn^{2+} (Tallant and Cheung, 1984a; Gupta *et al.*, 1984; Merat *et al.*, 1984). EDTA or EGTA decreased the activity of the proteolyzed phosphatase; the decrease was dependent upon both time of incubation with the chelator and concentration of the chelator. The EDTA-induced decrease could be restored by Mn^{2+} but not by other divalent cations, suggesting that a tightly bound metal ion, probably Mn^{2+}, is necessary for activity (Gupta *et al.*, 1984).

Calmodulin-stimulated phosphatase activity was not affected by Cu^{2+}, Be^{2+}, Cd^{2+}, Fe^{2+}, Al^{3+}, Fe^{3+}, or Pb^{2+} (Pallen and Wang, 1984; King and Huang, 1983; Gupta *et al.*, 1984). However, millimolar concentration of heavy metals such as Zn^{2+}, Fe^{2+}, Cu^{2+}, Pb^{2+}, or Cd^{2+} inhibited phosphatase activity (Gupta *et al.*, 1984); enzyme activity was also inhibited by F^- (Tallant and Cheung, 1984b; Klee *et al.*, 1983a). Both basal and calmodulin-dependent activities were inhibited by *p*-hydroxymercuribenzoate and *N*-ethylmaleimide; hence, the phosphatase may require sulfhydryl groups for activity (Tallant and Cheung, 1984b).

Atomic absorption spectrophotometric determinations on the phosphatase revealed almost stoichiometric amounts of bound Fe^{3+} and Zn^{2+} (King and Huang, 1984). However, no significant amounts of Mn^{2+}, Ni^{2+}, or Co^{2+} were detected. The phosphatase thus appears to be an Fe- and Zn-containing metalloenzyme, but one which requires an additional divalent cation for maximal activity. Activation by the metal ions could result from formation of a more stable conformation of the enzyme; alternatively, the metal ions could facilitate the interaction of the substrate with the enzyme. Although evidence *in vitro* suggests that Mn^{2+} satisfies the divalent cation requirement, the physiological metal cation has not been established.

XI. DEACTIVATION BY CALCIUM AND CALMODULIN

Although the activity of the phosphatase is stimulated by calmodulin in the presence of Ca^{2+}, a time-dependent deactivation of the phosphatase also appears to occur *in vitro*. The addition of the Ca^{2+}–calmodulin complex results in an initial burst of phosphatase activity that is followed by a slower deactivation (King and Huang, 1984; Pallen and Wang, 1984). Deactivation of the phosphatase requires the simultaneous presence of both Ca^{2+} and calmodulin. Stewart et al. (1982, 1983) first noted that the skeletal muscle phosphatase was converted to a relatively inactive form following chromatography on calmodulin-Sepharose; reactivation was observed after a preincubation with Mn^{2+}.

The deactivation process appeared to be accelerated by increasing the substrate concentration, using p-nitrophenyl phosphate as a substrate, and was also enhanced by pyrophosphate and various phospho compounds (King and Huang, 1984). The diphosphates were the most effective (ADP, GDP), and the purine nucleotides were more potent than pyrimidine nucleotides. The nucleotides and pyrophosphate not only accelerated the rate of deactivation but also competitively inhibited the phosphatase reaction; the nucleotides did not serve as alternative substrates.

King and Huang (1984) have investigated several possible mechanisms to explain the calmodulin-dependent deactivation of the phosphatase. They concluded that deactivation was not due to any of the following: (1) dissociation of calmodulin from the deactivated phosphatase, since the deactivation was not reversed by EGTA; (2) self-dephosphorylation of the phosphatase (although the enzyme does contain a small amount of endogenous phosphate, the same amount was present before and after deactivation); (3) removal of the tightly bound Zn^{2+} or Fe^{3+}, since no significant changes were found in their content; or (4) proteolysis of the enzyme, since the process was completely reversed by certain cations.

Deactivation of the phosphatase by Ca^{2+} and calmodulin could be reversed or prevented by divalent metal ions such as Mn^{2+} or Ni^{2+} (King and Huang, 1984; Pallen and Wang, 1984). Mg^{2+}, Co^{2+}, or Zn^{2+} were less effective, and Sr^{2+} had no effect (King and Huang, 1984). It is conceivable that deactivation involves a Ca^{2+}/calmodulin-induced removal of an essential metal ion. The importance, if any, of the Ca^{2+}/calmodulin-induced deactivation of the phosphatase in vivo is not known.

XII. SUBSTRATE SPECIFICITY

Calmodulin-dependent phosphatase from bovine brain has been shown to dephosphorylate various substrates, including the α-subunit of phosphorylase kinase, inhibitor-1 (Stewart et al., 1982, 1983), casein, histone (Yang et al., 1982), protamine, phosvitin, troponin I (Tallant and Cheung, 1984b), myosin light chain from both smooth and skeletal muscle (Manalan and Klee, 1983), the regulatory subunit of Type II cAMP-dependent protein kinase (Blumenthal and Krebs, 1983), MAP-2 (a microtubule-associated protein) (Murthy and Flavin,

1983), and four mammalian brain phosphoproteins—DARPP-32, G-substrate, Protein K.-F., and synapsin I (King *et al.*, 1984). However, comparison of the catalytic efficiency of the phosphatase toward some of these substrates (represented by the K_{cat}/K_m in Table VII) indicates that dephosphorylation of only DARPP-32, G-substrate, and Protein K.-F. occurs at significant rates *in vitro*.

DARPP-32 (dopamine- and cyclic AMP-regulated phosphoprotein of molecular weight 32,000) is a neuronal phosphoprotein which is highly concentrated in the neostriatum, predominantly in regions that are highly innervated by dopaminergic neurons (Hemmings *et al.*, 1984b; Walaas *et al.*, 1983; Walaas and Greengard, 1984; Ouimet *et al.*, 1984). The state of phosphorylation of DARPP-32 can be regulated by dopamine and by cyclic AMP in intact nerve cells (Walaas *et al.*, 1983; Walaas and Greengard, 1984). DARPP-32 is a potent inhibitor of the type 1 protein phosphatase. It inhibits the enzyme at nanomolar concentrations and is only active as a phosphatase inhibitor in its phosphorylated state (Hemmings *et al.*, 1984a). DARPP-32 is dephosphorylated by the calmodulin-dependent phosphatase at a significant rate *in vitro*, suggesting that it may represent a physiological substrate (King *et al.*, 1984). In addition, both the phosphatase and DARPP-32 are located at identical sites not only within the basal ganglia of the brain (Wallace *et al.*, 1980; Walaas *et al.*, 1984; Walaas and Greengard, 1984), but also within the cell (Wood *et al.*, 1980a; Ouimet *et al.*, 1984).

G-substrate is also dephosphorylated by the calmodulin-dependent phosphatase (King *et al.*, 1984). G-substrate is a soluble, 23,000-Da protein which is phosphorylated on two threonine residues by cGMP-dependent protein kinase. It is relatively enriched in the cerebellum, where it is located almost exclusively

TABLE VII

Catalytic Efficiency of Phosphatase

Substrate	K_m (μM)	K_{cat} (sec^{-1})	K_{cat}/K_m (sec^{-1}/M^{-1})
Casein[a]	2.1	0.014	6.7×10^3
Histone[b]	2.2	0.008	3.6×10^3
Synapsin (site I)[c]	7.0	0.053	7.6×10^3
Synapsin (site II)[c]	4.4	0.040	9.1×10^3
G-Substrate[c]	3.8	0.410	1.1×10^5
DARPP-32[c]	1.6	0.200	1.3×10^5
Protein K.-F.[c]	3	0.7	2×10^5

[a]Tallant and Cheung (1984b).
[b]King and Huang (1983).
[c]King *et al.* (1984).

within the Purkinje cells (Schlicter *et al.*, 1978; Aswad and Greengard, 1981; Detre *et al.*, 1984). G-substrate has also recently been identified as a phosphatase inhibitor in its phosphorylated state (Nestler and Greengard, 1984).

Dephosphorylation of Protein K.-F. by the calmodulin-dependent phosphatase also occurs at significant rates *in vitro* (King *et al.*, 1984). Protein K.-F. (M_r = 18,000) is a membrane-bound phosphoprotein which is phosphorylated on a serine residue by a cyclic nucleotide/Ca^{2+}-independent protein kinase present in the particulate fraction of bovine brain (Greengard and Chan, 1983). Protein K.-F. and myelin basic protein appear to be related proteins, since they have many properties in common (Greengard and Chan, 1983). In this respect, Gupta *et al.* (1984) have found that myelin basic protein appeared to be a good substrate for the phosphatase.

Although the phosphatase did not dephosphorylate synapsin I at significant rates *in vitro*, this does not entirely eliminate a physiological role for this reaction (King *et al.*, 1984). Synapsin I (M_r = 80,000 and 86,000 doublet) is phosphorylated at various sites by cAMP- and by Ca^{2+}/calmodulin-dependent protein kinases; it is present throughout the brain and is localized on synaptic vesicles (Nestler and Greengard, 1984). The phosphatase has also been found to dephosphorylate MAP-2, a microtubule-associated protein of molecular weight 270,000 (Murthy and Flavin, 1983). The dephosphorylation of MAP-2 was dependent upon calmodulin, but only to an extent that was not considered to be physiologically significant.

Although calmodulin-dependent phosphatase from bovine brain dephosphorylated casein, histone, troponin I, protamine, and phosvitin (Yang *et al.*, 1982; Tallant and Cheung, 1984b), the rates of dephosphorylation were fairly low, and their physiological relevance appears uncertain. The kinetic parameters for the dephosphorylation of inhibitor-1, phosphorylase kinase, and myosin light chain by the brain enzyme are not known; however, the skeletal muscle phosphatase does dephosphorylate these substrates at significant rates (Table VI), suggesting that the same may be true for the brain enzyme. Similarly, the regulatory subunit of Type II cAMP-dependent protein kinase may also represent a physiological substrate for the brain enzyme, since it was dephosphorylated at significant rates by the heart calmodulin-dependent phosphatase (Table VI) (Blumenthal and Krebs, 1983).

DARPP-32, G-substrate, and inhibitor-1 appear to be good substrates for the calmodulin-dependent phosphatase (King *et al.*, 1984; Stewart *et al.*, 1982, 1983), and they share certain common properties (Nimmo and Cohen, 1978; Aswad and Greengard, 1981; Hemmings *et al.*, 1984b). They are phosphorylated on threonine residues by cyclic nucleotide–dependent protein kinases, and the sequences around their phosphorylation sites are similar (Cohen *et al.*, 1977; Aitken *et al.*, 1981; Hemmings *et al.*, 1984c). In addition, their phosphorylated forms function as protein phosphatase inhibitors. The question arises as to whether they are members of a family of regulatory proteins involved in the control of phosphatase activity. The ability of the calmodulin-dependent phosphatase to dephosphorylate the three proteins suggests a mechanism by which Ca^{2+} antagonizes the effects of cyclic nucleotides. It also suggests the presence of a protein phosphatase cascade, in which the Ca^{2+}/calmodulin-dependent phosphatase regulates the activity of type 1 phosphatases by mediating the state of phosphorylation of various phosphatase inhibitors.

Calmodulin-dependent phosphatase can also dephosphorylate various low-molecular-weight, nonprotein phosphodiesters. The ability of the enzyme to dephosphorylate the chromogenic compound p-nitrophenyl phosphate, originally reported by Pallen and Wang (1983), has been useful in studying enzyme kinetics (Wang et al., 1984; King and Huang, 1984; Li, 1984). The phosphatase is also active toward β-naphthyl phosphate, α-naphthyl phosphate, phenylphthalein mono- and diphosphate, phenyl dihydrogen phosphate, and methylumbelliferyl phosphate (Li, 1984; Wang et al., 1984). The order of effectiveness of dephosphorylation was β-naphthyl phosphate > p-nitrophenyl phosphate > α-naphthyl phosphate. It seems that a hydrophobic group bulkier than the phenyl group might fit better near the active site, since β-naphthyl phosphate was a better substrate than was p-nitrophenyl phosphate (Li, 1984). The phosphatase also dephosphorylated phosphotyrosine in the presence of either Mn^{2+}, Co^{2+}, or Ni^{2+} (Pallen and Wang, 1984; Li, 1984); however, no significant activity was found toward phosphoserine, phosphothreonine, NADP, glucose 6-phosphate, $2'$-AMP, $3'$-AMP, $5'$-AMP, $5'$-IMP, $5'$-GMP, ADP, ATP, or β-glycerophosphate. In addition, the phosphatase dephosphorylated the phosphotyrosine residues of the EGF-receptor kinase (Pallen et al., 1984) and of phosphotyrosyl proteins phosphorylated by the EGF-receptor kinase (Chernoff et al., 1984); this activity required the presence of a divalent cation ($Ni^{2+} \gg Mn^{2+} > Mg^{2+} > Ca^{2+}$) and was stimulated by calmodulin.

Although the brain phosphatase dephosphorylates a wide range of substrates in vitro, its catalytic efficiency toward these proteins suggests that it may have a more limited specificity in vivo. Moreover, no physiological substrates for the brain phosphatase have yet been shown in vivo. In this respect, Famulski and Carafoli (1984) showed that, in an isolated microsomal fraction of rat liver, the addition of exogenous brain phosphatase in the presence of Ca^{2+} and calmodulin inhibited the transport of Ca^{2+} into the microsomes; thus, Ca^{2+}/calmodulin-dependent dephosphorylation may regulate or be involved in Ca^{2+} transport. In a rat brain synaptosomal fraction, Krueger et al. (1977) noted that upon incubation with agents known to increase Ca^{2+} permeability there was a decreased incorporation of ^{32}P into two protein bands of approximately 90,000 Da each; the decreased phosphorylation could have been due to a Ca^{2+}-dependent activation of the calmodulin-dependent phosphatase. The identification of substrates of the brain phosphatase *in vivo* should elucidate its physiological role in nervous tissue.

The skeletal muscle phosphatase exhibited a more restricted substrate specificity, since it dephosphorylated only the α-subunit of phosphorylase kinase, inhibitor-1, and myosin light chains from rabbit skeletal muscle (Stewart *et al.*, 1982, 1983). No significant activity was found toward glycogen phosphorylase, glycogen synthetase, histone H1 or H2B, eukaryotic initiation factor-2, or a variety of other protein substrates (Ingebritsen and Cohen, 1983b). When muscle extracts were incubated in the presence of Ca^{2+}, calmodulin-dependent phosphatase accounted for about 60% of the dephosphorylation of inhibitor-1 (Ingebritsen *et al.*, 1983). Inhibitor-1, phosphorylated by cAMP-dependent protein kinase, is a powerful inhibitor of the type-1 phosphatase; the dephosphorylation of inhibitor-1 by calmodulin-dependent phosphatase would release its inhibition of phosphatase 1 and may represent an example of a protein phosphatase cascade (Ingebritsen and Cohen, 1983b). Similarly, the calmodulin-dependent phosphatase appears to be the major enzyme catalyzing the dephosphorylation of the α-subunit of phosphorylase kinase. In contrast, the enzyme represented only 30% of the total potential dephosphorylation of myosin light chains in skeletal muscle extracts (Ingebritsen *et al.*, 1983) or of inhibitor-2 in contracting muscle (Tonks and Cohen, 1984).

XIII. CONCLUDING REMARKS

Phosphorylation of proteins by kinases and dephosphorylation by phosphatases constitute major cellular regulatory mechanisms. The activity of the kinases and the phosphatases is in turn controlled and coordinated by various external physiological stimuli. The identification of a calmodulin-dependent phosphatase, activated by micromolar concentrations of Ca^{2+}, suggests that this enzyme would be regulated by elevations of intracellular Ca^{2+}, such as after neural or hormonal stimulation. The steady state concentration of Ca^{2+} in the cytosol of mammalian cells is 10^{-8} to $10^{-7} M$; stimulation causes transient increases of intracellular Ca^{2+} to a micromolar level, which would be sufficient to allow calmodulin activation of the phosphatase. However, the phosphatase may be subject to a much more complex system of regulation. Not only is it activated by Ca^{2+}/calmodulin, but it also undergoes a deactivation process in the presence of the Ca^{2+}/calmodulin complex, a process which is enhanced by various nucleotides and pyrophosphate. The phosphatase is a Zn- and Fe-containing metalloenzyme, but its activity is also dependent upon an additional divalent metal ion, such as Mn^{2+} or Ni^{2+}. The divalent cation appears to regulate the catalytic activity of subunit A; subunit B, itself a metal-binding protein, may function in some additional regulatory role. Much more work is required before the complex nature of these observations can be unraveled at the molecular level.

The phosphatase is present in a variety of tissues and cells, but especially in skeletal muscle and brain. In skeletal muscle, it appears to be involved in glycogen metabolism, both in dephosphorylation of phosphorylase kinase and of inhibitor-1, which in turn regulates the state of phosphorylation of many of the glycolytic enzymes. The high level of the protein in brain tissue (particularly in the neostriatum), its immunohistochemical localization at synaptic sites and microtubules, and its apparent increase during synaptogenesis suggest that its role in brain tissue may be in the regulation of neurotransmission and microtubular function. The colocalization of certain phosphoproteins which are efficiently dephosphorylated by the phosphatase *in vitro* implies that the phosphatase may be involved in their regulation *in vivo*. The further identification of some of these substrates as region-specific phosphatase inhibitors may indicate an important role for the phosphatase and, thus, for Ca^{2+} in antagonizing the action of cyclic nucleotide-dependent and Ca^{2+}-independent protein kinases. The recent discovery that the phosphatase possesses phosphotyrosyl-protein phosphatase activity toward the EGF-receptor kinase and the proteins that it phosphorylates suggests a further possible role for the enzyme in growth regulation.

In conclusion, the calmodulin-dependent protein phosphatase appears to be an important regulatory enzyme in brain, skeletal muscle, and possibly various other tissues. Our understanding of its exact role in these tissues is dependent, to a large extent, upon identification of its endogenous substrates.

ACKNOWLEDGMENTS

The work from our laboratory was supported by Cancer Center Support (CORE) grant CA 21765, by project grants NS 08059 and GM 28178, and by American–Lebanese–Syrian Associated Charities. The authors would also like to thank Dr. Paul Greengard, Dr. Philip Cohen, and Dr. C. Y. Huang for making manuscripts available prior to their publication.

REFERENCES

Aitken, A., Bilham, T., Cohen, P., Aswad, D. W., and Greengard, P. (1981). A specific substrate from cerebellum for guanosine 3′ : 5′-monophosphate-dependent protein kinase. III. Amino acid sequences at the two phosphorylated sites. *J. Biol. Chem.* **256**, 3501–3506.

Aitken, A., Cohen, P., Santikarn, S., Williams, D. H., Calder, A. G., Smith, A., and Klee, C. B. (1982). Identification of the NH_2-terminal blocking group of calcineurin B as myristic acid. *FEBS Lett.* **150**, 314–318.

Aitken, A., Klee, C. B., and Cohen, P. (1984). The structure of subunit B of calcineurin. *Eur. J. Biochem.* **139**, 663–671.

Aswad, D. W., and Greengard, P. (1981). A specific substrate from rabbit cerebellum for guanosine 3′ : 5′-monophosphate-dependent protein kinase. I. Purification and characterization. *J. Biol. Chem.* **256,** 3487–3493.

Berridge, M. J. (1982). A novel cellular signaling system based on the integration of phospholipid and calcium metabolism. *In* "Calcium and Cell Function" (W. Y. Cheung, ed.), Vol. 3, pp. 1–36. Academic Press, New York.

Berridge, M. J. (1984). Inositol triphosphate and diacylglycerol as second messengers. *Biochem. J.* **220,** 345–360.

Billingsley, M., Kuhn, D., Velletri, P. A., Kincaid, R., and Lovenberg, W. (1984). Carboxylmethylation of phosphodiesterase attenuates its activation by Ca^{2+}-calmodulin. *J. Biol. Chem.* **259,** 6630–6635.

Blumenthal, D. K., and Krebs, E. G. (1983). Dephosphorylation of cAMP-dependent protein kinase regulatory subunit (Type II) by calcineurin (protein phosphatase 2B). *Biophys. J.* **4,** 409a (abstract).

Blumenthal, D. K., and Stull, J. T. (1980). Activation of skeletal muscle myosin light chain kinase by calcium (Ca^{2+}) and calmodulin. *Biochemistry* **19,** 5608–5614.

Brissette, R. E., Cunningham, E. B., and Swislocki, N. I. (1983). A Ca^{2+}-dependent phosphoprotein phosphatase of the erythrocyte membrane. *Fed. Proc., Fed. Am. Soc. Exp. Biol.* **42,** 2030 (abstr.).

Burgess, W. H., Jemiolo, D. K., and Kretsinger, R. H. (1980). Interaction of calcium and calmodulin in the presence of sodium dodecyl sulfate. *Biochim. Biophys. Acta* **623,** 257–270.

Carlin, R. K., Grab, D. J., and Siekevitz, P. (1981). Function of calmodulin in postsynaptic densities. III. Calmodulin-binding proteins of the postsynaptic density. *J. Cell Biol.* **89,** 449–455.

Carr, S. A., Biemann, K., Shoji, S., Parmelee, D. C., and Titani, K. (1982). n-Tetradecanoyl is the NH_2-terminal blocking group of the catalytic subunit of cyclic-AMP-dependent protein kinase from bovine cardiac muscle. *Proc. Natl. Acad. Sci. U.S.A.* **79,** 6128–6131.

Chernoff, J., and Li, H.-C. (1983). Multiple forms of phosphotyrosyl- and phosphoseryl-protein phosphatase from cardiac muscle: Partial purification and characterization of an EDTA-stimulated phosphotyrosyl-protein phosphatase. *Arch. Biochem. Biophys.* **226,** 517–530.

Chernoff, J., Sells, M. A., and Li, H.-C. (1984). Characterization of phosphotyrosyl-protein phosphatase activity associated with calcineurin. *Biochem. Biophys. Res. Commun.* **121,** 141–148.

Cheung, W. Y. (1967). Cyclic 3′,5′-nucleotide phosphodiesterase: pronounced stimulation by snake venom. *Biochem. Biophys. Res. Commun.* **29,** 478–482.

Cheung, W. Y. (1971). Cyclic 3′,5′-nucleotide phosphodiesterase. Evidence for and properties of a protein activator. *J. Biol. Chem.* **246,** 2859–2869.

Cheung, W. Y. (1980). Calmodulin plays a pivotal role in cellular regulation. *Science* **207,** 19–27.

Cohen, P. (1973). The subunit structure of rabbit skeletal-muscle phosphorylase kinase, and the molecular basis of its activation reactions. *Eur. J. Biochem.* **34,** 1–14.

Cohen, P. (1980). The role of calmodulin and troponin in the regulation of phosphorylase kinase from mammalian skeletal muscle. *In* "Calcium and Cell Function" (W. Y. Cheung, ed.), Vol. 1, pp. 184–198. Academic Press, New York.

Cohen, P. (1982). The role of protein phosphorylation in neural and hormonal control of cellular activity. *Nature (London)* **296,** 613–620.

Cohen, P., Watson, D. C., and Dixon, G. H. (1975). The hormonal control of activity of skeletal muscle phosphorylase kinase. Amino acid sequences at the two sites of action of adenosine-3′ : 5′-monophosphate-dependent protein kinase. *Eur. J. Biochem.* **51,** 79–92.

Cohen, P., Rylatt, D. B., and Nimmo, G. A. (1977). The hormonal control of glycogen metabolism: The amino acid sequence at the phosphorylation site of protein phosphatase inhibitor-1. *FEBS Lett.* **76,** 182–186.

Cohen, P., Burchell, A., Foulkes, J. G., Cohen, P. T. W., Vanaman, T. C., and Nairn, A. L. (1978). Identification of the Ca^{2+}-dependent modulator protein as the fourth subunit of rabbit skeletal muscle phosphorylase kinase. *FEBS Lett.* **92,** 287–293.

Cohen, P., Picton, C., and Klee, C. B. (1979). Activation of phosphorylase kinase from rabbit skeletal muscle by calmodulin and troponin. *FEBS Lett.* **104,** 25–29.

Cohen, P., Foulkes, J. G., Goris, J., Hemmings, B. A., Ingebritsen, T. S., Stewart, A. A., and Strada, S. T. (1981). Classification of protein phosphatases involved in cellular regulation. *In* "Metabolic Interconversion of Enzymes 1980" (H. Holzer, ed.), pp. 28–43. Springer-Verlag, Berlin and New York.

Cohen, S., Chinkers, M., and Ushiro, H. (1981). EGF-receptor-protein kinase phosphorylates tyrosine and may be related to the transforming kinase of Rous sarcoma virus. *Cold Spring Harbor Conf. Cell Proliferation* **8,** 801–808.

Collett, M., Purchio, A., and Erikson, R. (1980). Avian sarcoma virus-transforming protein, pp60scr, shows protein kinase activity specific for tyrosine. *Nature (London)* **285,** 167–169.

Cooper, N. G. F., McLaughlin, B. J., Wallace, R. W., Tallant, E. A., and Cheung, W. Y. (1982). Immunocytochemical localization of calcineurin in chick retina. *Invest. Ophthalmol. Visual Sci.* **22,** 246 (abstr.).

Cooper, N. G. F., McLaughlin, B. J., Tallant, E. A., and Cheung, W. Y. (1985). Calmodulin-dependent protein phosphatase: immunocytochemical localization in chick retina. *J. Cell Biol.* **101,** 1212–1218.

Depaoli-Roach, A. A., Gibbs, J. B., and Roach, P. J. (1979). Calcium and calmodulin activation of muscle phosphorylase kinase. Effect of tryptic proteolysis. *FEBS Lett.* **105,** 321–324.

Detre, J. A., Nairn, A. C., Aswad, D. W., and Greengard, P. (1984) Localization in mammalian brain of G-substrate, a specific substrate for guanosine 3′:5′-cyclic monophosphate-dependent protein kinase. *J. Neurosci.* **4,** 2843–2849.

Ek, B., Westermark, B., Wasteson, A., and Heldin, C.-H. (1982). Stimulation of tyrosine-specific phosphorylation by platelet-derived growth factor. *Nature (London)* **295,** 419–420.

Famulski, K. S., and Carafoli, E. (1984). Calmodulin-dependent protein phosphorylation and calcium uptake in rat liver microsomes. *Eur. J. Biochem.* **141,** 15–29.

Foulkes, G., Ernst, V., and Levin, D. (1982). Separation and identification of Type 1 and Type 2 protein phosphatases from rabbit reticulocyte lysates. *Fed. Proc., Fed. Am. Soc. Exp. Biol.* **41,** 648 (abstr.).

Foulkes, J. G., and Maller, J. L. (1982). In vivo actions of protein phosphatase inhibitor-2 in Xenopus oocytes. *FEBS Lett.* **150,** 155–159.

Foulkes, J. G., Erikson, E., and Erikson, R. L. (1983). Separation of multiple phosphotyrosyl-and phosphoseryl-protein phosphatases from chicken brain. *J. Biol. Chem.* **258,** 431–438.

Gagnon, C. (1983). Enzymatic carboxyl methylation of calcium-binding proteins. *Can. J. Cell Biol.* **61,** 921–926.

Gagnon, C., Kelley, S., Manganiello, V., Vaughan, M., Odya, C., Strittmatter, W., Hoffman, A., and Hirata, F. (1981). Modification of calmodulin function by enzymatic carboxyl methylation. *Nature (London)* **291,** 515–516.

Gopalakrishna, R., and Anderson, W. B. (1982). Ca^{2+}-induced hydrophobic site on calmodulin: Application for purification of calmodulin by phenyl-Sepharose affinity chromatography. *Biochem. Biophys. Res. Commun.* **104,** 830–836.

Gopalakrishna, R., and Anderson, W. B. (1983). Calmodulin interacts with cyclic nucleotide phosphodiesterase and calcineurin by binding to a metal ion-independent hydrophobic region on these proteins. *J. Biol. Chem.* **258,** 2405–2409.

Greengard, P. (1978). Phosphorylated proteins as physiological effectors. *Science* **199,** 146–152.

Greengard, P., and Chan, K.-F. J. (1983). Identification, purification, and partial characterization of a membrane-bound phosphoprotein from bovine brain. *Fed. Proc., Fed. Am. Soc. Exp. Biol.* **42,** 2048. (abstract).

Gupta, R. C., Khandelwal, R. L., and Sulakhe, P. V. (1984). Intrinsic phosphatase activity of bovine brain calcineurin requires a tightly bound trace metal. *FEBS Lett.* **169,** 251–255.

Hayakawa, T., Perkins, J. P., and Krebs, E. G. (1973). Studies on the subunit structure of rabbit skeletal muscle phosphorylase kinase. *Biochemistry* **12,** 574–580.

Hemmings, H. C., Jr., Greengard, P., Tung, H. Y. L., and Cohen, P. (1984a). DARPP-32, a dopamine-regulated neuronal phosphoprotein, is a potent inhibitor of protein phosphatase-1. *Nature (London)* **310,** 503–505.

Hemmings, H. C., Jr., Nairn, A. C., Aswad, D. W., and Greengard, P. (1984b). DARPP-32, a dopamine and adenosine 3′:5′-monophosphate-regulated phosphoprotein enriched in dopamine-innervated brain regions. II. Purification and characterization of the phosphoprotein from bovine caudate nucleus. *J. Neurosci.* **4,** 99–110.

Hemmings, H. C., Jr., Williams, K. R., Konigsberg, W. H., and Greengard, P. (1984c). DARPP-32, a dopamine- and adenosine-3′:5′-monophosphate-regulated neuronal phosphoprotein. I. Amino acid sequence around the phosphorylated threonine. *J. Biol. Chem.* **259,** 14486–14490.

Henderson, L. E., Krutzsch, H. C., and Oroszlan, S. (1983). Myristyl amino-terminal acylation of murine retrovirus proteins: An unusual post-translational protein modification. *Proc. Natl. Acad. Sci. U.S.A.* **80,** 339–343.

Huang, C. Y., Chau, V., Chock, P. B., Wang, J. H., and Sharma, R. K. (1981). Mechanism of activation of cyclic nucleotide phosphodiesterase: Requirement of the binding of four Ca^{2+} to calmodulin for activation. *Proc. Natl. Acad. Sci. U.S.A.* **78,** 871–874.

Huang, F. L., and Glinsmann, W. H. (1975). Inactivation of rabbit muscle phosphorylase phosphatase by cyclic AMP-dependent protein kinase. *Proc. Natl. Acad. Sci. U.S.A.* **72,** 3004–3008.

Huang, F. L., and Glinsmann, W. H. (1976). A second heat-labile protein inhibitor of phosphorylase phosphatase from rabbit muscle. *FEBS Lett.* **62,** 326–329.

Hunter, T., and Sefton, B. (1980). Transforming gene product of Rous sarcoma virus phosphorylates tyrosine. *Proc. Natl. Acad. Sci. U.S.A.* **77,** 1311–1315.

Ingebritsen, T. S., and Cohen, P. (1983a). The protein phosphatases involved in cellular regulation. 1. Classification and substrate specificities. *Eur. J. Biochem.* **132,** 255–261.

Ingebritsen, T. S., and Cohen, P. (1983b). Protein phosphatases: Properties and role in cellular regulation. *Science* **221,** 331–338.

Ingebritsen, T. S., Stewart, A. A., and Cohen, P. (1983). The protein phosphatases involved in cellular regulation. 6. Measurement of type-1 and type-2 protein phosphatases in extracts of mammalian tissues: An assessment of their physiological roles. *Eur. J. Biochem.* **132,** 297–307.

Jamieson, G. A., Jr., and Vanaman, T. C. (1979). Calcium-dependent affinity chromatography of calmodulin on an immobilized phenothiazine. *Biochem. Biophys. Res. Commun.* **90,** 1048–1096.

Kasuga, M., Zick, Y., Blithe, D. L., Crettaz, M., and Kahn, C. R. (1982). Insulin stimulates tyrosine phosphorylation of the insulin receptor in a cell-free system. *Nature (London)* **298,** 667–669.

Keller, C. H., LaPorte, D. C., Toscano, D. C., Jr., Storm, D. R., and Westcott, K. R. (1980). Ca^{2+} regulation of cyclic nucleotide metabolism. *Ann. N.Y. Acad. Sci.* **356,** 205–219.

Kincaid, R. L., Vaughan, M., Osborne, J. C., Jr., and Tkachuk, V. A. (1982). Ca^{2+}-dependent interaction of 5-dimethyl-aminonaphthalene-1-sulfonyl-calmodulin with cyclic nucleotide phosphodiesterase, calcineurin, and troponin I. *J. Biol. Chem.* **257,** 10638–10643.

Kincaid, R. L., Manganiello, V. C., Odya, C. E., Osborne, J. C., Jr., Stith-Coleman, I. E.,

Danello, M. A., and Vaughan, M. (1984). Purification and properties of calmodulin-stimulated phosphodiesterase from mammalian brain. *J. Biol. Chem.* **259**, 5158–5166.

King, M. M., and Huang, C. Y. (1983). Activation of calcineurin by nickel ions. *Biochem. Biophys. Res. Commun.* **114**, 955–961.

King, M. M., and Huang, C. Y. (1984). The calmodulin-dependent activation and deactivation of the phosphoprotein phosphatase, calcineurin, and the effect of nucleotides, pyrophosphate and divalent metal ions. *J. Biol. Chem.* **259**, 8847–8856.

King, M. M., Huang, C. Y., Chock, P. B., Nairn, A. C., Hemmings, H. C., Jr., Chan, K.-F. J., and Greengard, P. (1984). Mammalian brain phosphoproteins as substrates for calcineurin. *J. Biol. Chem.* **259**, 8080–8083.

King, M. M., Lynn, K. K., and Huang, C. Y. (1985). Activation of the calmodulin-dependent phosphoprotein phosphatase by nickel ions. *In* "Progress in Nickel Toxicology" (S. S. Brown and F. W. Sunderman, Jr., eds.). Blackwell, Oxford (in press).

Klee, C. B., and Haiech, J. (1980). Concerted role of calmodulin and calcineurin in calcium regulation. *Ann. N.Y. Acad. Sci.* **356**, 43–54.

Klee, C. B., and Krinks, M. H. (1978). Purification of cyclic 3′,5′-nucleotide phosphodiesterase inhibitory protein by affinity chromatography on activator protein coupled to Sepharose. *Biochemistry* **17**, 120–126.

Klee, C. B., Crouch, T. H., and Krinks, M. H. (1979). Calcineurin: A calcium- and calmodulin-binding protein of the nervous system. *Proc. Natl. Acad. Sci. U.S.A.* **79**, 6270–6273.

Klee, C. B., Oldewurtel, M. D., Williams, J. F., and Lee, J. W. (1981). Analysis of Ca^{2+}-binding proteins by high performance liquid chromatography. *Biochem. Int.* **2**, 485–493.

Klee, C. B., Krinks, M. H., Manalan, A. S., Cohen, P., and Stewart, A. A. (1983a). Isolation and characterization of bovine brain calcineurin: A calmodulin-stimulated protein phosphatase. *In* "Methods in Enzymology" (A. R. Means and B. W. O'Malley, eds.), Vol. 102, pp. 227–244. Academic Press, New York.

Klee, C. B., Newton, D. L., and Krinks, M. (1983b). Versatility of calmodulin as a cytosolic regulator of cellular function. *In* "Affinity Chromatography and Biological Recognition" (I. M. Chaiken, M. Wilchek, and I. Parikh, eds.), pp. 55–67. Academic Press, New York.

Krebs, E. G., and Beavo, J. A. (1979). Phosphorylation-dephosphorylation of enzymes. *Annu. Rev. Biochem.* **48**, 923–959.

Kretsinger, R. H. (1980). Structure and evolution of calcium-modulated proteins. *CRC Crit. Rev. Biochem.* **8**, 119–174.

Krinks, M. H., Haiech, J., Rhoads, A., and Klee, C. B. (1984). Reversible and irreversible activation of cyclic nucleotide phosphodiesterase: Separation of the regulatory and catalytic domains by limited proteolysis. *Adv. Cyclic Nucleotide Protein Phosphorylation Res.* **16**, 31–47.

Krueger, B. K., Forn, J., and Greengard, P. (1977). Depolarization-induced phosphorylation of specific proteins, mediated by calcium ion influx, in rat brain synaptosomes. *J. Biol. Chem.* **252**, 2764–2773.

Lamy, F. (1982). Regulation of biochemical processes through protein phosphorylation and dephosphorylation: Several important examples. *In* "Cell Regulation by Intracellular Signals" (S. Swillers and J. F. Dumont, eds.), pp. 173–193. Plenum, New York.

LaPorte, D. C., Wierman, B. M., and Storm, D. R. (1980). Calcium-induced exposure of a hydrophobic surface on calmodulin. *Biochemistry* **19**, 3814–3819.

Larsen, F. F., Raess, B. U., Hinds, T. R., and Vincenzi, F. F. (1978). Modulator binding protein antagonizes activation of $(Ca^{2+}-Mg^{2+})$-ATPase and Ca^{2+} transport of red blood cell membranes. *J. Supramol. Struct.* **9**, 269–274.

Lee, E. Y. C., Brandt, H., Capulong, Z. L., and Killilea, S. D. (1976). Properties and regulation of liver phosphorylase phosphatase. *Adv. Enzyme Regul.* **14**, 467–490.

Lee, E. Y. C., Silberman, S. R., Ganapathi, S. P., and Paris, H. (1980). The phosphoprotein phosphatases: Properties of the enzymes involved in the regulation of glycogen metabolism. *Adv. Cyclic Nucleotide Res.* **13**, 95–131.

Levin, R. M., and Weiss, B. (1977). Binding of trifluoperazine to the calcium-dependent activator of cyclic nucleotide phosphodiesterase. *Mol. Pharmacol.* **13**, 690–697.

Li, H.-C. (1982). Phosphoprotein phosphatases. *Curr. Top. Cell. Regul.* **21**, 129–174.

Li, H.-C. (1984). Activation of brain calcineurin phosphatase towards nonprotein phosphoesters by Ca^{2+}, calmodulin and Mg^{2+}. *J. Biol. Chem.* **259**, 8801–8807.

Lin, Y. M., and Cheung, W. Y. (1980). Ca^{2+}-dependent cyclic nucleotide phosphodiesterase. *In* "Calcium and Cell Function" (W. Y. Cheung, ed.), Vol. 1, pp. 79–104. Academic Press, New York.

Lin, Y. M., Liu, Y. P., and Cheung, W. Y. (1974). Cyclic 3′,5′-nucleotide phosphodiesterase. Purification, characterization, and active form of the protein activator from bovine brain. *J. Biol. Chem.* **249**, 4943–4954.

Liu, Y. P., and Cheung, W. Y. (1976). Cyclic 3′:5′-nucleotide phosphodiesterase: Ca^{2+} confers more helical conformation to the protein activator. *J. Biol. Chem.* **251**, 4193–4198.

Lynch, T. J., and Cheung, W. Y. (1979). Human erythrocyte Ca^{2+}-Mg^{2+}-ATPase: Mechanism of stimulation by Ca^{2+}. *Arch. Biochem. Biophys.* **182**, 124–133.

Manalan, A. S., and Klee, C. B. (1983). Activation of calcineurin by limited proteolysis. *Proc. Natl. Acad. Sci. U.S.A.* **80**, 4291–4295.

Meijer, L., and Guerrier, R. (1982). Activation of calmodulin dependent NAD^+ kinase by trypsin. *Biochim. Biophys. Acta* **702**, 143–146.

Merat, D. L., Hu, Z. Y., Carter, T. E., and Cheung, W. Y. (1984). Subunit A of calmodulin-dependent phosphatase requires Mn^{2+} for activity. *Biochem. Biophys. Res. Commun.* **122**, 1389–1397.

Merat, D. L., Hu, Z. Y., Carter, T. E., and Cheung, W. Y. (1985). Bovine brain calmodulin-dependent protein phosphatase: regulation of subunit A activity by calmodulin and subunit B. *J. Biol. Chem.* **260**, 11053–11059.

Miyake, M., Daly, J. W., and Creveling, C. R. (1977). Purification of calcium-dependent phosphodiesterases from rat cerebrum by affinity chromatography on activator protein-Sepharose. *Arch. Biochem. Biophys.* **181**, 39–45.

Murthy, A. S. N., and Flavin, M. (1983). Microtubule assembly using the microtubule-associated protein MAP-2 prepared in defined states of phosphorylation with protein kinase and phosphatase. *Eur. J. Biochem.* **137**, 37–46.

Nelson, R. L., and Branton, P. E. (1984). Identification, purification, and characterization of phosphotyrosine-specific protein phosphatases from cultured chicken embryo fibroblasts. *Mol. Cell. Biol.* **4**, 1003–1012.

Nestler, E. J., and Greengard, P. (1984). Protein phosphorylation in nervous tissue. *In* "Catecholamines, Part A: Basic and Peripheral Mechanisms," (E. Usdin, A. Carlsson, A. Dahlström and J. Engel, eds.), pp. 9–22. Alan R. Liss, Inc., New York.

Niggli, U., Adunyah, E. S., and Carafoli, E. (1981). Acidic phospholipids, unsaturated fatty acids, and limited proteolysis mimic the effect of calmodulin on purified erythrocyte Ca^{2+}-ATPase. *J. Biol. Chem.* **256**, 8588–8592.

Nimmo, G. A., and Cohen, P. (1978). The regulation of glycogen metabolism. Phosphorylation of inhibitor-1 from rabbit skeletal muscle, and its interaction with protein phosphatase-III and -II. *Eur. J. Biochem.* **87**, 353–367.

Ouimet, C. C., Miller, P. E., Hemmings, H. C., Jr., Walaas, S. I., and Greengard, P. (1984).

DARPP-32, a dopamine and adenosine $3':5'$-monophosphate regulated phosphoprotein enriched in dopamine-innervated brain regions. III. Immunocytochemical localization. *J. Neurosci.* **4**, 111–124.

Pallen, C. J., and Wang, J. H. (1983). Calmodulin-stimulated dephosphorylation of p-nitrophenyl phosphate and free phosphotyrosine by calcineurin. *J. Biol. Chem.* **258**, 8550–8553.

Pallen, C. J., and Wang, J. H. (1984). Regulation of calcineurin by metal ions. Mechanism of activation by Ni^{2+} and an enhanced response to Ca^{2+}/calmodulin. *J. Biol. Chem.* **259**, 6134–6141.

Pallen, C. J., Valentine, K. A., Wang, J. H., and Hollenberg, M. D. (1984). Calcineurin can dephosphorylate the epidermal growth factor receptor of human placental membrane. *Fed. Proc., Fed. Am. Soc. Exp. Biol.* **43**, 1519 (abstract).

Pato, M. D., and Adelstein, R. S. (1980). Dephosphorylation of the 20,000-dalton light chain of myosin by two different phosphatases from smooth muscle. *J. Biol. Chem.* **255**, 6535–6538.

Reed, L. J., and Pettit, F. (1981). Phosphorylation and dephosphorylation of pyruvate dehydrogenase. *Cold Spring Harbor Conf. Cell Proliferation* **8**, 701–711.

Resink, T. J., Hemmings, B. A., Tung, H. Y. L., and Cohen, P. (1983). Characterication of a reconstituted Mg-ATP-dependent protein phosphatase. *Eur. J. Biochem.* **133**, 455–461.

Richman, P. G., and Klee, C. B. (1978). Interaction of ^{125}I-labeled Ca^{2+}-dependent regulator protein with cyclic nucleotide phosphodiesterase and its inhibitory protein. *J. Biol. Chem.* **253**, 6323–6326.

Roufogalis, B. D. (1982). Specificity of trifluoperazine and related phenothiazines for calcium-binding proteins. *In* "Calcium and Cell Function" (W. Y. Cheung, ed.), Vol. 3, pp. 129–159. Academic Press, New York.

Schlicter, D. J., Casnellie, J. E., and Greengard, P. (1978). An endogenous substrate for cGMP-dependent protein kinase in mammalian cerebellum. *Nature (London)* **273**, 61–62.

Sharma, R. K., Desai, R., Waisman, D. M., and Wang, J. H. (1979). Purification and subunit structure of bovine brain modulator binding protein. *J. Biol. Chem.* **254**, 4276–4282.

Sharma, R. K., Taylor, W. A., and Wang, J. H. (1983). Use of calmodulin affinity chromatography for purification of specific calmodulin-dependent enzymes. *In* "Methods in Enzymology" (A. R. Means and B. W. O'Malley, eds.), Vol. 102, pp. 210–219. Academic Press, New York.

Sibley, D. R., Peters, J. R., Nambi, P., Caron, M. G., and Lefkowitz, R. J. (1984). Desensitization of turkey erythrocyte adenylate cyclase. β-adrenergic receptor phosphorylation is correlated with attenuation of adenylate cyclase activity. *J. Biol. Chem.* **259**, 9742–9749.

Stewart, A. A., Ingebritsen, T. S., Manalan, A., Klee, C. B., and Cohen, P. (1982). Discovery of a Ca^{2+}- and calmodulin-dependent protein phosphatase. Probable identity with calcineurin (CaM-BP$_{80}$). *FEBS Lett.* **137**, 80–84.

Stewart, A. A., Ingebritsen, T. S., and Cohen, P. (1983). The protein phosphatases involved in cellular regulation. 5. Purification and properties of a Ca^{2+}/calmodulin-dependent protein phosphatase (2B) from rabbit skeletal muscle. *Eur. J. Biochem.* **132**, 289–295.

Tallant, E. A. (1983). Purification and characterization of a calmodulin-dependent protein phosphatase from bovine brain. Ph.D. Dissertation, University of Tennessee Center for the Health Sciences, Memphis.

Tallant, E. A., and Cheung, W. Y. (1983). Calmodulin-dependent protein phosphatase: A developmental study. *Biochemistry* **22**, 3630–3635.

Tallant, E. A., and Cheung, W. Y. (1984a). Activation of bovine brain calmodulin-dependent protein phosphatase by limited trypsinization. *Biochemistry* **23**, 973–979.

Tallant, E. A., and Cheung, W. Y. (1984b). Characterization of bovine brain calmodulin-dependent protein phosphatase. *Arch. Biochem. Biophys.* **232**, 260–279.

Tallant, E. A., and Wallace, R. W. (1985). Characterization of a calmodulin-dependent phosphatase from human platelets. *J. Biol. Chem.* **260,** 7744–7751.

Tallant, E. A., Wallace, R. W., and Cheung, W. Y. (1983). Purification and radioimmunoassay of calmodulin-dependent protein phosphatase from bovine brain. *In* "Methods in Enzymology" (A. R. Means and B. W. O'Malley, eds.), Vol. 102, pp. 244–256. Academic Press, New York.

Teo, T. S., and Wang, J. H. (1973). Mechanism of activation of a cyclic adenosine 3′:5′-monophosphate phosphodiesterase from bovine heart by calcium ions. *J. Biol. Chem.* **248,** 5950–5955.

Tonks, N. K., and Cohen, P. (1983). Calcineurin is a calcium ion-dependent, calmodulin-stimulated protein phosphatase. *Biochim. Biophys. Acta* **747,** 191–193.

Tonks, N. K., and Cohen, P. (1984). The protein phosphatases involved in cellular regulation. Identification of the inhibitor-2 phosphatases in rabbit skeletal muscle. *Eur. J. Biochem.* **145,** 65–70.

Vandenheede, J. R., Yang, S.-D., Goris, J., and Merlevede, W. (1980). ATP-Mg-dependent protein phosphatase from rabbit skeletal muscle. II. Purification of the activating factor and its characterization as a bifunctional protein also displaying synthetase kinase activity. *J. Biol. Chem.* **255,** 11768–11774.

Walaas, S. I., and Greengard, P. (1984). DARPP-32, a dopamine and adenosine 3′:5′-monophosphate-regulated phosphoprotein enriched in dopamine-innervated brain regions. I. Regional and cellular distribution in the rat brain. *J. Neurosci.* **4,** 84–98.

Walaas, S. I., Aswad, D. W., and Greengard, P. (1983). A dopamine- and cyclic AMP-regulated phosphoprotein enriched in dopamine-innervated brain regions. *Nature (London)* **301,** 69–71.

Wallace, R. W., Lynch, T. J., Tallant, E. A., and Cheung, W. Y. (1978). An endogenous inhibitor protein of brain adenylate cyclase and cyclic nucleotide phosphodiesterase. *Arch. Biochem. Biophys.* **187,** 328–334.

Wallace, R. W., Lynch, T. J., Tallant, E. A., and Cheung, W. Y. (1979). Purification and characterization of an inhibitor protein of brain adenylate cyclase and cyclic nucleotide phosphodiesterase. *J. Biol. Chem.* **254,** 377–382.

Wallace, R. W., Tallant, E. A., and Cheung, W. Y. (1980). High levels of a heat-labile calmodulin-binding protein (CaM-BP$_{80}$) in bovine neostriatum. *Biochemistry* **19,** 1831–1837.

Wallace, R. W., Tallant, E. A., Dockter, M. E., and Cheung, W. Y. (1982). Calcium binding domains of calmodulin. *J. Biol. Chem.* **257,** 1845–1854.

Walsh, M. P., Dabrowska, R., Hinkins, S., and Hartshorne, D. J. (1982). Calcium-independent myosin light chain kinase of smooth muscle. Preparation by limited chymotryptic digestion of the calcium ion dependent enzyme, purification and characterization. *Biochemistry* **21,** 1919–1925.

Wang, J. H., and Desai, R. (1976). A brain protein and its effect on the Ca^{2+}- and protein modulator-activated cyclic nucleotide phosphodiesterase. *Biochem. Biophys. Res. Commun.* **72,** 926–932.

Wang, J. H., and Desai, R. (1977). Modulator binding protein: Bovine brain protein exhibiting the Ca^{2+}-dependent association with the protein modulator of cyclic nucleotide phosphodiesterase. *J. Biol. Chem.* **252,** 4175–4184.

Wang, J. H., Sharma, R. K., and Tam, S. W. (1980). Calmodulin-binding proteins. *In* "Calcium and Cell Function" (W. Y. Cheung, ed.), Vol. 1, pp. 305–328. Academic Press, New York.

Wang, J. H., Pallen, C. J., Brown, M. L., and Mitchell, K. J. (1984). A survey of calcineurin activity toward non-protein substrates. *Fed. Proc., Fed. Am. Soc. Exp. Biol.* **43,** 1897 (abstr.).

Watterson, D. M., and Vanaman, T. C. (1976). Affinity chromatography purification of a cyclic nucleotide phosphodiesterase using immobilized modulator protein, a troponin C-like protein from brain. *Biochem. Biophys. Res. Commun.* **73,** 40–46.

Watterson, D. M., Sharief, F., and Vanaman, T. C. (1980). The complete amino acid sequence of the Ca^{2+}-dependent modulator protein (calmodulin) from bovine brain. *J. Biol. Chem.* **255,** 962–975.

Winkler, M. A., Merat, D. L., Tallant, E. A., Hawkins, S., and Cheung, W. Y. (1984). The catalytic subunit of calmodulin-dependent protein phosphatase from bovine brain resides in subunit A. *Proc. Natl. Acad. Sci. U.S.A.* **81,** 3054–3058.

Witte, O. N., Dasgupta, A., and Baltimore, D. (1980). Abelson murine leukemia virus protein is phosphorylated in vitro to form phosphotyrosine. *Nature (London)* **283,** 826–831.

Wolf, H., and Hofmann, F. (1980). Purification of myosin light chain kinase from bovine cardiac muscle. *Proc. Natl. Acad. Sci. U.S.A.* **77,** 5852–5855.

Wolff, D. J., Poirier, P. G., Brostrom, C. O., and Brostrom, M. A. (1977). Divalent cation binding properties of bovine brain Ca^{2+}-dependent regulatory protein. *J. Biol. Chem.* **252,** 4108–4117.

Wood, J. G., Wallace, R. W., Whitaker, J. N., and Cheung, W. Y. (1980a). Immunocytochemical localization of calmodulin and a heat-labile calmodulin-binding protein ($CaM-BP_{80}$) in basal ganglia of mouse brain. *J. Cell Biol.* **84,** 66–76.

Wood, J. G., Wallace, R. W., Whitaker, J. N., and Cheung, W. Y. (1980b). Immunocytochemical localization of calmodulin in regions of rodent brain. *Ann. N.Y. Acad. Sci.* **356,** 75–82.

Yang, S.-D., Vandenheede, J. R., Goris, J., and Merlevede, W. (1980). ATP-Mg-dependent protein phosphatase from rabbit skeletal muscle. I. Purification of the enzyme and its regulation by the interaction with an activating protein factor. *J. Biol. Chem.* **255,** 11759–11767.

Yang, S.-D., Tallant, E. A., and Cheung, W. Y. (1982). Calcineurin is a calmodulin-dependent protein phosphatase. *Biochem. Biophys. Res. Commun.* **106,** 1419–1425.

Chapter 4

Biophysical Studies of Calmodulin

STURE FORSÉN
HANS J. VOGEL[1]
TORBJÖRN DRAKENBERG

Physical Chemistry, Chemical Centre
Lund University
Lund, Sweden

[1] Present address: Division of Biochemistry, Department of Chemistry, University of Calgary, Calgary, Alberta, Canada T2N 4NI

113

CALCIUM AND CELL FUNCTION, VOL. VI
Copyright © 1986 by Academic Press, Inc.
All rights of reproduction in any form reserved.

I. INTRODUCTION

Calmodulin (CaM) has been found in all eukaryotic cells. Upon binding of Ca^{2+} ions, the protein undergoes conformational changes that expose sites on the protein's surface that allow CaM to interact with a wide variety of enzymes and other proteins (Cheung, 1980; Krebs, 1981; Klee and Vanaman, 1982; Means *et al.*, 1982). The formation of such ternary complexes often leads to a modulation of the activities of these target systems. Thus, CaM's biological function is apparently to translate variations in the level of intracellular messenger Ca^{2+} into a metabolic response. Because of its manifold interactions and functions, CaM has captured the attention of many researchers. Presently it definitely is the most intensely studied member of the troponin-C superfamily of Ca^{2+}-binding proteins. The proteins in this class all contain several well-defined α helix–calcium binding loop–α helix regions in their amino acid sequences. This unit, often referred to as the EF-hand, was observed for the first time in the crystal structure of carp-parvalbumin (Kretsinger and Nockolds, 1973). Similar structures have since been observed in bovine intestinal calcium-binding protein (Szebenyi *et al.*, 1981) and turkey skeletal muscle troponin C (Herzberg and James, 1985). It is generally assumed that CaM's structure will not deviate much from these structures (Kretsinger, 1980). However, crystallographic studies are under way in several laboratories.

Referring to the EF-hand as the structural unit is an oversimplification. It is actually pairs of EF-hands, rather than single isolated sites, that form the building blocks of the proteins of the troponin C superfamily. The interaction between two sites in one pair of EF hands—mainly via hydrogen bonding arranged in a β-structure fashion—is probably pertinent to the functioning of these proteins, as will be discussed in this chapter in more detail.

After this chapter had been written the structure of CaM was determined crystallographically at 3.0 Å resolution (Babu *et al.*, 1985). The molecule is found to have a dumbbell shape with a long central helix that connects two remarkably similar globular lobes each containing two Ca^{2+} ions about 11.3 Å apart. There are no direct contacts between the two lobes. The helix–loop–helix structural pattern postulated for EF-type Ca^{2+}-binding loops is confirmed.

It is indeed encouraging that the body of biophysical data obtained on solutions of CaM and reviewed in this chapter have strongly indicated the protein molecule to be made up of two relatively independent domains of roughly equal size comprising the N-terminal and C-terminal halves of the molecule (see Section IX).

Calmodulin has been remarkably preserved throughout evolution. Amino acid sequences have been reported for CaM purified from bovine brain (Watterson *et al.*, 1980), rat testis (Dedman *et al.*, 1978), scallop, rabbit and chicken skeletal muscle (Toda *et al.*, 1981; Grand *et al.*, 1981; Putkey *et al.*, 1983), human brain (Sasagawa *et al.*, 1982), electric eel (Lagacé *et al.*, 1983), spinach (Burgess *et al.*, 1983), *Metridium senide* (Takagai *et al.*, 1980), *Tetrahymena pyriformis* (Yazawa *et al.*, 1981), *Renilla reniformis* (Jamieson *et al.*, 1980), and *Dictyostelium discoideum* (Marshak *et al.*, 1984). With the exception of some disputes on amide assignments, all vertebrate CaMs appear to be identical. Several amino acid substitutions have, however, been observed in plant, protozoan, and invertebrate CaMs. *Dictyostelium discoideum* CaM shows more differences from bovine CaM than does any other organism. Still, only 14 amino acids are different, including an additional C-terminal residue, and six of these 14 changes are conservative.

Means and co-workers have studied CaM at the level of gene organization and mRNA structure in electric eel and chicken. Their results can be summarized as follows: CaM is apparently not synthesized in a precursor form. This finding is in agreement with recent cell-free synthesis experiments (Walker *et al.*, 1984). The introns found in the chicken calmodulin gene were not in positions corresponding to the domain structure of CaM (Putkey *et al.*, 1983), where they would have been found if Gilbert's 1978 hypothesis concerning intron location and protein domains were equally applicable to all proteins. mRNAs have been found of differing lengths, apparently because of different polyadenylation sites in the CaM gene (Lagacé *et al.*, 1983). Most interesting of all was the isolation of a pseudogene from a chicken library (Stein *et al.*, 1983). This gene was completely devoid of intervening sequences, and it comprised 19 amino acid substitutions with respect to CaM. A large part of this gene has been cloned into, expressed in, and isolated from *Escherichia coli* (Putkey *et al.*, 1985). The protein produced in this fashion still contains 16 of the 19 amino acid changes. Nevertheless, it has retained many of CaMs functional properties, such as the Ca^{2+}-dependent binding to phenothiazines and phenyl-Sepharose, the Ca^{2+}-dependent change in electrophoretic mobility, and the Ca^{2+}-dependent activation of cyclic nucleotide phosphodiesterase (Putkey *et al.*, 1985). Since it is apparently transcribed in muscle (Stein *et al.*, 1983), it will be of importance to determine the physiological role of this CaM-like protein.

Several reviews have already dealt in detail with various aspects of CaM (see, for example, Cheung, 1980; Krebs, 1981; Klee and Vanaman, 1982; Means *et al.*, 1982). Our intent is to try to avoid duplicating these earlier accounts. Also, we will pay very little attention to the forces governing the interactions between CaM and its target systems, since this topic has been expertly dealt with in another review (Keller *et al.*, 1983; see also Olwin *et al.*, 1984). Instead, we will focus our attention on some biophysical studies that have been reported during the last few years and, using these results, attempt to present a more detailed picture of CaM's physical mode of action than was hitherto possible.

II. PROTEOLYTIC AND CHEMICAL MODIFICATION OF CaM

One of the lines of CaM research that has proved to be particularly useful in the past is the study of proteolytic fragments. This area of research was initiated by the work of the late Witold Drabikowski and colleagues at the Nencki Institute in Warsaw, Poland. They reported that controlled proteolysis of Ca^{2+}-saturated CaM by trypsin gives rise to two fragments of about equal size (Drabikowski *et al.*, 1977). Amino acid analysis of these peptides suggested that the cleavage took place at Lys 77 (Walsh *et al.*, 1977). Later, more detailed studies in our laboratory confirmed that cleavage mainly takes place at Lys 77 but that considerable amounts of hydrolysis can occur at Lys 74 or Arg 75 instead (Thulin *et al.*, 1984). Cleavage with thrombin can also be performed in a controlled manner; in the presence of EDTA two fragments, 1–106 and 107–148, are produced (Wall *et al.*, 1981; Andersson *et al.*, 1983b). The proteolytic fragmentation scheme is depicted in Fig. 1. Although the tryptic cleavage of CaM may occur after position 74, 75, or 77, in all cases the protein is essentially broken into two halves. The resultant mixture of fragments can be readily separated using ion-exchange chromatography, making use of the fact that the three different TR_2C fragments differ by only one positive charge (Thulin *et al.*, 1984) (Fig. 2). Some of these fragments retain certain properties of CaM, such as the Ca^{2+}-dependent binding to phenyl-Sepharose (Vogel *et al.*, 1983a). This binding provides a convenient means of purifying CaM (Gopalakrishna and Anderson, 1982) and other calcium-binding proteins, as well as some of their proteolytic fragments (Vogel *et al.*, 1983a; Brzeska *et al.*, 1983) (Fig. 3).

The proteolytic fragments described above have been studied for their capacity to substitute for CaM in activating different enzymes (Walsh *et al.*, 1977, Wall *et al.*, 1981; Kuznicki *et al.*, 1981, Newton *et al.* 1984, Guerini *et al.*, 1984). The outcome of these experiments has shown that only in a few instances could

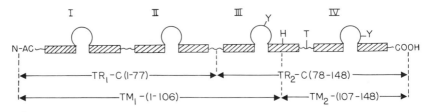

Fig. 1. Schematic representation of the proposed structure of calmodulin outlining the four domains, each of which comprises a calcium-binding loop and two flanking helix regions (hatched areas). Also indicated are the positions of the two tyrosine residues at positions 99 and 138 (Y) and the positions of the single histidine-106 (H) and trimethyllysine-115 (T) residues. Proteolytic fragmentation as induced by tryptic (TR) or thrombic (TM) cleavage is also indicated. Reprinted with permission from Andersson, A., Forsén, S., Thulin, E., and Vogel, H. J. (1983b), *Biochemistry* **22,** 2309. Copyright (1983) American Chemical Society.

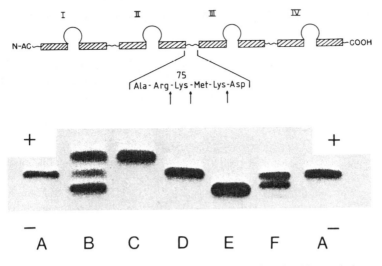

Fig. 2. Agarose gels (run in the presence of 1m*M* EDTA) of CaM and its tryptic fragments: (A) CaM; (B) mixture of $TR_2C_{a,b,c}$; (C) TR_2C_a (1–74); (D) TR_2C_b (1–75); (E) TR_2C_c (1–77); (F) TR_1C. The positions of the cathode and anode are indicated in the figure. Reprinted with permission from Thulin, E., Andersson, A., Drakenberg, T., Forsén, S., and Vogel, H. J. (1984) *Biochemistry* **23,** 1862. Copyright (1984) American Chemical Society.

fragments substitute in part for the intact protein. Since it has been difficult to point out specific interaction sites on CaM for different target enzymes, the idea that there would be different sites for the different target proteins on calmodulin had to be abandoned. The cooperativity between two halves is necessary to accomplish full activation. The activation that had been obtained in some early reports was only obtained at high ratios of peptide to enzyme. Such ratios give rise to the possibility that traces of contaminating nonhydrolysed CaM are responsible for the observed effects. Adequate control experiments have been introduced recently to allow the rigorous exclusion of such a possibility (Newton *et al.,* 1984, Guerini *et al.,* 1984).

Another explanation for the failure of the fragments to substitute for the intact protein could be that they have lost their tertiary structure. However, a wide variety of physical measurements—which will be discussed in detail later—have demonstrated that the two tryptic fragments encompassing the N- and C-terminal half of the protein retained not only their structure but also their affinity for Ca^{2+}. In contrast, the structure and Ca^{2+} affinity were perturbed in protein fragments containing less than two EF-hands. These results—as well as those obtained with synthetic peptide fragments—provided strong hints that the pairs of EF-hands, and not the isolated EF-hands, form the building blocks of the troponin C superfamily of proteins (for more detailed reviews, see Vogel *et al.,*

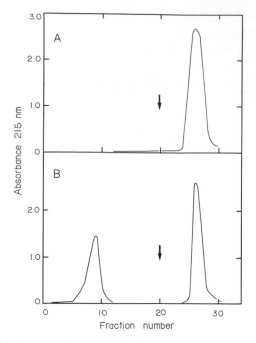

Fig. 3. (A) Ca^{2+}-dependent chromatography of CaM (1.1 mg) on a phenyl-Sepharose column (0.7×20 cm). The arrow indicates the time when 1 mM Ca^{2+} in the buffer was replaced by 1 mM EGTA. (B) Chromatography (as in Fig. 3A) of a mixture (1.8 mg) of the thrombic fragments TM_1 (1–106) and TM_2 (107–148). TM_2 is not retained on the column, whereas TM_1 binds in the presence of Ca^{2+}. From Vogel *et al.* (1983a).

1983b; Seamon and Kretsinger, 1983). Apparently the small piece of β structure (Szebenyi *et al.*, 1981) between the two Ca^{2+}-binding loops which can be detected spectroscopically both by ^{1}H NMR (Dalgarno *et al.*, 1983) and resonance raman studies (Seaton *et al.*, 1983) is of great importance for stabilizing the structure. Likewise, this structural feature may be responsible for the positive cooperativity that is built into this unit and which appears to be partially responsible for the increase in Ca^{2+} affinity over that in an isolated single EF-hand.

To date, relatively few reports are available concerning the chemical modification of CaM. The reason for this is simple: CaM does not contain many groups that can easily be modified in a selective nonperturbing manner. Except for the single His residue in the bovine CaM (Watterson *et al.*, 1980) or the single Cys residue in the plant CaM (Burgess *et al.*, 1983), the protein contains a blocked amino terminal and many acidic and basic amino acid side chains and methionines. Early studies by Walsh and Stevens (1977) demonstrated that one could never selectively modify one amino acid, but usually several at once. Chemical modification of lysines, methionines, and carboxyl groups resulted in a strong decrease or complete loss of activity (Walsh and Stevens, 1977, 1978), whereas O-carboxymethylation of CaM by enzymatic means produces only a small decrease in the ability of CaM to activate cyclic nucleotide phosphodiesterase (Billingsley et al., 1984).

Further selective modifications have been reported by Richman and Klee (1978). They demonstrated that both tyrosines could be nitrated by tetranitromethane in the presence of Ca^{2+}. However, in the presence of EDTA only Tyr 99 was modified. Interestingly, acetylation of the tyrosine with N-acetylimidazole only modified Tyr 99 in the presence of Ca^{2+}, and both tyrosines were modified by this reagent in the presence of EDTA. This differential reactivity created the possibility of selectively nitrating Tyr 138 by first acetylating Tyr 99 (Richman, 1978). These strategies are schematically depicted in Fig. 4. The nitrotyrosine chromophores are not only sensitive spectroscopic probes for studying, for example, metal ion binding to CaM (Richman and Klee, 1979; McCubbin et al., 1979), but they also open the way to some further chemistry. Addition of sodium dithionite quickly reduces nitrotyrosine to an amino tyrosine (Sokolovsky et al., 1967). The rather low pK_a (4–5) of the amino group on the amino tyrosine makes it ideally suited for specific attachments of various modification agents. This strategy has been used to date to attach cross-linking reagents (Klevitt and Vanaman, 1983), fluorescent labels (Lambooy et al., 1983), and ESR spin labels (H. J. Vogel and H. Wickström, unpublished results) to either one of the two tyrosines.

Several groups have prepared bifunctional-affinity and photoaffinity adjuncts of calmodulin. For example a preparation of azido-calmodulin labeled several

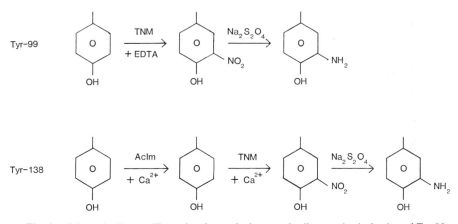

Fig. 4. Schematic diagram illustrating the synthetic routes leading to selectively nitrated Tyr 99 and Tyr 138 CaM. TNM, tetranitromethane; AcIm, acetylimidazole. Note that treatment with AcIm in the presence of Ca^{2+} modifies only Tyr-99 and not Tyr-138. Adapted from Richman and Klee (1978) and Richman (1978).

target proteins such as myosin light chain kinase and cyclic nucleotide phosphodiesterase and formed 1 : 1 complexes (Andreasen *et al.*, 1981). Such complexes should be useful for localization and purification of the CaM binding site in these proteins. CaM has also been used for cross-linking studies with membrane proteins as targets (Goewert *et al.*, 1982; Louis and Jarvis, 1982). Similarily, Kincaid (1984) reported the formation of a reversible 1 : 1 complex of 3-(2-pyridyldithio)propionyl-substituted CaM (PDP CaM) and brain cyclic-nucleotide phosphodiesterase. The covalent complex was active and was less dependent on Ca^{2+} for activity. Moreover it was not inhibited by typical inhibitors of this reaction such as TFP or W-7. A similar strategy was also used earlier where PDP CaM was coupled covalently to thiol-Sepharose. This column retained cyclic nucleotide phosphodiesterase which was subsequently eluted as the CaM complex merely by adding reducing agents. The authors suggest that this experimental strategy may be particularly advantageous in instances where the target protein is rather unstable in the absence of CaM (Kincaid and Vaughan, 1983).

III. CALCIUM-BINDING STUDIES

The binding of Ca^{2+} to CaM in solution has been investigated by a number of workers (Dedman *et al.*, 1977; Jarret and Kyte, 1979; Crouch and Klee, 1980; Cox *et al.*, 1981; Haiech *et al.*, 1981; Keller *et al.*, 1982; Burger *et al.*, 1984). Binding parameters have been obtained mainly from two types of measurements. The first, which we will call the "solution chemical approach," involves the measurement of the concentration or activity of free Ca^{2+} in equilibrium with CaM under varying degrees of saturation. Equilibrium or flow dialysis cells have usually been employed and the Ca^{2+} concentrations determined by means of atomic absorption or through the specific activity of the radioactive isotope ^{45}Ca. In a few cases Ca^{2+}-selective electrodes have been used. The second type involves the observation—usually spectroscopical—of changes, as a function of Ca^{2+} concentration, in some molecular property assumed to reflect the state of ligation. Let us call this the "protein conformation approach."

Before we discuss some recent experimental results on CaM and some of its proteolytic fragments it may be instructive to discuss briefly some general results concerning multiple ligand binding. In line with the primary sequence of CaM, let us assume CaM to have four sites of ligand binding. With no restrictive assumptions, a general four-site system is completely defined by 15 independent site-binding constants (k_i, k_{ij}, k_{ijk} or k_{ijkl}) and 4 stoichiometric constants (K_i) defined in Fig. 5 (Klotz and Hunston, 1979).

The number of parameters needed for a complete characterization of a four-site system is obviously far greater than one can reasonably hope to determine by

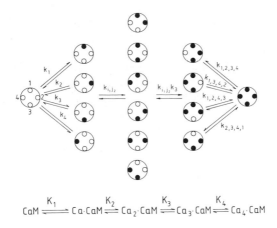

$$\text{CaM} \xrightleftharpoons{K_1} \text{Ca·CaM} \xrightleftharpoons{K_2} \text{Ca}_2\text{·CaM} \xrightleftharpoons{K_3} \text{Ca}_3\text{·CaM} \xrightleftharpoons{K_4} \text{Ca}_4\text{·CaM}$$

Fig. 5. The site-binding constants (k) and stoichiometric binding constants (K) in CaM treated as a general four-site system.

any technique. A complete characterization may in fact not be very meaningful. We would be content with a simplified description that captures the major properties of our system. The solution–chemical approach usually allows only the determination of the average ligand number, $\bar{\nu}$, as a function of the ligand concentration. In the four-site system, $\bar{\nu}$ can be completely specified by the stoichiometric constants K_1, K_2, K_3, and K_4 [the Adair equation (Adair, 1925)]:

$$\bar{\nu} = \frac{(K_1[\text{Ca}] + 2K_1K_2[\text{Ca}]^2 + 3K_1K_2K_3[\text{Ca}]^3 + 4K_1K_2K_3K_4[\text{Ca}]^4)}{1 + K_1[\text{Ca}] + K_1K_2[\text{Ca}]^2 + K_1K_2K_3[\text{Ca}]^3 + K_1K_2K_3K_4[\text{Ca}]^4} \quad (1)$$

The binding constants K_1 through K_4 may be obtained from the experimental $\bar{\nu}$ values through a nonlinear least squares fit. Needless to say, a high experimental precision is required in order to yield reliable K values. As a limiting case let us consider the four sites to be equal in their affinity for Ca^{2+} and completely independent. Let the site-binding constant be k. In this case the stoichiometric binding constants are related through the equation $K_i = k[(n - i + 1)/i]$ (Tanford, 1961) and Eq. (1) reduces to

$$\bar{\nu} = (4k[\text{Ca}])/(1 + k[\text{Ca}]) \quad (2)$$

All Ca^{2+}-binding data obtained so far by solution–chemical techniques have been analyzed using either Eq. (1) or (2). It is instructive to present the resultant binding-constant data using the graphic ''affinity profile'' approach of Klotz (1974) and Klotz and Hunston (1979). The affinity profile is simply a plot of $i \cdot K_i$ versus i. If the individual sites are independent and equivalent in their affinity, the plot should be a straight line with a certain "ideal" slope. If the slope of the line connecting any actual K_{i-1} with K_i is greater than the slope of the ideal line, the binding is positively cooperative, and, conversely, negative cooperativity is characterized by a slope less than that of the ideal line. The affinity profile for Ca^{2+} binding to CaM is given in Fig. 6. The plot is quite revealing. In all cases in which the experimental data have been analyzed by means of Eq. (1), the binding of the first two Ca^{2+} ions to CaM is seen to be positively cooperative. (The scatter in the individual binding data may partly be a result of the somewhat different experimental conditions employed—see legend to Fig. 6.) The same conclusion has also been drawn by some of the originators of the data. The

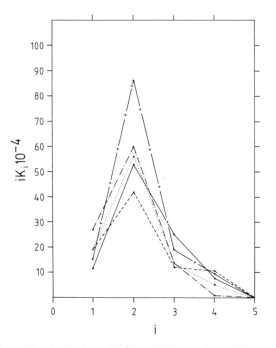

Fig. 6. Affinity profiles for binding of Ca^{2+} to CaM according to Klotz (1974) and Klotz and Hunston (1979). K_i represents the stoichiometric binding constant for the ith Ca^{2+} ion. If the binding sites are completely independent and equivalent, successive K_i values should be linearly related in a graph of iK_i versus i. Stoichiometric binding constants have been taken from binding data reported in the literature, and the data have been evaluated using an Adair-type Eq. (1). Data obtained under the explicit or implicit assumption of identical and independent sites have been omitted. Experimental conditions and references are as follows: ——— (Burger *et al.,* 1984 : 60 mM TES, pH 7.0, 150 mM NaCl); - - - - - - (Crouch and Klee, 1980 : 10 mM HEPES, pH 7.5, 100 mM KCl, 3 mM MgCl$_2$); .... (Keller, *et al.,* 1982 : 10 mM MOPS, pH 7.2, 150 mM KCl, 1 mM MgCl$_1$);—.—.— .—.— (Haiech *et al.,* 1981 : 10 mM Tris, pH 7.55, 200 mM KCl) -+-+-+ (Crouch and Klee, 1980 : 10 mM HEPES, pH 7.5, 100 mM KCl).

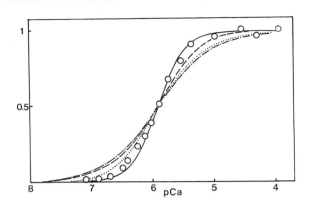

Fig. 7. Experimental points ($\bigcirc$) for the tyrosine fluorescence change in TR_2C as a function of calcium concentration (Drabikowski *et al.*, 1982) and theoretical curves calculated under various assumptions. _____ Total cooperativity in the binding, $K = 7.1 \times 10^5 M^{-1}$. _____ No cooperativity; binding to either site results in total change in fluorescence, $K = 3 \times 10^5 M^{-1}$. _____·_____ No cooperativity; binding to the first site has no effect; binding to the second site results in total change in flourescence, $K = 7.1 \times 10^5 M^{-1}$. ..... No cooperativity; binding to the first site has no effect; binding to the second site results in total change in fluorescence, $K = 1.7 \times 10^6 M^{-1}$.

relation between the other binding constants (for example K_3 and K_4) is less clear.

Stoichiometric binding constants provide little direct insight into the molecular details of the binding process. Such information may in principle be obtained through a variety of spectroscopic methods. A number of studies using UV spectroscopic fluorescence and circular dicroism have examined CaM as a function of Ca^{2+} concentration. We shall briefly discuss the claims by Burger *et al.* (1984) that the structural changes in CaM are sequential and ordered. A comparison between the Ca^{2+}-dependent changes in ellipticity at 222 nm and the intrinsic tyrosyl fluorescence of CaM and some of its tryptic fragments, notably TR_1C (aa 1–77) and TR_2C (aa 78–148), has been made by Drabikowski *et al.* (1982). The changes in tyrosyl fluorescence ($\lambda_{exc} = 286$ nm; $\lambda_{emiss} = 306$ nm) as a function of Ca^{2+} concentration are nearly identical for intact CaM and the fragment TR_2C (also J. D. Johnson, personal communication). It is therefore of some interest to analyze the spectral changes of TR_2C as a protein with only two potential Ca^{2+} binding sites. Figure 7 shows the experimental fluorescence changes and the results of some model calculations—normalized so as to coincide with the midpoint of the experimental curve. In the first model it is assumed that Ca^{2+} binding to the two sites in TR_2C is infinitely cooperative and that the fluorescence changes are concomitant with the appearance of the species Ca_2CaM. It is apparent that this model fits the experimental data reasonably well but that it predicts a somewhat more rapid change with $[Ca^{2+}]$ than is experimentally observed. Implicit in this model is that it is irrelevant if the structural changes causing the increased fluorescence occur after the binding of the first or the second Ca^{2+}; since the fraction of singly occupied molecules is always negligible, the result would remain the same. In the second model it is assumed that the two Ca^{2+} binding sites of TR_2C are completely independent and equal. In this case the two stoichiometric binding constants K_1 and K_2 are related to the site-binding constant K, through the equation $K_1 = 2k$ and $K_2 = \frac{1}{2}k$. If we assume that the fluorescence changes parallel the fraction of sites occupied (i.e., $\bar{v}/2$), the changes should follow Curve 2, in obvious disagreement with the experimental data. Next we assume that the two sites are completely independent and equal, but that the fluorescence changes are the result of binding of the second Ca^{2+}. Again the agreement with the experiment is poor. In the next model the sites are again assumed to be completely independent and equal, but the fluorescence changes are the result of the binding of the first Ca^{2+}, and the second Ca^{2+} has no additional effect. The slope of this curve clearly is greater than in any of the other noncooperative models, but still far less than required to give a reasonable fit to the experimental data. It seems fairly obvious that a good fit to the experiments on TR_2C can only be reached with a model that involves positive cooperative binding of Ca^{2+}. We may note that a calculation of the pertinent Ca^{2+}-binding constants for TR_2C from the experimental fluorescence data is not straightforward, since it is first necessary to assume a model in which the various possible liganded species are assigned a certain spectral change.

The studies of the TR_2C fragment of CaM are important because they show many of the characteristic behaviors of the native CaM. The total change in fluorescence intensity and in the transition midpoint in going from Ca^{2+}-free to Ca^{2+}-saturated CaM is very close to that found for TR_2C (Drabikowski et al., 1982). Buffer concentration (HEPES 20/50 100 mM) has little effect on the fluorescence curve, and addition of 100 mM KCl mainly shifts the transition midpoint. Such a shift is expected since the activity coefficient for Ca^{2+} will change considerably with the ionic strength, I. [When going from $I = 0$ to $I = 0.15$ the Ca^{2+}-activity coefficient is calculated to change from unity to about 0.33 (Thomas, 1982)]. The sum of the CD spectra of the fragments, TR_1C and TR_2C equals that of CaM under a variety of conditions (Drabikowski et al., 1982), indicating that the cleavage of the 77–78 peptide bond has little effect on the secondary structure on the NH_2- and COOH-terminal halves. As will be discussed in other sections, this conclusion may also be extended to the tertiary structures. It is therefore likely that data obtained for the fragments provides information on the NH_2- and COOH-terminal halves of the intact CaM molecule.

In our opinion experimental binding studies strongly suggest that the binding of at least the first two Ca^{2+} ions to CaM is cooperative under a variety of conditions, including the type of buffers used and the solution ionic strength. The experimental data allow an estimate of the energy coupling associated with the binding. The free energy of interaction between two sites, i and j, $\Delta G_{I,ij}$, in a system displaying cooperativity may be calculated as follows (Cantor and Schimmel, 1980):

$$\Delta G_{I,ij} = RT \ln(K_i K_j) - RT \ln\left(\frac{\Omega_{n,i-1}/\Omega_{n,i}}{\Omega_{n,j-1}/\Omega_{n,j}}\right) \quad (3)$$

where K_i and K_j are stoichiometric binding constants and $\Omega_{n,i}$, etc., are the statistical factors relating the stoichiometric to site-binding constants for a system with a total of n sites. Depending on which experimental values are used and which model is used for the purely statistical factors (a 2-site or 4-site model) a free energy of interaction, $\Delta G_{I,1,2}$ (where 1 and 2 stand for sites III and IV of CaM) is calculated to be in the range -0.7 to -1.3 kcal/mol.

IV. THERMOCHEMICAL STUDIES

Thermochemical studies of Ca^{2+} binding to CaM and to the tryptic fragments TR_1C and TR_2C have recently been performed by two groups. Tanokura and Yamada (1983, 1984) studied the enthalpy changes accompanying Ca^{2+} or Mg^{2+} binding to bovine brain CaM in a solution containing 20 mM PIPES/NaOH (pH 7.0) and 100 mM KCl. The effect of Mg^{2+} or Ca^{2+} on the reaction enthalpy upon binding of the other ion was also studied. The results obtained were surprising: the binding of 4 Ca^{2+} is endothermic in the absence of Mg^{2+} ($\Delta H_{tot} = 3.8$ kcal/mol at 25°C) as is the binding of Mg^{2+} in the absence of Ca^{2+} ($\Delta H_{tot} \approx 9.1$ kcal/mol at 25°C). This result is contrary to what has been found for Ca^{2+} binding to homologous proteins under similar conditions: parvalbumin, $\Delta H_{tot} \approx -17.9$ kcal/mol (Moeschler et al., 1980; Filimonov et al., 1978); STnC, $\Delta H_{tot} = -30.8$ kcal/mol (Potter et al., 1977) or -13.3 kcal/mol (Yamada and Kometani, 1982); CTnC, $\Delta H_{tot} = -4.6$ kcal/mol (Kometani and Yamada, 1983). The thermochemical results of the Japanese workers are also in disagreement with the results obtained in a recent Swedish study (Laynez et al., 1983; Sellers et al., 1985). CaM from both bovine testes and brain was employed with essentially the same result. In addition, Ca^{2+} binding to the fragments TR_1C and TR_2C was studied. The enthalpy change at different temperatures as a function of molar ratio of Ca^{2+} to protein is shown in Fig. 8. (Under the conditions of the experiment, this ratio is essentially equal to the number of Ca^{2+} bound to the protein.) The enthalpy change is negative at all temperatures, in line with the results for homologous Ca^{2+}-binding proteins.

One striking feature of the data in Fig. 8 is the approximately linear dependence of the enthalpy change on the Ca^{2+}/CaM ratio in the range 0–2. This linear behavior was also observed by Tanokura and Yamada (1983, 1984). Since

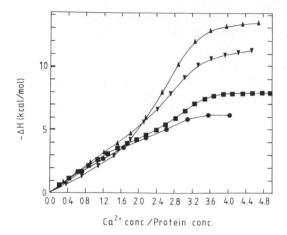

Ca^{2+} conc./Protein conc.

Fig. 8. Observed enthalpy change for binding of Ca^{2+} to bovine testes calmodulin. Observed enthalpy changes at four different temperatures (Laynez *et al.*, 1983; Sellers *et al.*, 1985). Conditions of the experiment were as follows: HEPES buffer 50 mM, pH 7.0, and calmodulin concentration 0.3 mM. Symbols: 37°C (▲), 30°C (▼), 25°C (■), 21°C(▨). The experiments were performed using an LKB Batch Titration Microcalorimeter (see Chen and Wadsö, 1982).

the Ca^{2+}/CaM ratio is almost identical to the number of moles of Ca^{2+} bound per mole of CaM, this kind of behavior is what one would expect if the binding of the first two Ca^{2+} was cooperative during a Ca^{2+} titration. In such a case the solution should mainly contain protein molecules with zero or two Ca^{2+} ions bonded.

Also striking is how little the temperature influences the enthalpy change during this first binding step in contrast with the second step involving the third and fourth Ca^{2+} ion. One interpretation of this difference would be, that the structural rearrangements of the protein in the two binding steps have different signatures (e.g., that mainly hydrophobic amino acid residues are exposed to the solvent in the first step).

Preliminary work has been done on the enthalpy changes for the binding of Ca^{2+} to the fragments TR_1C and TR_2C (Sellers *et al.*, 1985). At 25°C the following values were determined for the binding of two Ca^{2+} ions under the same solution conditions as used in the study of the parent protein: $\Delta H_{TR_1C} \approx -3.8K$ cal/mol; $\Delta H_{TR_2C} \approx -3.5$ kcal/mol. The sum of the enthalpy changes, $\Delta H_{TR_1C} + \Delta H_{TR_2C} = -7.3$ kcal/mol, is very nearly equal to the total enthalpy change accompanying binding of four Ca^{2+} ions to CaM, $\Delta H_{tot} = -7.9$ kcal/mol. This preliminary result is clearly compatible with a model of CaM as consisting of two nearly independent domains.

In order to throw some light on the apparent discrepancy between the thermochemical results obtained by the Japanese group and those obtained by the Swedish group, Sellers *et al* (1985) have very recently supplemented their earlier study with an investigation into how the details of the solution composition do affect the thermochemical parameters. The same type of microcalorimetric equipment was used in both studies. This new investigation has revealed that the ionic strength of the medium has a marked effect on the thermochemical parameters in contrast to what has been reported for TnC (Yamada *et al.* 1983). At high ionic strength (0.15 M KCl) the enthalpy changes upon binding of Ca^{2+} have changed sign and are now positive. It thus presently appears that the discrepency has largely been resolved—further studies are, however, certainly necessary to quantify the medium effects.

The enthalpy changes also allow for reasonably accurate calculations of the entropy changes associated with Ca^{2+} binding. Rather than trying to dissect the entropy changes characterizing the individual single ion-binding steps, we will aim for average values per mole Ca^{2+}. The binding of one Ca^{2+} ion to TR_2C— the tryptic fragment with the two high-affinity sites—is associated with a free energy change ΔG_{TR_2C} of about -8.25 kcal/mol ($K_{ass} \approx 10^6 M^{-1}$) (Drabikowski *et al.*, 1982). Using the corresponding enthalpy change per bonded Ca^{2+} ion as measured by Sellers *et al.*, (1985) (i.e., -1.75 kcal/mol) (we consider the slight differences in solution conditions to be of minor importance), the entropy change is obtained as $\Delta S_{TR_2C} \approx +21.7$ cal/mol, °K.

A similar calculation for the binding of Ca^{2+} ions to the low-affinity fragment TR_1C will suffer in accuracy due to the less well known value for the change in free energy. Using $\Delta G_{TR_1C} \approx -6.8$ kcal/mol (corresponding to $K_{ass} \approx 10^5 M^{-1}$) and with $\Delta H_{TR_1C} = -1.9$ kcal/mol, we obtain $\Delta S_{TR_1C} = 16.3$ cal/mol, °K.

For a calculation of the average entropy change associated with the binding of Ca^{2+} to the whole CaM molecule, we must decide on what value to choose for the average free-energy change $\overline{\Delta G}_{CaM}$. Noting that $\overline{\Delta G}_{CaM}$ is related to the stoichiometric binding constants K_i through the equation

$$4 \, \overline{\Delta G}_{CaM} = - \, RT \ln (K_1 K_2 K_3 K_4) \qquad (4)$$

we obtain $\overline{\Delta G}_{CaM} = -6.8 \pm 0.5$ kcal/mol as an average from binding constants obtained in six studies (data summarized by Burger *et al.*, 1984). With the average enthalpy change of -1.97 kcal/mol, the average entropy change $\overline{\Delta S}_{CaM}$ is obtained as 16.1 cal/mol, °K.

It is of some interest to compare the above data with those pertinent to Ca^{2+} binding in small molecule complexes which have some general structural resemblance to the CaM binding sites. For EDTA in aqueous solutions with an ionic strength of about 0.1, the binding of Ca^{2+} is characterized by the following approximate thermodynamic parameters (20–25°C): $\Delta G_{EDTA} = -14.5$ kcal/mol; $\Delta H_{EDTA} = -6.6$ kcal/mol; $\Delta S_{EDTA} = 26$ cal/mol, °K [see critical compilations by Christensen and Izatt (1970) and Martell and Smith (1982)]. The binding of Ca^{2+} to CaM and its tryptic fragments TR_1C and TR_2C is partly an entropy-driven process, as it is for binding to EDTA. The positive entropies of binding strongly point to the importance of water of solvation—both that of the cation and that of the negatively charged or polar ligands will be "released" in the binding process. Inasmuch as the spectroscopically observed conformational changes in CaM and its tryptic fragments upon binding of Ca^{2+} lead to an "ordering" of the protein structure, one would expect this effect to be associated with an entropy decrease. Judging from the ΔS changes given above, these conformational entropy changes are of secondary importance, in particular that of the fragment TR_2C. It remains a possibility, however, that the less positive value of ΔS_{TR_1C} by comparison to ΔS_{TR_2C} is a result of more extensive structural rearrangements occurring upon Ca^{2+} binding in fragment TR_1C than in TR_2C. It is also of considerable interest to compare the thermochemical data obtained for CaM with those recently obtained for an extracellular Ca^{2+}-binding enzyme (i.e., bovine pancreatic phospholipase A_2). This enzyme has one high-affinity Ca^{2+} site at the active site $(K \approx 3.5 \times 10^3 \, M^{-1})$ (Drakenberg et al., 1984). The thermochemical data pertaining to Ca^{2+} binding to this site (Hedwig and Biltonen, 1984) are: $\Delta G = -5.0 \pm 0.1$ kcal/mol, $\Delta H = -3.9 \pm 0.3$ kcal/mol, and $\Delta S = 3.6 \pm 1.4$ cal/mol, °K. It is apparent from the small entropy term that Ca^{2+} complex formation with the enzyme does not follow the pattern established for CaM and small carboxylic ligands. Hedwig and Biltonen (1984) have suggested that a large part of this entropic difference can be attributed to a low degree of hydration of the ligand groups already in the Ca^{2+}-free enzyme. If this explanation is valid, one would expect to observe similar small entropy changes associated with Ca^{2+} binding to other enzymes where the liganding groups have limited accessibility to the solvent. Studies in this area are presently in progress in our laboratory.

A microcalorimetric study of intramolecular melting of CaM and proteolytic fragments at various concentrations of divalent and monovalent cations has been presented (Tsalkova and Privalov, 1985). A thermodynamic analysis of the experimentally determined excess heat capacities does not fit with the model in which the four calcium-binding domains are integrated into a single cooperative system. Different parts of the molecule, tentatively assigned to pairs of domains, appear to melt separately. The data also indicate interactions between domains I and II, as well as III and IV, and from the point of view of melting these two domain pairs appear as single cooperative blocks—the former more markedly than the latter. Interaction between the two domain pairs is by contrast small and negative.

V. STOPPED-FLOW KINETIC STUDIES OF CaM AND ITS PROTEOLYTIC FRAGMENTS

The kinetics of the binding of Ca^{2+} to CaM and of the structural changes of the protein associated with Ca^{2+} binding or dissociation have been studied by a variety of physical techniques. Results from ^1H and ^{43}Ca NMR studies are discussed in Section VI. In the present section we will deal only with results from stopped-flow studies. Since a discussion of the developments of the field up to the end of 1982 has been published (Seamon and Kretsinger, 1983), we will mainly report on results of the last few years.

Since a distinction has not always been clearly made in the literature, we should perhaps start by recalling that different kinetic methods monitor different time-dependent phenomena that may or may not be related. Structural transitions of CaM accompanying the dissociation or association of Ca^{2+} may, for example, be studied by monitoring fluorescence changes of either "intrinsic" tyrosine or fluorescent probes attached to the protein. In a stopped-flow study, Malencik et al. (1981) followed the changes in tyrosine fluorescence after mixing $(Ca^{2+})_4$ CaM with EGTA. A single first-order process with a rate constant $k_{diss} = 10.5 \pm 1.5$ sec^{-1} was observed. In an analogous experiment, fluorescence changes in dansylated CaM were studied. In this one, two time-dependent processes were resolved, one with a rate identical with that found for the native protein ($k_{diss} = 10.4 \pm 1.0$ sec^{-1}) and one with a slower rate ($k_{diss} = 0.39 \pm 0.05$ sec^{-1}). It is not yet clear what kind of molecular process is responsible for the latter rate. Malencik et al. also attempted to use tyrosine fluorescence to follow the rate of the structural changes accompanying Ca^{2+}-binding to Ca^{2+}-free CaM. The fluorescence change was, however, complete within the instrumental dead-time, and it was estimated that the rate constant for the association step must exceed 3×10^3 sec^{-1}. In subsequent work from the same group (Schimerlik et al., 1982), 9-anthroylcholine was employed as an extrinsic fluorescent probe. This molecule binds to calcium-saturated calmodulin with fluorescence enhancement. From the time course of the fluorescence decrease observed when the $(Ca^{2+})_4$–CaM complex was mixed with EGTA at constant anthroyl concentration, it was concluded that the rate of structural changes following Ca^{2+} dissociation was $k_{diss} = 10.1 \pm 0.7$ sec^{-1} at room temperature—the same rate obtained from the kinetic studies involving intrinsic tyrosine fluorescence. Chan et al. (1982) have used both tyrosine fluorescence and an indicator system employing the chelator BAPTA (1) to obtain a value of k_{diss} of approximately 17.5 sec^{-1}. They also suggested that faster processes could be present but were unable to resolve them. The observation that the value of k_{diss} is independent of the monitoring system used suggests that the slow dissociation of Ca^{2+} directly corresponds to the change in CaM conformation that results in the change in tyrosine fluorescence.

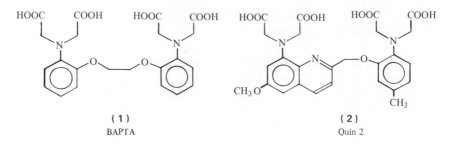

(1)

BAPTA

(2)

Quin 2

The development of new fluorescent Ca^{2+} chelators like Quin 2 (**2**) [a methoxyquinoline derivative of BAPTA (Tsien *et al.*, 1982)] has greatly facilitated a direct determination of the rate(s) of dissociation of Ca^{2+} from Ca^{2+}-binding proteins. This approach was recently used by Bayley *et al.* (1984) to study the rates of dissociation of Ca^{2+} from the Ca–CaM complex at a range of temperatures. At room temperature (28°C) only one single rate process was observed with a rate constant ($k_{obs} = 9.1 \pm 1$ sec^{-1}) closely similar to that obtained in the earlier studies just discussed. At lower temperatures (19 and 11°C) a faster process ($k_{obs} = 550 \pm 175$ sec^{-1} at 19°C) was clearly resolvable (Fig. 9). Both the fast and the slow processes correspond to the release of two Ca^{2+} ions. Activation parameters for the slow process were determined as $\Delta H^{\neq} = 14.1 \pm 2.4$ kcal/mol and $\Delta S^{\neq} = -7 \pm 7$ cal/mol, °K. The Anglo–Swedish group has very recently (Martin *et al.*, 1985) extended their kinetic studies to include ionic strength effects on the rates of Ca^{2+} dissociation not only from the Ca–CaM complex but also from the Ca^{2+} complexes formed by the two tryptic fragments

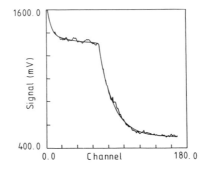

Fig. 9. Ca^{2+} dissociation from the Ca^{2+}–CaM complex. CaM plus Ca^{2+} mixed with Quin 2 (final concentrations: 6.25 μM CaM, 62.5 μM Ca^{2+}, 100 μM Quin 2). The time scale is 0.5 msec/channel (channels 0–75) and 25 ms/channel (channels 76–end). The line drawn through the curve represents the fit to the data with $k^{fast}_{obs} = 356$ sec^{-1} and $k^{slow}_{obs} = 2.19$ s^{-1}. When corrected for the finite dead time of the apparatus, the amplitude changes in both the fast and slow processes are the same and correspond to the release of two Ca^{2+} each. The temperature was 11°C. From Bayley *et al.* (1984).

TR_1C (aa 1–77) and TR_2C (aa 78–148). Again rates were studied at different temperatures in order to obtain estimates of activation parameters. In addition a study was also undertaken of the kinetic characteristics of the release of Ca^{2+} from the Ca–Quin 2 complex—a study made possible by the use of the stronger Ca^{2+} chelator EDTA. Some of the new results obtained are summarized in Table I together with the earlier results of Bayley et al. (1984).

The stopped-flow results collected in Table I have many interesting aspects. To begin with they are in good agreement with rates of dissociation obtained from earlier ^{43}Ca NMR studies on Ca–CaM complexes—these data are further discussed in Section VI. One of the major results of the NMR studies—the existence in $(Ca^{2+})_4$–CaM solutions of a slow and a fast dissociative process differing in rate by almost two orders of magnitude—is confirmed. That the fast and slow processes correspond to the release of two Ca^{2+} ions each is convincingly shown by amplitude factors that are proportional to the Quin 2 fluorescence intensity. The difference in rate between the slow and fast dissociation processes of the Ca–CaM complex is most marked in the absence of KCl. The effect of 100 mM KCl is that the rate of the fast process is decreased while that of the slow process is increased. It is apparent that a simple explanation in terms of kinetic salt effects is not adequate. A complicating factor is that K^+ may compete for the Ca^{2+} binding sites (Haiech et al., 1981) and that the effect of KCl addition is not entirely a result of increased ionic strength. Further studies with other salts are necessary to clarify this point. Nevertheless the fact remains that the dissociation of Ca^{2+} from the Ca–CaM complex apparently takes place in two fairly distinct steps, and it seems intuitively obvious that k^{slow} should correspond to the rate of dissociation of Ca^{2+} from two strong binding sites and k^{fast}, to the rate of dissociation from two weaker sites. This idea is strongly supported by the results on the tryptic fragments TR_1C and TR_2C. Unless some other structural rearrangements occur upon cleavage, it seems evident that the "slow" sites reside in fragment TR_2C containing binding sites III and IV and the "fast" sites in TR_1C containing sites I and II.

Again these inferences are in complete agreement with results from NMR studies (see Section VI). It is indeed striking how little the cleavage of the native CaM molecule into two halves affects the rates of dissociation of Ca^{2+}. In terms of free energy of activation, a change in rate constant by a factor of 2 corresponds to only 0.4 kcal/mol. The notion of CaM as being constructed from two largely autonomous domains coinciding with the fragments TR_1C and TR_2C gains support.

An estimate of the rate of Ca^{2+} association to the tryptic fragment TR_2C can be made with reasonable confidence. In the presence of 100 mM KCl, the off rate is about 40 sec^{-1} at 25°C, and the effective Ca^{2+} binding constant is close to 10^6 M^{-1} (Drabikowski et al., 1982). Thus, k_{on} can be calculated to 4×10^7 $M^{-1}sec^{-1}$. This value may be compared with the rate constant for the binding of

TABLE I

Kinetic Parameters for the Dissociation of Ca^{2+} from the $(Ca^{2+})_4CaM$, $(Ca^{2+})_2TR_1C$, and $(Ca^{2+})_2TR_2C$ Complexes as Induced by the Fluorescent Chelator Quin 2[a]

	Temperature (°C)	k_{off}[a] (s^{-1})	Activation[b] parameters (kcal/mol or cal/mol, °K)	References
	A. The release of Ca^{2+} from $(Ca^{2+})_4$ CaM			
[KCl] = 0	11	293 ± 92[c]		Bayley et al., 1984
		2.1 ± 0.4[c]		
	19	550 ± 175[c]	$\Delta H^{\neq} = 14.1 \pm 2.4$	Bayley et al., 1984
		5.3 ± 0.9[c]	$\Delta S^{\neq} = -7 \pm 7$	
	11.0	303 ± 77[c]		Martin et al., 1985
		3.6 ± 1.2[c]		
[KCl] = 100 mM	11.3	90 ± 20[c]	$\Delta H^{\neq}_{fast} = 9.8 \pm 1.2$	Martin et al., 1985
		7.3 ± 2[c]	$\Delta S^{\neq}_{fast} = -15 \pm 4$	
	21.4	155 ± 36[c]	$\Delta H^{\neq}_{slow} = 15.2 \pm 1.3$	Martin et al., 1985
		16.6 ± 2.9[c]	$\Delta S^{\neq}_{slow} = -1.4 \pm 4.4$	
	B. The release of Ca^{2+} from $(Ca^{2+})_2$ TR_2C			
[KCl] = 0	13.3	4.5 ± 0.5	$\Delta H^{\neq} = 14.6 \pm 1$	Martin et al., 1985
	23.0	13.6 ± 0.5	$\Delta S^{\neq} = -4.3 \pm 3.9$	
[KCl] = 100 mM	13.9	13.5 ± 3.1	$\Delta H^{\neq} = 12.5 \pm 1.2$	Martin et al., 1985
	20.9	23.9 ± 3.1	$\Delta S^{\neq} = -9.4 \pm 4.1$	
	C. The release of Ca^{2+} from $(Ca^{2+})_2$ TR_1C			
[KCl] = 0	13.0	313 ± 21	$\Delta H^{\neq} = 6.7 \pm 0.9$	Martin et al., 1985
	20.0	399 ± 23	$\Delta S^{\neq} = -23.7 \pm 4.5$	
[KCl] = 100 mM	12.8	175 ± 3.9	$\Delta H^{\neq} = 5.8 \pm 0.7$	Martin et al., 1985
	22.0	237 ± 38	$\Delta S^{\neq} = -28.6 \pm 2.2$	
	D. The release of Ca^{2+} from Ca^{2+} (Quin 2)[a]			
[KCl] = 0	13.6	32.3 ± 0.5	$\Delta H^{\neq} = 6.7 \pm 0.2$	Martin et al., 1985
	19.4	41.9 ± 0.7	$\Delta S^{\neq} = -30 \pm 2.5$	

[a]All solutions were prepared in 20 mM/KOH pipes buffer (pH 7.0). Quin 2 was 200 μM, CaM or tryptic fragment TR_1C or TR_2C was 12.5–15.0 μM, and Ca^{2+} was in about 5- to 10-fold excess. As indicated, 100 mM KCl was also present in some solutions.

[b]The errors given are standard deviations. With the exception of the data for TR_1C, all activation parameters have been determined from more than three temperatures. The standard deviations given probably underestimate the error associated with the activation parameters.

[c]The fast and slow rate processes have equal amplitudes within the experimental error.

[d]The release of Ca^{2+} from the Ca^{2+}-(Quin 2) complex was induced by EDTA. In this experiment 50 μM Quin 2 and 100 μM Ca^{2+} were in one syringe and 10 mM EDTA in the other.

Ca^{2+} to Quin 2 calculated to be $6 \times 10^8 M^{-1}sec^{-1}$ from the experimental data in Table I (D) and the reported Ca^{2+} binding constant of Quin 2 ($1.25 \times 10^7 M^{-1}$) (Tsien et al., 1982). Although the protein is not quite as effective in capturing Ca^{2+} ions as the low-molecular-weight chelator is, it apparently has a flexible binding region that allows the ligands to wrap swiftly around Ca^{2+} ions yet is sufficiently restricted in its mobility to discriminate between Ca^{2+} and other ions like Mg^{2+}.

Finally we may ponder the fact that *two* Ca^{2+} ions are released in a dynamic process that can be accurately described by a *single* rate constant. If we take the Ca^{2+} complexes with TR_1C of TR_2C as example, the release of Ca^{2+} from $(Ca^{2+})_2 TR_1C$ or $(Ca^{2+})_2 TR_2C$ should occur in two steps. A simple model for the dissociation as observed in the stopped-flow experiments should be a consecutive reaction that schematically may be written as shown in Eq. (5)

$$(Ca)_2P \xrightarrow[-Ca^{2+}]{k_{21}} (Ca)P \xrightarrow[-Ca^{2+}]{k_{10}} P \tag{5}$$

where P symbolizes the protein. The back reactions may be neglected in presence of an excess Ca^{2+} chelator as in the actual experiments. An analysis of the kinetic scheme above (Frost and Pearson, 1961) shows that at a time t, the concentration of Ca^{2+} released (*i.e.*, bound by the chelator) may be written as shown in Eq. (6)

$$[Ca^{2+}] = [(Ca)_2P]_0 \left[2 + \left(\frac{e^{-x\eta} - (2\eta - 1)e^{-x}}{\eta - 1} \right) \right] \tag{6}$$

where $[(Ca)_2P]_0$ is the concentration of $(Ca)_2$-P at $t=0$, and we have introduced the dimensionless parameters $x = k_{21} t$ and $\eta = k_{10}/k_{21}$. It is easily shown that in the case when $k_{10} >> k_{21}$ (*i.e.*, $\eta >> 1$) Eq. (6) reduces to

$$[Ca^{2+}] = 2 [(Ca)_2P]_0 (1 - e^{-x}) \tag{7}$$

and the observed rate of release of Ca^{2+} is simply the rate of dissociation in the first step of Eq. (5). When k_{21} and k_{10} are equal, then $\eta = 1$ Eq. (6) reduces to

$$[Ca^{2+}] = 2[Ca_2P]_0 \left[1 - e^{-x}\left(1 + \frac{x}{2}\right) \right] \tag{8}$$

in which the expression in brackets on the right hand side for small x reduces to $(1 - e^{-0.5x})$.

Although the resolution of the experimental data obtained in the Anglo–Swedish study clearly favors a scheme in which $\eta > 1$, a more detailed analysis of the actual curve fitting problem would be needed in order to establish an upper limit for the value of η.

A comparison of the "true" Ca^{2+} dissociation rates obtained by ^{43}Ca NMR or the Quin 2 stopped-flow studies with the rates obtained using intrinsic protein fluorescent probes indicates that there is no experimental basis for making a distinction in kinetic terms between conformation change and Ca^{2+} dissociation from the Ca–CaM complex. Available data are consistent with the idea that the conformational change occurs synchronously with Ca^{2+} binding or dissociation (Bayley et al., 1984).

VI. NMR AND ESR STUDIES

NMR studies of calmodulin and its proteolytic fragments provide a good illustration of the versatility of the NMR method and the complementary information that can be obtained through studies of different magnetic nuclei. ^1H NMR has mainly been used to follow the conformational changes in the protein that are induced by metal-ion binding. The kinetics of Ca^{2+}-ion binding has been studied with ^{43}Ca NMR to give data complementary to stopped-flow results. ^{113}Cd NMR is uniquely suited to determining metal-ion populations at individual binding sites and has given strong evidence for cooperative ion binding in calmodulin as well as in some of its fragments. ^{113}Cd NMR has also been used to monitor conformational changes in the protein and its fragments as a result of drug binding.

A. ^1H NMR

Figure 10 shows ^1H NMR spectra of CaM with and without saturating amounts of calcium ions added, displaying the dramatic changes in the ^1H NMR spectrum as a result of Ca^{2+}-ion binding. The effects are most pronounced in the aromatic region (6–8 ppm) and in the high field region below 1 ppm. When metal-ion binding to CaM is studied with proton NMR, one observes two phases. Binding of the first two Ca^{2+} causes a conformational change in the protein. The conformations for apoCaM and CaM with two Ca^{2+} ions bound are in slow exchange with each other on the NMR time scale (i.e., $k_{exch} < 10$ sec^{-1}). During the addition of Ca^{2+} up to a Ca^{2+}/CaM ratio of two, the proton NMR spectrum is a superposition of two spectra with varying relative intensities, one increasing as the other decreases (Fig. 11) (Seamon, 1980; Ikura et al., 1983a,b, 1984; Andersson et al., 1983b; Thulin et al., 1984; Klevitt et al., 1984; Aulabaugh et al., 1984a; Dalgarno et al., 1984). There is no indication of the existence of a third component in these spectra, and the obvious conclusion is therefore that only CaM and $(Ca^{2+})_2CaM$ coexist to a measurable extent. Similarly, it has been observed for the fragment TR_2C that the Ca^{2+} dependence of the proton NMR spectrum is best explained by assuming that only TR_2C and $(Ca^{2+})_2TR_2C$ coexist to a measurable extent. This is, however, equivalent to saying that there is a strong positive cooperativity between the two binding sites. The same conclusion can be drawn from ^{113}Cd NMR spectra for CaM as well as for TR_2C (Andersson et al., 1983b; Thulin et al., 1984) and from tyrosine fluorescence measurements on TR_2C (Drabikowski et al., 1982).

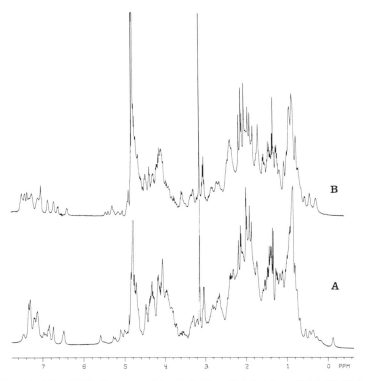

Fig. 10. 360 MHz ^{1}H NMR spectra of 1 mM CaM at pH 7.5 : (A) apoCaM, (B) (Ca)$_4$CaM.

The second phase in the Ca^{2+} binding to CaM comprises the binding of the third and fourth metal-ion equivalents. During this phase a single spectrum is observed with some resonances shifting continuously with added metal ions from a Ca^{2+}CaM ratio of 2 to 4. These metal ions are thus in fast exchange on the proton NMR time scale (i.e., $k_{exch} > 500$ sec^{-1}). The same behavior was found for the first two metal ions added to TR$_1$C.

Since the binding of calcium ions causes dramatic changes in the proton NMR spectrum of CaM and its fragments TR$_1$C and TR$_2$C, it is likely that proton NMR should also be sensitive to differences in the conformation of the proteins when different metal ions are bound to them. It has thus been shown that the proton NMR spectra from (Ca)$_2$CaM and (Cd)$_2$CaM are almost identical, and the same is true for (Ca)$_4$CaM and (Cd)$_4$CaM as well as for the tryptic fragments TR$_1$C and TR$_2$C (Andersson *et al.*, 1983b; Klevit *et al.*, 1984; Thulin *et al.*, 1984). These results justify the use of Cd^{2+} as a probe ion for Ca^{2+}. Similar experiments using La^{3+} as a replacement for Ca^{2+} resulted in spectra for (La)$_2$CaM which differed significantly from those for (Ca)$_2$CaM (W. P. Niemczura personal communication). This observation casts some doubt on the conclusions drawn from experiments using lanthanide ions as Ca^{2+} probes.

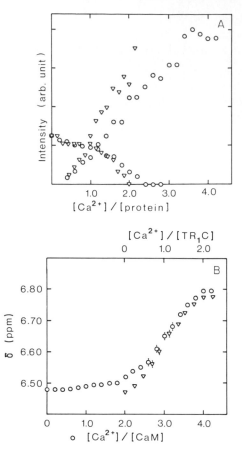

Fig. 11. Changes in the ¹H NMR spectra of CaM and its fragments upon Ca²⁺ addition: (A) intensities of the His-107 resonance in the spectra of TR₂C (○) and CaM (▽); (B) chemical shift of one of the upfield shifted phenylalanine resonances in the spectra of TR₁C (▽) and CaM (○). Reprinted with permission from Thulin, E., Andersson, A., Drakenberg, T., Forsén, S., and Vogel, H. J. (1984) *Biochemistry* **23,** 1862. Copyright (1984) American Chemical Society.

Figure 12 shows a part of the proton NMR spectra from CaM, TR₁C and TR₂C, and the difference spectrum (CaM–TR₁C–TR₂C). These spectra show convincingly that the CaM structure is well conserved in the tryptic fragments in apo- as well as in Ca²⁺-saturated forms, and they give strong support to the notion that CaMs structure is composed of two independently folded domains that are retained in the fragments TR₁C and TR₂C which are connected by a flexible loop. The two domains independently undergo Ca²⁺-dependent conformational changes.

Recently, more advanced ¹H NMR techniques have been used to study CaM,

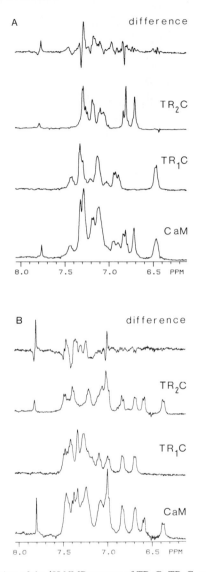

Fig. 12. Aromatic region of the ^{1}H NMR spectra of TR_1C, TR_2C, and CaM and the difference spectra of CaM ($TR_1C + TR_2C$): (A) apo forms, (B) Ca^{2+}-saturated forms. The solid dots indicate the position of the His-107 resonances, which are very sensitive to pH changes and hence show up in the difference spectra. Reprinted with permission from Thulin, E., Andersson, A., Drakenberg, T., Forsén, S., and Vogel, H. J. (1984). *Biochemistry* **23,** 1862. Copyright (1984) American Chemical Society.

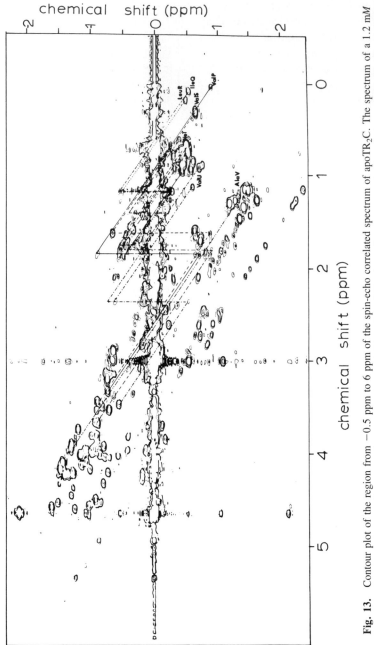

Fig. 13. Contour plot of the region from −0.5 ppm to 6 ppm of the spin-echo correlated spectrum of apoTR₂C. The spectrum of a 1.2 m*M* protein solution was measured in ²H₂O at 26°C, pH 7.5. Connectivities of the following spin-systems are indicated: (——) ValP, (..-..) IleQ, (.....) LeuR, (- - -) ValS, (- — -) IleT, (.-.-) ValU, and (.-.-) AlaV. From Aulabaugh *et al.* (1984b).

and especially the fragments. Figure 13 shows the 2D SECSY spectrum from TR_2C. In this spectrum several correlation peaks are indicated. These correlations are due to spin coupling interactions among protons within the protein and can in principle be used to elucidate the three-dimensional structure of the protein. This work has only recently been started and will probably take a few years to finish. However, the results obtained to date (Aulabaugh et al., (1984b) suggest that the solution structure of CaM strongly resembles that predicted earlier based on a parvalbumin structure (Kretsinger, 1980).

B. ^{43}Ca NMR

The only calcium isotope with a nonzero spin is ^{43}Ca, with a spin of 7/2 and a natural abundance of 0.14%. In order to observe the ^{43}Ca NMR signal at millimolar or submillimolar concentrations, it is necessary to use enriched isotopes—in our studies about 60%. ^{43}Ca is a quadrupolar nucleus ($I = 7/2$) which results in relatively broad resonances. Therefore, in most cases NMR signals from ions bound to the different protein binding sites are not resolved, in contrast to ^{113}Cd NMR signals, which are usually well resolved. ^{43}Ca NMR is, however, a very attractive method for measuring of Ca^{2+} ion-exchange rates under equilibrium conditions in the range of 10^2 to 10^4 sec^{-1}. For the upper range it is probably the only general method, whereas for the lower range stopped-flow techniques can be used as well (see Section V).

At less-than-saturating amounts of Ca^{2+} added to either calmodulin or its fragments TR_1C and TR_2C, the ^{43}Ca NMR spectrum shows a single, broad resonance with a width of 500–700 Hz. The half-width of the signal, in combination with the longitudinal relaxation rate, has been used to calculate the correlation time of the bound calcium ions. Such calculations have resulted in correlation times, τ_C, of approximately 8 ns for calmodulin (Andersson et al., 1983a) and approximately 5 ns for the fragment TR_2C (Andersson Teleman et al., 1985). These τ_C values are in fair agreement with the overall rotational correlation time that can be calculated for the protein. Thus the calcium ions do not appear to have any mobility with a correlation time in the ns range within the sites. However, there may well be faster, nonisotropic motions within the protein which would have the effect of reducing the quadrupole coupling constant from its value in a static site. Furthermore, the relaxation times would not be sensitive to motions much slower than the correlation time of the protein.

When more than saturating amounts of calcium are added (4 eq for CaM and 2 eq for TR_1C and TR_2C) only the TR_2C fragment appears to give rise to a spectrum of two overlapping signals, indicative of slow exchange between free and bound Ca^{2+} ions in a low-salt medium (Fig. 14). For CaM and TR_1C single, broad signals are observed at all calcium concentrations. The width of this signal has a temperature dependence typical for intermediate exchange. Detailed analysis of this temperature dependence has resulted in well determined values of k_{off}

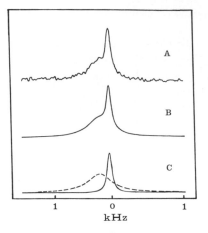

Fig. 14. 24 MHz ^{43}Ca NMR spectrum at 50°C of 0.8 mM TR$_2$C and 2.2 mM Ca^{2+} (60% enrichment in ^{43}Ca). (A) Experimental spectrum. (B) Simulated spectrum using two Lorentzian lines with 88 and 476 Hz line widths, respectively, and with a chemical shift difference between the lines of 160 Hz. (C) The individual lines used in (B).

for the two weakly bound calcium ions and estimates of k_{off} for the two strongly bound calcium ions in CaM (Andersson *et al.*, 1982; Drakenberg *et al.*, 1983). Also, good values for the exchange rate of the two ions bound to TR$_1$C have been obtained (Andersson Teleman *et al.*, 1985). The off-rate for the fast-exchanging Ca^{2+} ions was for both proteins found to be around 10^3 sec^{-1} at 25°C. Both exchange processes also have negative entropies of activation.

When the medium also contained 0.1 M KCl, the CaM ^{43}Ca NMR spectra for Ca/CaM ratios above 4 showed strongly nonLorentzian lines which could alternatively be seen as two overlapping Lorentzian lines with slightly different chemical shifts and different line-widths. The Ca^{2+} off-rate for the rapidly exchanging sites has been estimated to be 300 sec^{-1} at 25°C, i.e., a reduction in rate as compared to the salt-free medium (Andersson Teleman *et al.*, 1985).

The kinetics in the Ca^{2+} binding to calmodulin is also strongly affected by added hydrophobic drugs like TFP, chlorpromazine, and calmidazolium. The ^{43}Ca line width decreases when TFP is added to a sample containing Ca^{2+} in excess of possible binding sites in calmodulin. This effect was first interpreted in terms of a retarded calcium exchange rate (T. Andersson, 1981; Thulin *et al.*, 1980), and later, as due to a Ca^{2+}–TFP competition (Shimizu *et al.*, 1982). Shift reagents have been used to shift the ^{43}Ca NMR signal from ''free'' Ca^{2+} ions away from the signal from CaM-bound Ca^{2+} ions. In this way it has been shown unequivocally that there is no competition between Ca^{2+} and TFP bound to the protein, but the calcium exchange has been slowed down by the addition of TFP (Fig. 15) (Vogel *et al.*, 1984).

C. ^{113}Cd NMR

For cadmium there are two magnetic isotopes (^{111}Cd and ^{113}Cd) which can be used for NMR studies with equal success. However, since the natural abundance is about 10% for both isotopes, it is highly advantageous to use enrichments in experimental studies of Cd^{2+} ions bound to proteins at sub-mM concentration. Only ^{113}Cd has been used in studies of metal binding to CaM.

When ^{113}Cd^{2+} is added to either CaM or TR$_2$C, two well-resolved ^{113}Cd signals at high field are observed (Fig. 16; Andersson *et al.*, 1983a; Thulin *et al.*, 1984). The signals increase in intensity nearly in parallel up to a Cd/CaM (or TR$_2$C) ratio of 2. The same behavior has been observed for skeletal muscle (Forsén *et al.*, 1979) and heart muscle troponin C (Teleman *et al.*, 1983), i.e., calcium-binding proteins also containing a pair of strong "EF-hand" sites. We have taken this as strong indication of cooperative metal-ion binding to these pairs of sites. This interpretation is corroborated by detailed ^{1}H NMR studies

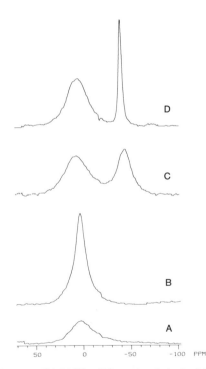

Fig. 15. ^{43}Ca NMR spectra (24.3 MHz, 10^6 transients) obtained for a sample containing 1 mM bovine testis CaM, 100 mM NaClO$_4$, pH 7.0 with the following additions: (A) 2 equivalents of Ca^{2+}, (B) 6 equivalents of Ca^{2+}, (C) 1.2 mM Dy(PPP)$_2^{7-}$ (i.e., a mixture of DyCl$_3$ and Na$_5$P$_3$O$_{10}$), and (D) 3 mM TFP. From Vogel *et al.* (1984).

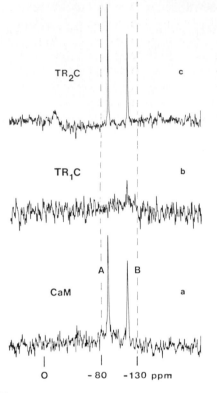

Fig. 16. 56.5 MHz ^{113}Cd NMR spectra from 1 to 2 mM protein solutions (pH 7.5). 2.5 10^4 transients were accumulated, and a weight function resulting in a line broadening of 30 Hz was applied. (a) (Cd)$_4$CaM, (b) (Cd)$_2$TR$_1$C, (c) (Cd)$_2$TR$_2$C.

(Teleman *et al.*, 1983). The only other possibility would be that for all these pairs of sites the two binding constants have to be virtually identical–not a very likely situation. That the chemical shifts of the two ^{113}Cd signals are identical within the error of measurement in CaM and TR$_2$C (Fig. 16) provides additional evidence for the structural integrity of the C-terminal half of the CaM molecule.

When more than two equivalents of Cd^{2+} ions are added to the tryptic fragment TR$_2$C, a ^{113}Cd resonance from "free" ions appears near $\delta = 0$ ppm, whereas for CaM four equivalents are needed. Thus, the third and fourth equivalents of ^{113}Cd^{2+} produce no apparent change in the ^{113}Cd NMR spectrum. This would at first seem to be in reasonable agreement with the ^{43}Ca NMR and stopped-flow studies showing that the off-rate for Ca^{2+} bound to the two weaker sites is of the order of 10^3 sec^{-1} at room temperature (Andersson *et al.*, 1982). However, in the presence of 0.1 M KCl the off-rate for Ca^{2+} is about 300 sec^{-1}

(Andersson Teleman *et al.*, 1985), which should result, if we assume the same exchange rate for Cd^{2+}, in a broadening of the ^{113}Cd NMR resonances of no more than 100 Hz. Such a slight broadening would still permit well-resolved ^{113}Cd NMR resonances. Furthermore, the line width of the resonance from "free" $^{113}Cd^{2+}$ ions in a low-salt medium is in good agreement with an off-rate of the order of 10^3 sec^{-1}, and certainly not with off-rate an order of magnitude larger. Therefore, we have to postulate the presence of another dynamic process which is faster than the metal exchange and which will alter the environment of the less strongly bound metal ions. Model calculations (Fig. 17) have shown that a conformational change with a rate constant of 10^4 sec^{-1} and with concomitant chemical shift differences of at least 10 ppm between the two conformations for both ^{113}Cd ions at the weak sites will result in spectra similar to those observed.

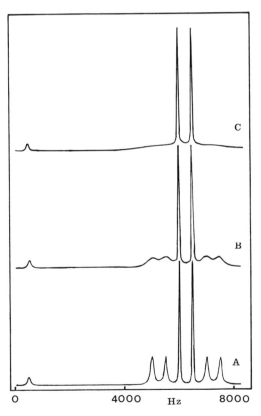

Fig. 17. Model calculation of ^{113}Cd NMR spectra assuming exchange between two conformations containing 4 metal binding sites each. Two of the sites are unaffected by the conformational change, whereas the other two have chemical shift differences of 2000 Hz between the two conformations. The exchange rates used in the calculation are (A) 10^2, (B) 10^3, and (C) 10^4 sec^{-1}.

This model is not in contradiction with any other available data sensitive to a conformational exchange with this rate. For example, in 360 MHz proton NMR spectra, two conformers with a shift difference of 0.1 ppm, exchanging with a rate of 10^4 sec^{-1}, would cause a resonance broadening of about 0.2 Hz. A conformational exchange similar to the one discussed for CaM has to be assumed for TR_1C to explain the lack of resonances for less than two equivalents of Cd^{2+} added (Fig. 16). In fact the same has to be assumed for TnC as well.

We may speculate that the effect of KCl, i.e., decreasing the off-rate for the fast exchanging sites, is in fact to change the populations among two or more conformations, as discussed previously. These have different Ca^{2+} off-rates. If the metal exchange is mainly from one conformation, then it should be sufficient to change the population ratio from 3 : 1 in favor of the conformer with the fast Ca^{2+} exchange to 1 : 3.

D. ESR Studies

Three principally different strategies are open to the experimenter who wants to study CaM using electron spin resonance techniques. First, one can use paramagnetic metal ions such as Mn^{2+} or Gd^{3+} as substitutes for Ca^{2+}. The second possibility is to selectively attach stable nitroxide spin labels to specific groups on the protein. Third, one can make use of ligands (drugs or proteins) that bind to CaM and that have been modified so that they carry the nitroxide group (Berliner, 1976). Such studies are of considerable potential value since distance information can be obtained from ESR studies combined with NMR studies. These may help delineate features of the solution structure of CaM. Although few ESR studies have been reported on CaM, all three experimental strategies have been used.

Hewgley and Puett (1980) utilized iodoacetate based spin labels that apparently modified two or more methionine residues. CaM modified in this manner still activated cyclic nucleotide phosphodiesterase, but with altered K_m. The ESR intensity was reduced dramatically upon Ca^{2+} addition, probably as a result of spin–spin interactions, i.e., addition of Ca^{2+} brings these labeled methionine residues in close proximity. Surprisingly, addition of TFP or cyclic nucleotide phosphodiesterase to spin-labeled CaM did not reduce the mobility of the labels.

Xü et al. (1983) repeated the experiment of Hewgley and Puett (1980), except that they prepared a modified CaM carrying only one iodoacetate nitroxide spin label per protein molecule. Determination of the modification sites would have added considerably to the scope of these studies. Xü et al. (1983) demonstrated that in their preparation, as expected, no reduction in ESR intensity took place upon Ca^{2+} addition. They observed, however, a slight decrease in correlation time τ_C of the labels. This decrease bears out the results of fluorescence studies (see below). The same authors also prepared a more stable CaM modified by lysine maleimide. While monitoring the free Mn^{2+} by ESR, Xü *et al.* (1983) observed two high- and two low-affinity sites for Mn^{2+} with K_d on the order of 10^{-6} and 10^{-5} M. Virtually identical results on Mn^{2+} binding have been reported by Yagi *et al.* (1982). Thus, binding of Mn^{2+} seems to resemble that of Ca^{2+} and Cd^{2+}, since it can be clearly separated into two low- and two high-affinity sites. Yagi *et al.* (1982) also used a modifying spin label which is suggested to be specific for tyrosine and found that the first two Mn^{2+} ions that bind to this modified CaM were located within 15 Å from the covalently attached spin labels. On the questionable assumption that this reagent modified only the two tyrosines and no other amino acids, they concluded that sites III and IV are the two strong sites for Mn^{2+}, a conclusion which is in agreement with NMR studies on the binding of Ca^{2+} and Cd^{2+}.

Al^{3+} binding to CaM has also been studied using ESR techniques (Siegel *et al.*, 1983; Siegel and Hang, 1983). The results obtained are consistent with the notion that binding of Al^{3+}, in contrast to binding of Ca^{2+}, gives rise to a more open structure of CaM. Furthermore, Rainteau *et al.* (1984) synthesized a spin-labeled TFP analog and found that the spin label was not strongly immobilized upon binding to CaM. They also found that four Ca^{2+} ions were needed to give complete binding of this phenothiazine analog. It would be of interest to use this analog in NMR studies in an attempt to delineate the location of the drug binding sites on CaM.

VII. FLUORESCENCE AND LUMINESCENCE STUDIES

Because of their attractive spectroscopic properties, lanthanides have been used as calcium probes in studies of calcium-binding proteins. Terbium luminescence in particular has been used to study metal-ion binding to calmodulin (Kilhoffer *et al.*, 1980a,b, 1981; Wallace *et al.*, 1982; Wang *et al.*, 1982). Direct excitation at 222 nm reveals binding of all terbium ions to a protein. Indirect excitation of terbium (via excitation of tyrosine at 280 nm) was observed for metal-ion binding sites where tyrosine is sufficiently close to terbium so that an energy transfer can take place. It was found in the majority of the studies that the tyrosine energy transfer appears only upon binding of the third and fourth equivalents of Tb^{3+}, and it was therefore concluded that Tb^{3+} binds first to sites I and II, where there are no tyrosines, and subsequently to sites III and IV, which both contain a tyrosine. Kilhoffer *et al.* (1983) argued strongly that Ca^{2+} would bind in the same order as Tb^{3+}. There are in our opinion few data to support the latter notion. On the contrary, the tyrosine fluorescence data can be interpreted to mean that sites III and IV are the two strong sites for Ca^{2+}, since the change in the tyrosine fluorescence quantum yield is completed after the binding of the second equivalent of calcium to calmodulin (Kilhoffer *et al.*, 1980b, 1981;

Wallace *et al.*, 1982; Wang *et al.*, 1982). The change in tyrosine fluorescence accompanying the binding of the first two Ca^{2+} ions as documented by these authors is parallelled by a similar if not identical change upon binding of two Ca^{2+} ions to the carboxy terminal fragment TR_2C (Drabikowski *et al.*, 1982). These observations are in agreement with the bulk of NMR data discussed in Section VI, which convincingly show that sites III and IV are the strong Ca^{2+} sites. The Tb^{3+}-luminescence results seem to present a warning against an uncritical use of lanthanides as probes for calcium binding in proteins that contain more than one metal binding site, since the available data suggest that Tb^{3+} binds in a different sequence to CaM than does Ca^{2+}. This hypothesis has recently been verified using stopped-flow measurements (Wang *et al.*, 1985) and ^1H NMR (W. P. Niemczura, personal communication). Wang *et al.* (1982) have shown that the change in tyrosine fluorescence as a function of free Ca^{2+} ions has a shape in agreement with a Hill coefficient of 2. This finding is supported by the cooperativity deduced from ^1H NMR and ^{113}Cd NMR studies (Andersson *et al.*, 1983b); Thulin *et al.*, 1984) and from fluorescence studies on the TR_2C fragment (Drabikowski *et al.*, 1982).

Tyrosine fluorescence has also been used to study temperature-dependent conformational changes. Gangola and Pant (1983) found that there is a conformational change in apocalmodulin between 20 and 26°C, which is also found in the thermochemical studies (see Section IV). Interestingly, temperature-dependent ^1H NMR studies have not provided any evidence for such a conformational change, perhaps because these measurements were performed mainly at temperatures higher than 25°C (Guerini and Krebs, 1983). It would be worthwhile to do further NMR studies using the calmodulin tryptic fragments. Using the assignments for the resonances in the aromatic part of the protein, one should be able to see in which part of the protein this structural transition occurs. The functional significance of the transition is not clear at present, but it does affect the calcium-binding properties of calmodulin (Gangola and Pant, 1983).

Pundak and Roche (1984) have studied the pH dependence of tyrosine fluorescence. With excitation at 278 nm, they found not only the typical tyrosine fluorescence at 305 nm but also a band characteristic of tyrosinate emission at 330–350 nm. The pH dependence of the excitation spectra from $(Ca)_4CaM$ at two emission wavelengths is shown in Fig. 18. The spectra both display pH maxima at 7 and 8.5, as has been observed for apoCaM. Pundak and Rodney (1984) offer some suggestions on the possible origin of the unusually low pK_a value for one of the tyrosines (Tyr 99). They argue that only the presence of a nucleophilic group close to tyrosyl, which is basic enough to accept the phenolic proton, could cause such ionization. This is an interesting idea which deserves further investigation.

Steiner and co-workers have studied the molecular dynamics of the tyrosine resonance of calmodulin under a variety of experimental conditions (Lambooy *et*

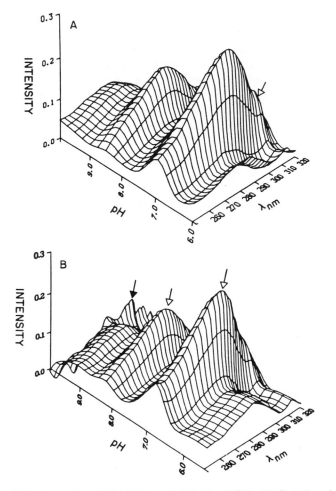

Fig. 18. Excitation surface of CaM (0.4 mg/ml) in 15 mM Tris. (A) Emission of holoCaM at 305 ± 4 nm. (B) Emission of holoCaM at 345 ± 4 nm. Reprinted with permission from Pundak, S., and Roche, R. S. (1984) *Biochemistry* **23**, 1549. Copyright (1984) American Chemical Society.

al., 1982; Steiner *et al.*, 1983). These authors studied static and dynamic fluorescence anisotropy using the intrinsic fluorescence of tyrosine groups. They also made extrinsic fluorochromes that were attached using the selective nitration technique discussed previously and succeeded in attaching fluorescent probes such as dansyl or fluorescein selectively to the tyrosine residues. All their results from the differnt approaches are consistent with the idea of considerable mobility for amino acid residues in apoCaM, which is somewhat reduced upon addition of Ca^{2+}. (Lambooy *et al.*, 1982). However, CaM does not appear to be a rigid protein in the Ca^{2+}-saturated form. Calmodulin structure and dynamic properties are dependent on electrostatic stress. This makes CaM different from typical globular proteins, such as ribonuclease whose structures are not very sensitive to a wide range of electrostatic stress. Nevertheless CaM does retain its α-helical structure under these conditions. Based on their results, the authors suggest that the inherent molecular flexibility of CaM may play a role in its function (Steiner et al., 1983).

VIII. LOCALIZATION OF INTERACTION SITES ON CALMODULIN

Several lines of evidence suggest that the interactions between CaM and its target proteins depend to a large extent on hydrophobicity. It has been known for some time that numerous hydrophobic substances bind to CaM in a Ca^{2+}-dependent manner. These include several classes of drugs such as phenothiazines (Levin and Weiss, 1977), naphthalene sulfonamides (Hidaka et al., 1981), imidazolines (Gietzen et al., 1981), and dihydropyridines (Boström et al., 1981). Also several typical hydrophobic probe molecules with fluorescent properties bind to CaM in a Ca^{2+}-dependent manner. These include 9AC, ANS (LaPorte et al., 1980), and TNS (Tanaka and Hidaka, 1980). These three compounds as well as the antihypertensive drug felodipine (Johnson et al., 1985) undergo an increase in fluorescence upon binding to CaM, whereas TFP undergoes a decrease in fluorescence (Johnson and Wittenauer, 1983). Binding of these substances is usually accompanied by a shift of the fluorescence optimum to lower wavelengths. These observations provide further support for the idea that binding takes place at a site which is mainly hydrophobic in nature. Furthermore, the observation that CaM and its two tryptic fragments interact only in the presence of Ca^{2+} with the hydrophobic support phenyl-Sepharose (Gopalakrisha and Anderson, 1982; Vogel et al., 1983a) and hydrophobic photoaffinity labels (Krebs et al., 1984) provides further support for this idea. Moreover, metal ion-independent hydrophobic sites have been found on some of the target enzymes (Gopalakrishna and Anderson, 1983) that are likely to be responsible for interaction with CaM.

The majority of the compounds that bind to CaMs surface also inhibit the CaM activation of target enzymes. Thus, it was quickly concluded that these hydrophobic substances interacted with the same CaM surfaces as the target enzymes. However, caution should be exercised in making such inferences since some of these drugs can also bind directly to the target enzymes and modify their activity in this manner (Schatzman et al., 1983; Zimmer and Hofmann, 1984; Tuana and MacLennan, 1984). Moreover, Roufogalis (1981) demonstrated that a range of phenothiazines which had widely different clinical potencies had comparable affinities for CaM. Nevertheless, these compounds have provided an extremely useful tool for studying CaM and its interactions. Most importantly, Ca^{2+}-dependent affinity chromatography on phenothiazine drugs coupled to inert resins provided a convenient method for CaM purification (Charbonneau and Cormier, 1978; Jamieson and Vanaman, 1979; Hart et al., 1983). This affinity ligand also proved useful for the purification of the related calcium-binding proteins, S100b and troponin C (Marshak et al., 1981). Other affinity ligands such as naphtalenesulfonamides (Endo et al., 1981), ω-aminoalkylagaroses (Tanaka et al., 1984), dihydropyridines (Boström et al., 1985), as well as the commercially available reagent phenyl-Sepharose (Gopalakrishna and Anderson, 1982), are as effective as phenothiazine columns for the purification of CaM. In fact, phenyl-Sepharose has recently been shown to be equally useful for the purification of other calcium-binding proteins such as S100 (Baudier et al., 1982), skeletal and cardiac muscle troponin C (Vogel et al., 1983a), α-lactalbumins (Lindahl and Vogel, 1984), and protein kinase C (Walsh et al., 1984).

In addition to hydrophobic organic substances, a wide series of small usually hydrophobic and/or basic peptides also bind to CaM in a Ca^{2+}-dependent manner. These include histones and myelin proteins (Grand and Perry, 1980); extracellular hormones, such as the vasoactive intestinal peptide VIP, glucagon and ACTH (Malencik and Anderson, 1983); bee-venom peptides, such as melittin (Maulet and Cox, 1983); and other peptides, such as β-endophins (Giedroc et al., 1983), dynorphins, and mastoparans (Malencik and Anderson, 1984). Some of these bind very strongly to CaM; the affinity for melittin is on the order of a nanomolar (Maulet and Cox, 1983; Malenick and Anderson, 1984). All of these peptides require the presence of Ca^{2+} for high-affinity binding.

Three different approaches have been used to localize the binding sites for the hydrophobic compounds on CaM. These include studies of proteolytic fragments of CaM (often combined with affinity chromatography), the use of affinity and photoaffinity labeling, and the use of spectroscopic distance measurements. The last approach is in principle nonperturbing and should therefore be preferred. Unfortunately only a few detailed studies have been reported using this strategy. Klevitt et al. (1981) and Krebs and Carafoli (1982) reported on the basis of ^{1}H NMR studies that methionines were involved in the binding of TFP. However, since the complete assignment of the methionine resonances has not been elucidated, this did not directly pinpoint the binding site. Steiner (1984) used fluorescence energy transfer to prove that the dominant binding site for the hydrophobic probe ANS is located in the N-terminal half of the molecule.

Head et al. (1982) reported that after cyanogen bromide digestion of CaM only the fragment stretching from 72 to 123 was retained on a phenothiazine column. This located one TFP binding site on this part of the protein. Although there is general agreement at present that there are two TFP binding sites with very similar

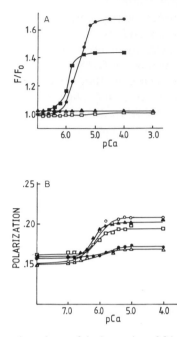

Fig. 19. (A) The calcium dependence of the interaction of felodipine with calmodulin and its fragments. Calcium titrations of 4 micromolar felodipine and 2 micromolar calmodulin (●); fragment Tr1-E, 1–106 (■); TR1-C, 1–77 (▲); and TR2-E, 78–148 (□) are shown. Fragments TR2-C, 1–90 and TR3-E, 107–148 produced no calcium dependent change in felodipine fluorescence. (B) Calcium dependence of felodipine fluorescence polarization in the presence of calmodulin and its fragments. Corrected polarization values are shown as a function of calcium for felodipine and whole calmodulin (○); fragment TR1-C, 1–77 (●); fragment TR2-C, 78–148 (△); TR1-E, 1–106 (▲); and TR2-E, 1–90 (□). Calcium titrations were conducted on 1 μm felodipine and 2 μm protein. Excitation was at 365 nm and emission was monitored at 445 nm. Redrawn from data by Johnson *et al.*, 1985.

but not identical affinities for CaM (see, for example, Levin and Weiss, 1977; Klee and Vanaman, 1982; Vogel *et al.*, 1984), surprisingly only one equivalent of a phenothiazine isothiocyanate affinity analog reacted with CaM (Newton *et al.*, 1983). However, since the binding of this phenothiazine analog was potentiated[2] by low levels of TFP, there obviously existed a second phenothiazine binding site which does not recognize the affinity analog. Studies using phenyl-Sepharose chromatography and hydrophobic-probe photoaffinity labeling (Vogel *et al.*, 1983a; Brzeska *et al.*, 1983; Krebs *et al.*, 1984) provided strong evidence for the

---

[2] This potentiating effect drew attention to the positive allosteric effects among the drug binding sites, which had earlier been demonstrated by the work of Johnson (1983). Thus, not only Ca^{2+} binding, but also drug binding can take place in a cooperative manner.

location of one Ca^{2+}-dependent hydrophobic surface on each half of the molecule. Further studies showed that the nonspecific hydrophobic-probe molecule
TNS also bound to both halves of the protein (Johnson et al., 1985), which
provided some idea about the location of the two TNS binding sites on CaM
(Follenius and Gérard, 1984). By comparing ^{113}Cd NMR spectra of CaM and its
two tryptic fragments in the presence and absence of TFP, it was concluded that
the strongest of the two TFP sites was located in the carboxy terminal half, and that
there was very little difference between the two TFP sites found in the two tryptic
fragments as compared to the intact protein (Forsén et al., 1979; Thulin et al.,
1984). However, when felodipine binding was studied using its intrinsic fluorescence, it did not bind to either half of the protein (Fig. 19) (Johnson et al., 1985).
Binding was observed to CaM, TM_1 (1–106), and TR_2E (1–90) (see Fig. 19b).
These data suggested that the stretch of amino acids between amino acid residues
70 and 90 is of particular relevance for the binding of this compound. This idea
found further support after we observed that felodipine offered significantly more
protection against tryptic cleavage at residues 74, 75, or 77 than did any of the
other drugs (see Fig. 20) (Johnson et al., 1985). Apart from the methionine
residues at position 71, 72, and 76, this stretch of amino acids is not particularly
hydrophobic. It is of considerable interest, however, that the major site for the
binding of the peptides appears to be between residues 72 and 106 (Malencik and
Anderson, 1984). These authors comment that the site does not necessarily need to
be hydrophobic, but that it could be rigidly hydrophilic instead.

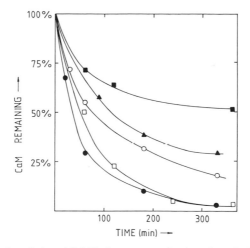

Fig. 20. Tryptic degradation of CaM in the presence of various drugs (ratio drug: CaM = 2 : 1).
Detailed experimental conditions are described elsewhere (Johnson et al., 1985). Note that felodipine
is better at protecting against proteolytic cleavage at positions 74, 75, and 77 than are the other
compounds. Symbols: Felodipine (■), W_7 (▲), TFP (○), TNS (□), Control (●).

Thus, two calcium-exposed hydrophobic surfaces are located in each half of the molecule, and they are apparently mainly responsible for the binding with TNS, TFP, phenyl-Sepharose, etc. The loop between the two halves of the CaM molecule appears to be involved in the binding of some drugs and peptides. This property of the connecting loop may explain why the structurally intact tryptic fragments can bind to target proteins without activating them, and the loop is probably important in providing specificity and fine-tuning for the interactions between CaM and its target systems.

IX. CONCLUDING REMARKS

The molecular properties and general characteristics of CaM in solution have now been investigated with the aid of a variety of biophysical methods. On the whole the results are generally in good accord. It seems that of the four Ca^{2+} binding sites, a distinction can be drawn between two with higher affinity and two with lower affinity—the difference being most pronounced at low ionic strength. Furthermore, the strong sites can be identified as sites III and IV residing in the C-terminal half of the molecule.

It also appears indisputable that under a variety of solution conditions the binding of Ca^{2+} ions to sites III and IV is a cooperative process, both in CaM and in the tryptic fragment TR_2C constituting the C-terminal half of the native molecule. It is even possible to estimate that the free energy of interaction between these sites is on the order of -1.0 kcal/mol. This value is in the range of coupling free energies found for other small ions or molecules interacting with a protein (Weber, 1975). Little can be said with certainty at present regarding the free energy of interaction characterizing Ca^{2+} binding to sites I and II of CaM or to the tryptic fragment TR_1C.

Perhaps the most remarkable finding concerning CaM obtained in recent years has followed from biophysical studies of the tryptic fragments TR_1C and TR_2C. It is apparent that the proteolytic cleavage of the native molecule at the bond joining amino acids 77 and 78 results in two peptides that seem to have largely retained the properties they had in the native molecule. CaM may therefore be considered as constructed from two domains comprising the N-terminal and the C-terminal halves of the molecule. This is not to say that the sum of the two halves equals the whole under all circumstances. In particular the interplay between CaM and other molecules, drugs, and/or target proteins depends on the interactions between different regions of the unbroken native protein. One manifestation of this interplay is the synergistic binding of drug molecules to CaM; another is the inability of the tryptic fragments to activate target proteins (Newton et al., 1984).

REFERENCES

Adair, G. S. (1925). *J. Biol. Chem.* **63**, 529–545.

Andersson, A., Drakenberg, T., Thulin, E., and Forsén, S. (1983a). *Eur. J. Biochem.* **134**, 459–465.

Andersson, A., Forsén, S., Thulin, E., and Vogel, H. J. (1983b). *Biochemistry* **22**, 2309–2313.

Andersson, T. (1981). Ph.D. Thesis, Univ. of Lund, Sweden.

Andersson, T., Drakenberg, T., Forsén, S., and Thulin, E. (1982). *Eur. J. Biochem.* **126**, 501–505.

Andersson Teleman, A., Drakenberg, T., Forsén, S., and Thulin, E. (1985). In preparation.

Andreasen, T. J., Keller, C. H., LaPorte, D. C., Edelman, A. M., and Storm, D. R. (1981). *Proc. Natl. Acad. Sci. U.S.A.* **78**, 2782–2785.

Aulabaugh, A., Niemczura, W. P., and Gibbons, W. A. (1984a). *Biochem. Biophys. Res. Commun.* **118**, 225–232.

Aulabaugh, A., Niemczura, W. P., Blundell, T. L., and Gibbons, W. A. (1984b). *Eur. J. Biochem.* **143**, 409–418.

Babu, Y. S., Sack, J. S., Greenhough, T. J., Bugg, C. E., Means, A. R., and Cook, W. J. (1985). *Nature (London)* **315**, 37–40.

Baudier, J., Hotzscherer, C., and Gérard, D. (1982). *FEBS Lett.* **148**, 231–234.

Bayley, P., Ahlström, P., Martin, S. R., and Forsén, S. (1984). *Biochem. Biophys. Res. Commun.* **120**, 185–191.

Berliner, L. J. (1976). "Spin Labeling." Academic Press, New York.

Billingsley, M., Kuhn, D., Velletri, P. A., Kincaid, R. A., and Lovenberg, W. (1984). *J. Biol. Chem.* **259**, 6630–6635.

Boström, S.-L., Ljung, B., Mårdh, S., Forsén, S., and Thulin, E. (1981). *Nature* **292**, 777–778.

Boström, S.-L., Ljung, B., Rochester, S., and Vogel, H. J. (1985). In preparation.

Brzeska, H., Szynkiewicz, J,, and Drabikowski, W. (1983). *Biochem. Biophys. Res. Commun.* **115**, 87–93.

Burger, D., Cox, J. A., Comte, M., and Stein, E. A. (1984). *Biochemistry* **23**, 1966–1971.

Burgess, W. H., Schleicher, M., van Eldik, L. J., and Watterson, D. M. (1983). *In* "Calcium and Cell Function" (W. Y. Cheung, ed.), Vol. 4, pp. 209–261. Academic Press, New York.

Cantor, C. R., and Schimmel, P. R. (1980). "Biophysical Chemistry," pp. 859–866. Freeman, San Francisco, California.

Chan, V., Huang, C. Y., Chock, P. B., Wang, J. H., and Sharma, R. K. (1982). *In* "Calmodulin and Intracellular Ca Receptors" (S. Kakiduchi, H. Hidaka, and A. R. Means, eds.), pp. 199–217. Plenum, New York.

Charbonneau, H., and Cormier, M. J. (1979). *Biochem. Biophys. Res. Commun.* **90**, 1040–1047.

Chen, A.-T., and Wadsö, I. (1982). *J. Biochem. Biophys. Methods* **6**, 307–316.

Cheung, W. Y. (1980). *Science* **207**, 19–27.

Christensen, J. J., and Izatt, R. M. (1970). "Handbook of Metal Ligand Heats and Related Thermodynamic Quantities," pp. 82–83. Dekker, New York.

Cox, J. A., Malnoë, A., and Stein, E. A. (1981). *J. Biol. Chem.* **256**, 3218–3222.

Crouch, T. H., and Klee, C. B. (1980). *Biochemistry* **19**, 3692–3698.

Dalgarno, D. C., Levine, B. A., and Williams, R. J. P. (1983). *Biosci. Rep.* **3**, 443–452.

Dalgarno, D. C., Klevitt, R. E., Levine, B. A., Williams, R. J. P., Dobrowolski, Z., and Drabikowski, W. (1984). *Eur. J. Biochem.* **138**, 281–289.

Dedman, J. R., Potter, J. D., Jackson, T. K., Johnson, J. D., and Means, A. R. (1977). *J. Biol. Chem.* **252**, 8415–8422.

Dedman, J. R., Jackson, R. L., Schreiber, W. C., and Means, A. R. (1978). *J. Biol. Chem.* **253**, 343–346.

Drabikowski, W., Kuznicki, J., and Grabarek, Z. (1977). *Biochim. Biophys. Acta* **485**, 124–133.
Drabikowski, W., Brzeska, H., and Venyaminov, S. Yu. (1982). *J. Biol. Chem.* **257**, 11584–11590.
Drakenberg, T., Forsén, S., and Lilja, H. (1983). *J. Magn. Reson.* **53**, 412–422.
Drakenberg, T., Andersson, T., Forsén, S., and Wieloch, T. (1984). *Biochemistry* **23**, 2387–2392.
Endo, T., Tanaka, T., Isobe, T., Kasai, H., Okuyama, T., and Hidaka, H. (1981). *J. Biol. Chem.* **256**, 12485–12489.
Filimonov, W. P., Tsalkova, T. N., and Privalov, P. L. (1978). *Biophys. Chem.* **8**, 117–122.
Follenius, A., and Gerard, D. (1984). *Biochem. Biophys. Res. Commun.* **119**, 1154–1160.
Forsén, S., Thulin, E., and Lilja, H. (1979). *FEBS LeH.* **104**, 123–126.
Forsén, S., Thulin, E., Drakenberg, T., Krebs, J., and Seamon, K. (1980) *FEBS Lett.* **117**, 189–194.
Frost, A. A., and Pearson, R. G. (1961). "Kinetics and Mechanism," pp. 166–169. Wiley, New York.
Gangola, P., and Pant, H. C. (1983). *Biochem. Biophys. Res. Commun.* **111**, 301–305.
Giedroc, D. P., Puett, D., Ling, N., and Staros, J. V. (1983). *J. Biol. Chem.* **258**, 16–19.
Gietzen, K., Wüthrich, A., and Bader, H. (1981). *Biochem. Biophys. Res. Commun.* **101**, 418–425.
Gilbert, W. (1978). *Nature (London)* **271**, 501–505.
Goewert, R. G., Landt, M., and McDonald, J. (1982). *Biochemistry* **21**, 5310–5315.
Gopalakrishna, R., and Anderson, W. B. (1982). *Biochem. Biophys. Res. Commun.* **104**, 830–836.
Gopalakrishna, R., and Anderson, W. B. (1983). *J. Biol. Chem.* **258**, 2405–2409.
Grand, R. J. A., and Perry, S. V. (1980). *Biochem. J.* **189**, 227–240.
Grand, R. J. A., Shenolikar, S., and Cohen, P. (1981). *Eur. J. Biochem.* **113**, 359–367.
Guerini, D., and Krebs, J. (1983). *FEBS Lett.* **164**, 105.
Guerini, D., Krebs, J., and Carafoli, E. (1984). *J. Biol. Chem.* **259**, 15172–15177.
Haiech, J., Klee, C. B., and Demaille, J. G. (1981). *Biochemistry* **20**, 3890–3897.
Hart, R. C., Hice, R. E., Charbonneau, H., Putnam-Evans, C., and Cormier, M. J. (1983). *Anal. Biochem.* **135**, 208–220.
Head, J. F., Masure, H. R., and Kaminer, B. (1982). *FEBS Lett.* **137**, 71–71.
Hedwig, G. R., and Biltonen, R. L., (1984). *Biophys. Chem.* **19**, 1–11.
Herzberg, O., and James, M. (1985). Submitted for publication.
Hewgley, P. B., and Puett, D. (1980). *Ann. N.Y. Acad. Sci.* **356**, 20–32.
Hidaka, H., Asano, M., and Tanaka, T. (1981). *Mol. Pharmacol.* **20**, 571–578.
Ikura, M., Kiraoki, T., Kikichi, K., Mikuni, T., Yazawa, M., and Yagi, K. (1983a). *Biochemistry* **22**, 2568–2572.
Ikura, M., Hiraoki, T., Hikichi, K., Mikuni, T., Yazawa, M., and Yagi, K. (1983b). *Biochemistry* **22**, 2573–2579.
Ikura, M., Hiraoki, T., Hikichi, K., Minowa, O., Yamaguchi, H., Yazawa, M., and Yagi, K. (1984). *Biochemistry* **23**, 3124–3128.
Jamieson, G. A., and Vanaman, T. C. (1979). *Biochem. Biophys. Res. Commun.* **90**, 1048–1056.
Jamieson, G. A., Bronson, D. D., Schachat, F. H., and Vanaman, T. C. (1980). *Ann. N.Y. Acad. Sci.* **356**, 1–13.
Jarrett, H. W., and Kyte, J. (1979). *J. Biol. Chem.* **254**, 8237–8244.
Johnson, J. D. (1983). *Biochem. Biophys. Res. Commun.* **112**, 787–793.
Johnson, J. D., and Wittenauer, L. A. (1983). *Biochem. J.* **211**, 473–479.
Johnson, J. D., Thulin, E., and Vogel, H. J. (1985). In preparation.
Keller, C. H., Olwin, B. B., LaPorte, D. C., and Storm, D. R. (1982). *Biochemistry* **21**, 156–162.
Keller, C. H., Olwin, B. B., Heideman, W., and Storm, D. R. (1983). *In* "Calcium and Cell Function" (W. Y. Cheung, ed.), Vol. 3, pp. 103–127. Academic Press, New York.

Kilhoffer, M.-C., Demaille, J. G., and Gerard, D. (1980a). *FEBS Lett.* **116**, 269–272.
Kilhoffer, M.-C., Gerard, D., and Demaille, J. G. (1980b). *FEBS Lett.* **120**, 99–103.
Kilhoffer, M.-C., Demaille, J. G., and Gerard, D. (1981). *Biochemistry* **20**, 4407–4414.
Kilhoffer, M.-C., Haiech, J., and Demaille, J. G. (1983). *Mol. Cell. Biochem.* **51**, 33–54.
Kincaid, R. L. (1984). *Biochemistry* **23**, 1143–1147.
Kincaid, R. L., and Vaughan, M. (1983). *Biochemistry* **22**, 826–830.
Klee, C. B. and Vanaman, T. C. (1982). *Adv. Protein Chem.* **35**, 213–321.
Klevitt, R. E., and Vanaman, T. C. (1983). *Fed. Proc., Fed. Am. Soc. Exp. Biol.* **46**, 2615.
Klevitt, R. E., Levine, B. A., and Williams, R. J. P. (1981). *FEBS Lett.* **123**, 25–29.
Klevitt, R. E., Dalgarno, D. C., Levine, B. A., and Williams, R. J. P. (1984). *Eur. J. Chem.* **139**, 109–114.
Klotz, I. M. (1974). *Acc. Chem. Res.* **7**, 162–168.
Klotz, I. M., and Hunston, D. L. (1979). *Arch. Biochem. Biophys.* **193**, 314–328.
Kometani, K., and Yamada, K. (1983). *Biochem. Biophys. Res. Commun.* **114**, 162–167.
Krebs, J. (1981). *Cell Calcium* **2**, 295–311.
Krebs, J., and Carafoli, E. (1982). *Eur. J. Biochem.* **124**, 619–627.
Krebs, J., Buerkler, J., Guerini, D., Brunner, J., and Carafoli, E. (1984). *Biochemistry* **23**, 400–403.
Kretsinger, R. H. (1980). *Ann. N.Y. Acad. Sci.* **356**, 14–19.
Kretsinger, R. H., and Nockolds, C. E. (1973). *J. Biol. Chem.* **248**, 3313–3326.
Kuznicki, J., Grabarek, Z., Brzeska, H., Drabikowski, W., and Cohen, P. (1981). *FEBS Lett.* **130**, 141–145.
Lagacé, L., Chandra, T., Woo, S. L., and Means, A. R. (1983). *J. Biol. Chem.* **258**, 1684–1688.
Lambooy, P. K., Steiner, R. F., and Sternberg, H. (1982). *Arch. Biochem. Biophys.* **217**, 517.
LaPorte, D. C., Wierman, B. M., and Storm, D. R. (1980). *Biochemistry* **19**, 3814–3819.
Laynez, J., Sellers, P., and Thulin, E. (1983). *Int. Symp. Thermodyn. Proteins Biol. Membr., 1983*, P-I-13.
Levin, R. M., and Weiss, B. (1977). *Mol. Pharmacol.* **13**, 690–697.
Lindahl, L., and Vogel, H. J. (1984). *Anal. Biochem.* **140**, 394–402.
Louis, C. F., and Jarvis, B. (1982). *J. Biol. Chem.* **257**, 15187–15191.
McCubbin, W. D., Hincke, M. T., and Kay, C. M. (1979). *Can. J. Biochem.* **57**, 15–20.
Malencik, D. A., and Anderson, R. R. (1983). *Biochemistry* **22**, 1995–2001.
Malencik, D. A., and Anderson, R. R. (1984). *Biochemistry* **23**, 2420–2428.
Malencik, D. A., Anderson, S. R., Shalitin, Y., and Schimerlik, M. I. (1981). *Biochem. Biophys. Res. Commun.* **101**, 390–395
Marshak, D. R., Watterson, D. M., and van Eldik, L. J. (1981). *Proc. Natl. Acad. Sci. U.S.A.* **78**, 6793–6797.
Marshak, D. R., Clarke, M., Roberts, D. M., and Watterson, D. M. (1984). *Biochemistry* **23**, 2891–2899.
Martell, A. E., and Smith, R. M. (1982). "Critical Stability Constants," Vol. 5, Suppl. 1. Plenum, New York.
Martin, S. R., Anderson, A., Bayley, P., Drakenburg, T., and Forsén, S. (1985). *Eur. J. Biochem.* (in press).
Maulet, Y., and Cox, J. A. (1983). *Biochemistry* **22**, 5680–5686.
Means, A. R., Tash, J. S., and Chafoulos, J. G. (1982). *Physiol. Rev.* **62**, 1–39.
Moeschler, H. J., Scharr, J.-J., and Cox, J. A. (1980). *Eur. J. Biochem.* **111**, 73–78.
Newton, D. L., Burke, T. R., Rice, K. C., and Klee, C. B. (1983). *Biochemistry* **22**, 5472–5476.
Newton, D. L., Oldewurtel, M. D., Krinks, M. H., Shiloach, J., and Klee, C. B. (1984). *J. Biol. Chem.* **259**, 4419–4431.

Olwin, B. D., Edelman, A. M., Krebs, E. G., and Storm, D. R. (1984). *J. Biol. Chem.* **259**, 10949–10955.

Potter, J. D., Hsu, F.-J., and Pownall, H. J. (1977). *J. Biol. Chem.* **252**, 2452–2454.

Pundak, S., and Roche, R. S. (1984). *Biochemistry* **23**, 1549–1555.

Putkey, J. A., Ts'ui, K. F., Tanaka, T., Lagacé, L., Stein, J. P., Lai, E. C., and Means, A. R. (1983). *J. Biol. Chem.* **258**, 11864–11870.

Putkey, J. A., Slaughter, G. R., and Means, A. R. (1985). *J. Biol. Chem.* (in press).

Rainteau, D., Wolf, C., Bereziat, G., and Polonovski, J. (1984). *Biochem. J.* **221**, 659–663.

Richman, P. G. (1978). *Biochemistry* **17**, 3001–3005.

Richman, P. G., and Klee, C. B. (1978). *Biochemistry* **17**, 928–935.

Richman, P. G., and Klee, C. B. (1979). *J. Biol. Chem.* **254**, 5372–5376.

Roufogalis, B. D. (1981). *Biochem. Biophys. Res. Commun.* **98**, 607–613.

Sasagawa, T., Ericsson, L. H., Walsh, K. A., Schreiber, W., Fisher, E. H., and Titani, K. A. (1982) *Biochemistry* **21**, 2565–2569.

Schatzman, R. C., Raynor, R. L., and Kuo, J. F. (1983). *Biochim. Biophys. Acta* **755**, 144–147.

Schimerlik, M. I., Malencik, D. A., Anderson, S. R., and Shalitin, Y. (1982). *Biophys. Biochem. Res. Commun.* **106**, 1331–1339.

Seamon, K. B. (1980). *Biochemistry* **19**, 207–215.

Seamon, K. B., and Kretsinger, R. H. (1983). *In* "Metal Ions in Biology" (T. G. Spiro, ed.), Vol. VI, pp. 1–51. Wiley, New York,

Seaton, B. A., Head, J. F., Lord, R. C., and Petsko, G. A. (1983). *Biochemistry* **22**, 973–978.

Sellers, P., Laynes, J., Thulin, E., and Forsén, S. (1985). In preparation.

Shimizu, T., Hatano, M., Nagao, S., and Nozawa, Y. (1982). *Biochem. Biophys. Res. Commun.* **106**, 1112–1118.

Siegel, N., and Hang, A. (1983). *Biochim. Biophys. Acta* **744**, 36–45.

Siegel, N., Coughlin, R., and Hang, A. (1983). *Biochem. Biophys. Res. Commun.* **115**, 512–517.

Sokolovsky, M., Riordan, J. F., and Vallee, B. L. (1967). *Biochem. Biophys. Res. Commun.* **27**, 20–27.

Stein, J. P., Munjaal, R. P., Lagacé, L., Lai, E. C., O'Malley, B. W., and Means, A. R. (1983). *Proc. Natl. Acad. Sci. U.S.A.* **80**, 6485–6489.

Steiner, R. F. (1984). *Arch. Biochem. Biophys.* **228**, 105–112.

Steiner, R. F., Lambooy, P. K., and Sternberg, H. (1983). *Arch. Biochem. Biophys.* **222**, 158.

Szebenyi, D. M. E., Obendorf, S. K., and Moffat, K. (1981). *Nature (London)* **294**, 327–332.

Takagi, T., Nemoto, T., Konishi, K., Yazawa, M., and Yagi, K. (1980). *Biochem. Biophys. Res. Commun.* **96**, 1493–1505.

Tanaka, T., and Hidaka, H. (1980). *J. Biol. Chem.* **255**, 11078–11080.

Tanaka, T., Umekawa, H., Ohmura, T., and Hidaka, H. (1984). *Biochim. Biophys. Acta* **787**, 158–164.

Tanford, C. (1961). "Physical Chemistry of Macromolecules." Wiley, New York.

Tanokura, M., and Yamada, K. (1983). *J. Biochem. (Tokyo)* **84**, 607–609.

Tanokura, M., and Yamada, K. (1984). *J. Biochem. (Tokyo)* **95**, 643–649.

Teleman, O., Drakenberg, T., Forsén, S., and Thulin, E. (1983). *Eur. J. Biochem.* **134**, 453–457.

Thomas, M. V. (1982). "Techniques in Calcium Research," pp. 25–32. Academic Press, London.

Thulin, E., Forsén, S., Drakenberg, T., and Andersson, T. (1980). *In* "Calcium Binding Proteins: Structure and Function" (F. L. Siegel, ed.), p. 243. Elsevier/North-Holland, Amsterdam.

Thulin, E., Andersson, A., Drakenberg, T., Forsén, S., and Vogel, H. J. (1984). *Biochemistry* **23**, 1862–1870.

Toda, H., Yazawa, M., Kondo, K., Henna, T., Narita, K., and Yagi, K. (1981). *J. Biochem. (Tokyo)* **90**, 1493–1505.

Tsalkova, T. N., and Privalov, P. L. (1985). *J. Mol. Biol.* **181**, 533–544.

Tsien, R. Y., Pozzan, T., and Rink, T. J. (1982). *J. Cell Biol.* **94**, 325–334.

Tuana, B. S., and MacLennan, D. H. (1984). *J. Biol. Chem.* **259**, 6979–6983.

Vogel, H. J., Lindahl, L., and Thulin, E. (1983a). *FEBS Lett.* **157**, 241–246.

Vogel, H. J., Drakenberg, T., and Forsén, S. (1983b). *In* "NMR of Newly Accessible Nuclei" (P. Laszlo, ed.), Vol. 1, pp. 157–192. Academic Press, New York.

Vogel, H. J. Andersson, T., Braunlin, W. H., Drakenberg, T., and Forsén, S. (1984). *Biochem. Biophys. Res. Commun.* **122**, 1350–1356.

Walker, S. W., Warla, J. D., MacNeil, S., Mellersh, H., Brown, B. L., and Tomlinson, S. (1984). *Biochem. J.* **217**, 827–832.

Wall, C. M., Grand, R. J., and Perry, S. V. (1981). *Biochem. J.* **195**, 307–316.

Wallace, R. W., Tallant, E. A., Dockter, M. E., and Cheung, W. Y. (1982). *J. Biol. Chem.* **257**, 1845–1854.

Walsh, M., and Stevens, F. C. (1977). *Biochemistry* **16**, 2742–2749.

Walsh, M., and Stevens, F. C. (1978). *Biochemistry* **17**, 3924–3929.

Walsh, M., Stevens, F. C., Kuznicki, J., and Drabikowski, W. (1977). *J. Biol. Chem.* **252**, 7440–7443.

Walsh, M., Valentine, K. A., Ngai, P. K., Carruthers, C. A., and Hollenberg, M. (1984). *Biochem. J.* **224**, 117–127.

Wang, C.-L. A., Aquaron, R. R., Leavis, P. C., and Gergely, J. (1982). *Eur. J. Biochem.* **124**, 7–12.

Wang, C.-L. A., Leavis, P. G., and Gergely, J. (1985). *Biochemistry* (submitted for publication).

Watterson, D. M., Sharief, F., and Vanaman, T. C. (1980). *J. Biol. Chem.* **255**, 962–975.

Weber, G. (1975). *Adv. Protein Chem.* **29**, 1–48.

Xü, Y. H., Gietzen, K., and Galla, H.-J. (1983). *Int. J. Biol. Macromol.* **5**, 154–158.

Yagi, K., Matsuda, S., Nagamoto, H., Mikuni, T., and Yazawa, M. (1982). *In* "Calmodulin and Intracellular Ca^{2+} Receptors" (S. Kakiuchi, H. Hidaka, and A. R. Means, eds.), pp. 75–91. Plenum, New York.

Yamada, K., and Kometani, K. (1982). *J. Biochem. (Tokyo)* **92**, 1505–1517.

Yamada, K., and Kometani, K. (1983). *Biochem. Biophys. Res. Commun.,* **114**, 162–167.

Yazawa, M., Yagi, K., Toda, H., Kondo, K., Narita, K., Yamazaki, R., Sobue, K., Kakiuchi, S., Nagao, S., and Nozawa, Y. (1981). *Biochem. Biophys. Res. Commun.* **99**, 1052–1057.

Zimmer, M., and Hofmann, F. (1984). *Eur. J. Biochem.* **142**, 393–397.

Chapter 5

Regulation of Ca^{2+}-Dependent Proteinase of Human Erythrocytes

SANDRO PONTREMOLI
EDON MELLONI

Institute of Biological Chemistry
University of Genoa
Genoa, Italy

159

CALCIUM AND CELL FUNCTION, VOL. VI

I. INTRODUCTION[1]

Calcium-dependent neutral proteinases indicated with the generic name of calpain have been studied extensively in recent years and found to be ubiquitously distributed in all mammalian cells so far examined (Murachi, 1983). The numerous reports on Ca^{2+} proteinases have revealed some basic properties of this class of hydrolases for all cell types. These include location in the cytosolic fraction, neutral pH activity dependent on the presence of calcium ions, cysteine type of proteinases, heterodimeric molecular structure, and specific inhibition by an endogenous inhibitor protein with a cellular location identical to that of calpain(s) (for a review, see Pontremoli et al., 1985a).

In spite of all these findings, uncertainties remain concerning calpain's Ca^{2+} requirement. There appear to be either two distinct classes of calpains, one with a low and the other with a high Ca^{2+} requirement, or only one class with a high Ca^{2+} requirement which can be converted to the low-Ca^{2+}-requiring form by an autoproteolytic digestion (Pontremoli et al., 1984a). This question bears directly on physiological function, since the level of calcium ions required for activity or for autoproteolytic conversion is much higher than the actual intracellular Ca^{2+} concentration (Wiley et al., 1982). Other questions include the specific functions of the two subunits which compose the molecular structure of calpain and the type and location of calpain's physiological substrate(s) in various cell types.

II. THE PROTEOLYTIC ENZYMES OF HUMAN ERYTHROCYTES

Human erythrocytes contain a number of proteolytic enzymes characterized by distinct properties, specificities, and cellular distributions (Pontremoli et al., 1979, 1980a). In the cytosolic compartment two amino peptidases and two dipeptidyl aminopeptidases are present in addition to a single Ca^{2+}-dependent neutral endopeptidase (calpain), first discovered by us (Pontremoli et al., 1980a) and more recently described by others (Murakami et al., 1981).

In the erythrocyte membrane three acid endopeptidases have been identified which have properties similar to those of Cathepsin D and are distinguishable from each other on the basis of different electrophoretic mobilities (Pontremoli et al., 1980b).

In spite of a report (Golovtchenko-Matsumoto et al., 1982) concerning the presence of a Ca^{2+}-requiring proteinase, no other endo- or exopeptidase activities have been detected in the erythrocyte membrane. Human erythrocytes thus appear to contain two distinct types of endopeptidases with different cellular

---

[1] Procalpain is the inactive proenzyme form of calpain; calpain, the active form of the Ca^{2+}-dependent neutral proteinase; CaP hemoglobin, β-hemoglobin modified by limited proteolysis with calpain.

locations: one type includes three acidic endopeptidases, similar to Cathepsin D, which are associated with the erythrocyte membrane; the other type consists of a single Ca^{2+}-dependent neutral proteinase located in the red cell cytoplasm. As will be discussed, the two types of endopeptidases, although different in location and in catalytic properties, seem to be functionally related (Melloni et al., 1982a). This chapter will consider the catalytic and regulatory properties of human erythrocyte calpain in an effort to understand how this enzyme participates in intracellular proteolysis.

III. PHYSICAL PROPERTIES

The molecular weight of calpain purified from human erythrocytes is approximately 110,000 (Melloni et al., 1982b). Analysis by SDS–polyacrylamide gel electrophoresis indicates that the enzyme is composed of two nonidentical subunits of 80,000 and 30,000 Da. The heterodimeric nature of calpain seems to be a general property in all organs and tissues with the exception of the liver, which has two distinct dimeric forms, each composed of equal molecular weight subunits (Melloni et al., 1984a). We found that the 80-kDa subunit carries the catalytic site and expresses, in its isolated form, full catalytic activity (Melloni et al., 1982c). The function of the 30-kDa subunit in human erythrocyte calpain seems to be related to the maintenance of the inactive proenzyme form. It dissociates from the catalytic subunit during the activation process (Melloni et al., 1984b). Additional investigation is needed on the exact relationship of the two subunits because of questions raised by recent data concerning the platelet (Yoshida et al., 1983) and the muscle enzyme (Ishiura et al., 1978).

Calpain appears to be a thiol proteinase because of its high sensitivity to leupeptin, an inhibitor of thiol proteinases (Melloni et al., 1982b), and because of its complete inactivation by alkylating agents (Melloni et al., 1982c).

IV. EFFECT OF HIGH CONCENTRATIONS OF CALCIUM IONS ON THE ACTIVITY OF CALPAIN

The proteolytic activity of purified calpain can only be detected in the presence of calcium ions. As shown in Fig. 1, half-maximum activity is observed at 40 μM Ca^{2+}, and full activity at 100 μM Ca^{2+}.

At 1–5 μM Ca^{2+} concentration no proteolytic activity was detected. It must be emphasized, however, that under routine assay conditions the incubation is carried out for several minutes during which calpain is converted to an active, low-Ca^{2+}-requiring form. This conclusion is supported by the observation that almost instantaneously following incubation in 1 mM Ca^{2+} (Fig. 2A), or in 0.1

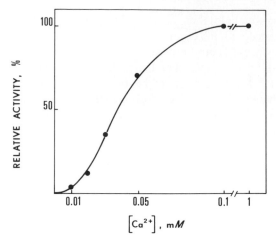

Fig. 1. Effect of Ca^{2+} on the activity of procalpain. For experimental details see Melloni *et al.* (1982b).

mM Ca^{2+} (Fig. 2B), the proteinase becomes fully active at micromolar Ca^{2+} and shows significant changes in its calcium requirement (Fig. 3). Thus, the high concentration of calcium ions apparently required for full activity is necessary for a preliminary conversion of an inactive procalpain form to a fully active calpain. We suggest that human erythrocyte calpain be considered a low-Ca^{2+}-requiring proteinase which actually expresses catalytic activity at low concentrations of the

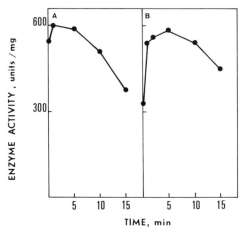

Fig. 2. Conversion of procalpain to the active low-calcium-requiring form. The purified pro-enzyme (30 μg) was incubated at room temperature in 0.5 ml of 50 mM sodium borate, pH 7.5, containing 1 mg of substrate and 1 mM Ca^{2+} (A), or 0.1 mM Ca^{2+} (B). At the times indicated, 0.025-ml aliquots of the incubation mixtures were collected and assayed for proteinase activity in the presence of 5 μM Ca^{2+}. For experimental details see Melloni *et al.* (1984b).

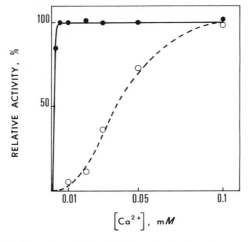

Fig. 3. Effect of Ca^{2+} on the activity of active calpain. (- - - -) represents procalpain's Ca^{2+} requirement (taken from Fig. 1). For experimental details see Melloni *et al.* (1984b).

metal ion. This conclusion receives support from the observation that following treatment of intact erythrocytes with ionophore A23187 and increasing concentrations of Ca^{2+}, the ratio of inactive procalpain to active calpain progressively increases and accounts for more than 50% of the total calpain activity in 50 μM Ca^{2+}, and for 100% in 0.1 mM Ca^{2+} (Fig. 4).

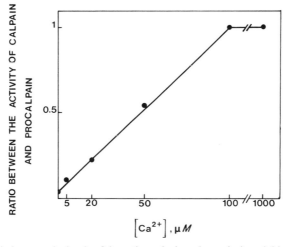

Fig. 4. Ratio between the levels of the active calpain and procalpain activities in intact erythrocytes treated with A23187 ionophore at increasing concentrations of Ca^{2+}. Human erythrocytes were incubated with A23187 ionophore and the indicated concentration of Ca^{2+} as described by Pontremoli *et al.* (1984b). The activities of calpain and procalpain were measured using 5 μM Ca^{2+} and 1 mM Ca^{2+}, respectively (Melloni *et al.*, 1984b).

These results confirm that in the intact red cell the proteinase exists in the procalpain inactive form and that it can be converted to active calpain when the intracellular calcium concentration is raised. The type of calpain (high- or low-Ca^{2+}-requiring form) that can be isolated from intact cells can thus be artifactually modified, which may explain why under some conditions only one form or the other has been isolated.

V. EFFECT OF SUBSTRATE ON THE CONVERSION OF PROCALPAIN TO ACTIVE CALPAIN AT MICROMOLAR Ca^{2+} CONCENTRATIONS

The conversion of human erythrocyte procalpain to active calpain is promoted, as previously discussed, by high concentrations of Ca^{2+} and does not require the presence of a substrate. Similar, though slower, conversion is observed when procalpain is incubated in the presence of 5 μM Ca^{2+} with an appropriate amount of a digestible substrate such as β-hemoglobin or heme-free β-globin chains (Melloni et al., 1984b). The conversion of procalpain at low concentrations of Ca^{2+} is dependent not only on the amount of substrate but also on the

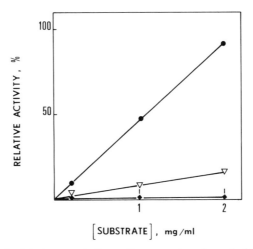

Fig. 5. Effect of substrate concentration on the conversion of procalpain to active calpain at 5 μM Ca^{2+}. The purified proenzyme (10 μg) was incubated at room temperature in 0.2 ml of sodium borate, pH 7.5, containing 5 μM Ca^{2+} and the indicated amounts of denatured globin (●), casein (▽), or bovine serum albumin (BSA) (◆). After 60 min, 0.025-ml aliquots of the mixtures were removed and assayed for proteinase activity in the presence of 5 μM Ca^{2+}. For experimental details see Melloni et al. (1984b).

particular substrate used (Fig. 5). Thus, we can conclude that proteolytic activity requires the conversion of inactive procalpain to active calpain, and that this conversion is promoted by 1 or 0.1 mM Ca^{2+}, or in a more physiological setting by micromolar Ca^{2+} in the presence of a digestible substrate. It is important to emphasize that, although there may be two mechanisms for the formation of active calpain, there is only one catalytically active form of the enzyme, which shows identical rates of hydrolysis of denatured globin at concentrations of Ca^{2+} from 1 μM to 1 mM.

VI. THE MOLECULAR BASIS OF THE CONVERSION OF PROCALPAIN TO ACTIVE CALPAIN

The conditions (1 mM Ca^{2+} or 5 μM Ca^{2+} plus substrate) which promote the formation of active calpain induce dissociation of procalpain and generate the free catalytic subunit. Neither conversion to active calpain nor any change in the molecular weight of procalpain was observed if incubation was carried out in either substrate or 5 μM Ca^{2+} alone.

The dissociation of procalpain into constituent subunits by Ca^{2+} precedes the activation. We demonstrated this requirement by using the carboxymethylated ^{14}C inactive proenzyme, which still retains its dimeric structure. The carboxymethylated procalpain (preincubated in 0.1 mM, 1 mM Ca^{2+}, or 5 μM Ca^{2+} plus saturating concentrations of substrate revealed a molecular weight of 80,000 both in coefficient distribution measurements (Table I) and gel filtration (Fig. 6).

TABLE I

Molecular Size of the Active Calpain Form[a]

Activating system	Enzyme form produced	K_D[b]	M_r[c]
None	Inactive	0.6	110,000
5 μM Ca^{2+}	Inactive	0.6	110,000
2 mg/ml globin	Inactive	0.6	110,000
5 μM Ca^{2+} + 2 mg/ml globin	Active	0.8	80,000
0.1 mM Ca^{2+}	Active	0.8	80,000
1 mM Ca^{2+}	Active	0.8	80,000

[a]For experimental details see Melloni *et al.* (1984b).

[b]K_D = distribution coefficient determined following the procedure described by MacGregor *et al.* (1980).

[c]M_r = molecular weight established from the K_D values of standard proteins.

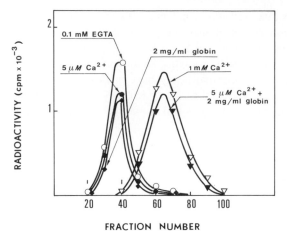

Fig. 6. Changes in the elution volume of [^{14}C]carboxymethylated procalpain induced by 1 mM Ca^{2+} (∇) and by 5 μM Ca^{2+} in the presence of the substrate ($\blacktriangledown$). Symbols: $\bigcirc$, 0.1 mM EGTA; $\bullet$, 5 μM Ca^{2+}; $\blacklozenge$, 2 mg/ml globin. Procalpain was carboxymethylated with [^{14}C]iodoacetate as described by Melloni *et al.* (1982c). The labeled enzyme was filtered on a Sephadex G-200 column previously equilibrated with 50 mM sodium borate, pH 7.5, containing the specified additions. Native procalpain (110,000 Da) was eluted in the same column at fraction 40; bovine serum albumin, at fraction 76.

VII. MECHANISM OF PROCALPAIN ACTIVATION

Although conversion of procalpain to active calpain is related to dissociation of the oligomeric structure, dissociation alone appears only to precede activation, and not to produce it. The 80-kDa subunit was in fact isolated from the calpain-inhibitor complex (Pontremoli *et al.*, 1985b) in an inactive form and shown to require, as the inactive procalpain, high concentrations of Ca^{2+} alone, or micromolar Ca^{2+} plus the substrate, to be converted to the active form (S. Pontremoli and E. Melloni, unpublished data). This activation process appears to involve a limited autocatalytic digestion of the 80-kDa subunit with a decrease in its molecular weight to approximately 75,000 (Fig. 7). The 75-kDa subunit retains its monomeric structure and does not reassociate with the 30-kDa subunit, whereas this reassociation process does occur in the 80-kDa inactive subunit (S. Pontremoli and E. Melloni, unpublished data).

The limited digestion which produces the active catalytic subunit appears to be an autocatalytic, intramolecular process, because active calpain failed to generate any additional proteinase activity when added to undissociated procalpain or

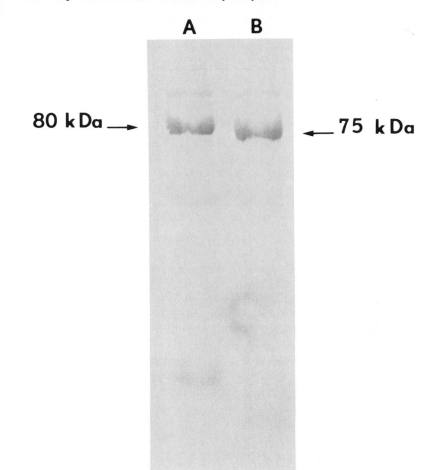

Fig. 7. SDS–polyacrylamide gel electrophoresis of procalpain (A) and of active calpain (B). For experimental details see Pontremoli *et al.* (1984a).

to the isolated, inactive 80-kDa subunit (Table II). The results so far discussed are consistent with the model presented in Fig. 8.

VIII. SPECIFICITY AND SELECTIVITY OF PROTEOLYTIC DIGESTION BY CALPAIN

Active calpain shows negligible hydrolyzing activity with commonly used synthetic substrates of cathepsin or with peptide substrates. It is most active with

TABLE II

Effect of Incubation of Procalpain or of the Single Inactive 80-kDa Subunit with Small Quantities of Active Calpain[a,b]

Mixture	Calpain activity (units)	
	0 Time	After 60 min
Procalpain (10 μg) + calpain (0.8 μg)	0.48	0.45
Single 80-kDa subunits (10 μg) + calpain (0.8 μg)	0.50	0.44

[a]Activation of procalpain by active calpain was carried out as described by Pontremoli et al. (1984a).

[b]Experimental details on the isolation and activation of the 80-kDa subunit will be reported elsewhere.

isolated heme-free α- or β-globin chains and less effective (50–60% of full activity) with α- or β-hemoglobin chains (Table III). This difference in the rate of digestion may result from the fact that heme-free globins are present as monomers, while native hemoglobins are present as tetramers (Pontremoli et al., 1984b). Such structural requirements apply not only to the expression of catalytic activity, but also to the promoting effect of the substrates in the conversion of procalpain to active calpain. We found β-hemoglobin to be 60% as effective as β-globin in the activation process (Pontremoli et al., 1984a).

Human erythrocyte calpain possesses a high degree of specificity toward globin chains. It produces a selective cleavage of a single peptide bond in a region close to the amino-terminal end of both globin molecules (Melloni et al., 1984c). In α-hemoglobin or α-globin, the cleavage occurs between Lys_{11} and Ala_{12}; in β-hemoglobin or β-globin, it occurs between Lys_8 and Ser_9 (Fig. 9). Two peptides, composed of 11 and 8 amino acids respectively, are thus produced, while no further degradation of the isolated peptides or of the digested proteins occurs (Melloni et al., 1984c). Apparently the modified β-hemoglobin and β-globin lose their structural properties as substrates and accordingly lose their ability to promote the conversion from procalpain to calpain (Table IV). The isolated octapeptide produced by digestion of β-heme or heme-free β-globins with calpain fails to promote such conversion (S. Pontremoli and E. Melloni, unpublished data). These observations indicate a high degree of specificity of calpain for soluble proteins. The amino terminus of the globin molecule is a recognition site of the proteinase, although activation appears to require the integrity of the globin molecule, and not just the amino terminus peptides.

One conclusion that can be drawn from these results concerns the stability of globin variants in the intact red cell. For instance, the rate of digestion of α-

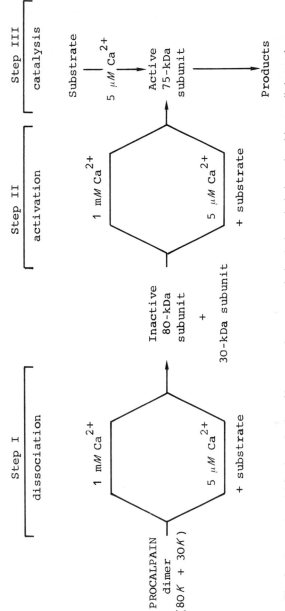

Fig. 8. Proposed model for the activation of human erythrocyte procalpain. Active calpain is produced by two distinct pathways: one requiring a high (mM) concentration of Ca^{2+}, and a second requiring a more nearly physiological concentration (μM) of Ca^{2+} and the presence of a digestible substrate. The first step of the process is the dissociation of the native oligomeric proenzyme into its constituent subunits, followed by an autoproteolytic conversion of the native 80,000-Da catalytic subunit to the active 75,000-Da subunit. The active subunit expresses full catalytic activity at concentrations of Ca^{2+} ranging from 1 μM to 1 mM.

TABLE III

**Effect of Native or Denatured Hemoglobin Chains on the Conversion
of Procalpain to Active Calpain**[a]

Substrate	Concentration (mg/ml)	Procalpain to calpain (%)
Denatured α-globin	2	100
Denatured β-globin	2	100
Native α-hemoglobin	2	18
Native β-hemoglobin	2	24
Native α-hemoglobin	8	48
Native β-hemoglobin	8	61

[a]10 μg of purified procalpain was incubated as described by Pontremoli *et al.* (1984b) with the indicated amount of substrate in the presence of 5 μ*M* Ca^{2+}. After 60 min the amount of active calpain produced was determined.

Hasharon (which carries a mutation Asp-His at position 47) is higher than that of normal α-globin, while the β-San José (which carries a mutation Glu-Gly in position 7—very close to the site of cleavage) was almost completely resistant to digestion (Fig. 10).

IX. REGULATION OF HUMAN ERYTHROCYTE CALPAIN

A. Regulation by the Endogenous Inhibitor

The generally accepted regulatory mechanism of the activity of human erythrocyte calpain is, as for all known Ca^{2+}-dependent neutral proteinases, based on the presence of an endogenous inhibitor which is also present in the cytosolic fraction (Murakami *et al.*, 1981; Murachi, 1983). The inhibitor is a tetrameric

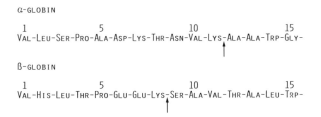

Fig. 9.   Site of cleavage of α- and β-hemoglobin chains by human erythrocyte calpain. For experimental details see Melloni *et al.* (1984c).

TABLE IV

Effect of Digested β-Hemoglobin Chains on the Conversion of Procalpain to Active Calpain[a]

Addition	Procalpain to calpain (%)
5 μM Ca²⁺ + β-globin (2 mg/ml)	100
5 μM Ca²⁺ + CaP–β-globin (2 mg/ml)	5
5 μM Ca²⁺ + β-hemoglobin (8 mg/ml)	61
5 μM Ca²⁺ + CaP–β-hemoglobin (8 mg/ml)	2

[a]The modified proteins (CaP–β-globin and CaP–β-hemoglobin) were prepared as described by Pontremoli *et al.* (1984a). 10 μg of purified procalpain was incubated with the indicated concentration of substrate at 25°C, and after 60 min the amount of calpain produced was determined.

protein of 240 kDa composed of 4 subunits of 60 kDa each (Melloni *et al.*, 1982c). Inhibition of calpain activity requires the presence of calcium ions and is nearly complete when either procalpain or active calpain is incubated with the inhibitor and Ca²⁺ ranging from 0.1 to 1 mM. No inhibition of previously activated calpain is observed at low concentrations of Ca²⁺ (Table V).

When the conversion of the proenzyme to active proteinase was promoted by high concentrations of Ca²⁺, the inhibitor was effective, but when conversion was carried out at low Ca concentrations in the presence of a substrate, the inhibitor was ineffective (Fig. 11). To explore this discrepancy we isolated the

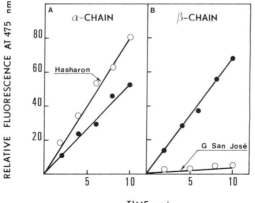

Fig. 10. Digestibility of normal (●) or variant (○) human globin chains by active calpain. For experimental details see Melloni *et al.* (1984c).

TABLE V

Effect of Ca²⁺ on the Inhibition of Procalpain and of Active Calpain by the Endogenous Inhibitor[a]

Enzyme form	[Ca²⁺]	Proteinase activity		
		− Inhibitor	+ Inhibitor	Inhibition
Procalpain	1 mM	625	54	92
Procalpain	5 μM	10	9	10
Calpain	5 μM	585	560	5
Calpain	1 mM	585	49	91

[a]For experimental details see Melloni *et al.* (1984b).

calpain–inhibitor complex formed at 1 mM Ca²⁺ (Fig. 12). The complex showed an apparent molecular mass of 140 kDa corresponding to the molecular weight of a dimer containing 1 subunit each of inhibitor and proteinase. These results can only be explained by the assumption that high concentrations of Ca²⁺, in addition to dissociating procalpain, also dissociated the inhibitor into single subunits of 80 kDa and 60 kDa, respectively. These changes in the inhibitor's molecular size occur at high concentrations of Ca²⁺, but not in the presence of low concentrations of Ca²⁺ plus substrate (Table VI). Since dissociation of the inhibitor is required to provide the interacting inhibitor subunits, it becomes clear why inhibition is not observed at micromolar Ca concentrations.

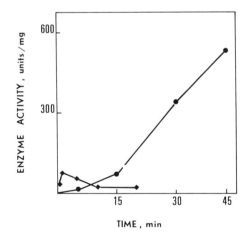

Fig. 11.   Effect of the endogenous inhibitor on the conversion of procalpain to active calpain. Symbols: ●, 5 μM Ca²⁺ + inhibitor; ◆, 1 mM Ca²⁺ + inhibitor. For experimental details see Melloni *et al.* (1984b).

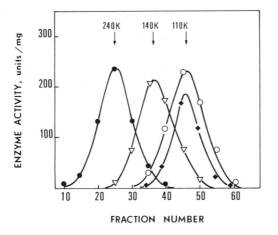

Fig. 12. Isolation of procalpain–inhibitor complex by gel chromatography. Symbols: ●, endogenous inhibitor alone; ○, procalpain alone; ▽, procalpain + inhibitor filtered in the presence of 1 mM Ca^{2+}; ◆, procalpain + inhibitor filtered in the presence of 5 μM Ca^{2+} and 2 mg/ml globin. For experimental details see Melloni *et al.* (1982c, 1984b).

The calpain–inhibitor complex dissociates when Ca^{2+} is removed. Because of this property we were able to reisolate the catalytic subunit which showed activity only when incubated with high concentrations of Ca^{2+} (Fig. 13). Thus, the inhibitor interacts, following its dissociation, with either inactive, single 80-kDa subunits or with active 75-kDa subunits (see Table V).

TABLE VI

Effect of Ca^{2+} Concentration on the Molecular Size of the Endogenous Inhibitor^a

Addition	$K_D{}^b$	Apparent $M_r{}^c$
0.1 mM EGTA	0.2	240,000
5 μM Ca^{2+}	0.25	230,000
5 μM Ca^{2+} + β-globin (2 mg/ml)	0.22	230,000
1 mM Ca^{2+}	0.91	60,000

^{a}For experimental details see Melloni *et al.* (1984b) and Pontremoli *et al.* (1985a).

$^b K_D$ = distribution coefficient following the procedure described by MacGregor *et al.* (1980).

$^c M_r$ = molecular weight established from the K_D values of standard proteins.

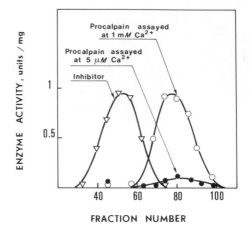

Fig. 13. Dissociation of the calpain-inhibitor complex by treatment with EGTA and ion exchange chromatography. Symbols: ○, procalpain assayed at 1 mM Ca^{2+}; ●, procalpain assayed at 5 μM Ca^{2+}; ▽, inhibitor. For experimental details see Melloni *et al.* (1984b).

B. Regulation by Autoproteolysis of the Active Calpain

Termination of the catalytic activity of calpain seems to be produced by an autoproteolytic degradation of the 75-kDa subunit into smaller fragments (Fig. 14). The rate of proteolytic inactivation appears to be related to the concentrations of Ca^{2+} and of an appropriate substrate (Melloni *et al.*, 1984b), but the molecular and physiological significance of these requirements remains unclear.

C. Regulation by the Substrate at Low Concentrations of Ca^{2+}

All our findings on the conversion of procalpain to active calpain indicate that modulation of the proteinase activity at low concentrations of Ca^{2+} is dependent on the amount and specificity of a digestible substrate. This conversion represents therefore another type of regulatory mechanism.

D. General Remarks on the Mechanism(s) for Regulation of Human Erythrocyte Calpain

We have discussed so far the regulation of calpain activity, taking into consideration the roles of the endogenous inhibitor, of the digestible substrate(s), and of autoproteolytic degradation and inactivation. In all of these factors the concentration of calcium ions is crucial. Thus, changes in the intracellular concentration of Ca^{2+} as a result of membrane damage (Eaton *et al.*, 1973; Palek, 1973; Clark *et al.*, 1981; Wiley *et al.*, 1982) or in response to external cellular effectors

TIME OF INCUBATION, min

CALPAIN 2   5   10   15   30

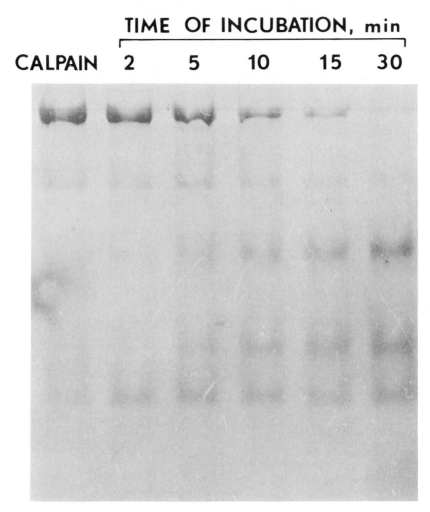

Fig. 14. SDS–polyacrylamide gel electrophoresis of active calpain and autodigested inactive calpain. For experimental details see Pontremoli *et al.* (1984b).

(Michell, 1983) may contribute a great deal to the activation or inactivation of calpain. Calcium therefore must be considered of primary importance in the modulation of calpain activity.

Regulation of calpain may occur at initial or later steps of its catalytic activity. The initial step involves dissociation of the 110-kDa procalpain to produce a single 80-kDa subunit; dissociation must then be followed by autoproteolytic digestion, which converts the inactive 80-kDa subunit to an active 75-kDa subunit. These two consecutive events can occur at high concentrations of Ca²⁺, in

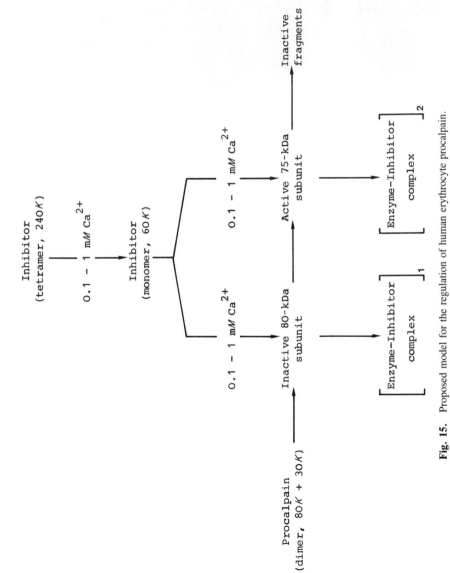

Fig. 15. Proposed model for the regulation of human erythrocyte procalpain.

which case regulation is accomplished by the endogenous inhibitor, or at low concentrations of Ca^{2+}, in which case a digestible substrate plays the regulatory role (see Fig. 8). Finally, calpain activity is terminated in the cell by some form of autoproteolysis. These integrated mechanisms of regulation at high concentrations of Ca^{2+} are schematically represented in Fig. 15.

X. PHYSIOLOGICAL FUNCTION OF CALPAIN IN HUMAN ERYTHROCYTES

In order to study calpain's physiological role, we have studied the proteolytic degradation of ^{3}H-labeled β-hemoglobin entrapped by osmotic shock in human erythrocytes. The cells were then incubated for various times, and the radioactivity was evaluated in the different cell fractions (Melloni et al., 1982a). The distribution of the radioactivity indicated that in addition to the acid-soluble material recovered in the extracellular compartment, a large fraction had been associated with the membranes, and the remaining radioactivity was present in the cytosol as acid-precipitable material (Table VII). Chromatographic analysis of the radioactive material recovered from the membrane fractions revealed a partially degraded form of β-hemoglobin (CaP-hemoglobin) identical in chromatographic behavior to that obtained by digestion with purified calpain (Melloni et al., 1982a). The acid-precipitated material recovered from the cytosol also contained partially degraded β-hemoglobin. We postulated that complete degradation of "excess" β-hemoglobin in intact human erythrocytes was accomplished through a multistep proteolytic process initiated in the cytosol by a limited proteolytic modification of the globin chains produced by calpain. When the CaP-hemoglobin was associated with the membranes, it became susceptible

TABLE VII

Distribution of Radioactivity Entrapped as Labeled β-Hemoglobin in Erythrocytes after 6 and 15 hr of Incubation[a]

Time (hr)	Total counts (cpm)	Extracellular medium		Membrane fraction		Cytosol fraction	
		Acid soluble (cpm)	Acid insoluble (cpm)	Acid soluble (cpm)	Acid insoluble (cpm)	Acid soluble (cpm)	Acid insoluble (cpm)
0	20,000	0	0	0	0	0	20,000
6	20,000	640	0	0	4630	780	14,000
15	20,000	1100	0	0	6000	1070	11,860

[a]For experimental details see Melloni et al. (1982a).

of complete degradation by the acidic endopeptidases of the erythrocyte membrane (Melloni *et al.*, 1982a).

More recently, taking advantage of new information concerning the catalytic and regulatory properties of calpain, we further investigated the postulated three-step process required for complete proteolytic degradation of β-hemoglobin chains. We designed experiments using inside-out vesicles to analyze the requirements for the binding of CaP-hemoglobin to the cytosolic domain of the erythrocyte membrane. This binding appeared to require not only a partial modification of the hemoglobin chains, but also a limited proteolytic digestion of the inner surface of the membranes (Pontremoli *et al.*, 1984b). These requirements were analyzed in a series of multicontrol experiments, all of which are reported in Table VIII.

We further examined the requirement for a proteolytic modification of the cytosolic domain of the erythrocyte membrane by evaluating the binding of CaP-hemoglobin to inside-out vesicles prepared from Ca^{2+}-enriched erythrocytes (Pontremoli *et al.*, 1984b). We observed maximum binding to membranes prepared from erythrocytes incubated with the ionophore and 5 μM Ca^{2+}. A progressive decrease in binding capacity was observed at higher concentrations of Ca^{2+} (Fig. 16). In the presence of leupeptin, a known inhibitor of calpain, the binding showed a significant reduction which provided additional evidence for the requirement of a proteolytic modification. Thus, degradation of the inner surface protein(s) of the erythrocyte membrane produced in the intact erythrocyte by activation of procalpain was as efficient in binding CaP-hemoglobin as was digestion with calpain of untreated inside-out vesicles.

TABLE VIII

Binding of Native and Modified β-Hemoglobin Chains to Inside-Out Erythrocyte Vesicles[a]

Additions		Binding
β-Hemoglobin form	Proteinase form	(μg/mg membrane protein)
Modified	None	38
Modified	Calpain	255
Modified	Procalpain	38
Native	None	nd[b]
Native	Calpain	200
Native	Procalpain	213

[a]The procedure for the modification of β-hemoglobin (CaP–β-hemoglobin) by digestion with calpain, as well as all other experimental details, is described in Pontremoli *et al.* (1984a).

[b]Abbreviation: nd, not detectable.

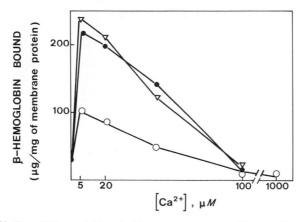

Fig. 16. Binding of β-hemoglobin to inside-out vesicles prepared from erythrocytes treated with Ca²⁺ and ionophore A23187. Symbols: △, + pepstatin; ●, control; ○, + leupeptin. For experimental details see Pontremoli *et al.* (1984b).

That the proteolytic modifications required for maximum binding were not detected by gel electrophoretic analysis of the membrane proteins implies a limited proteolytic digestion. As the concentration of Ca²⁺ increases above 5 μM, the binding capacity progressively decreases, in proportion to the degradation of several transmembrane proteins and notably of a band 4.1 (Pontremoli *et*

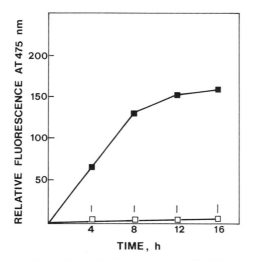

Fig. 17. Release of acid-soluble peptides from calpain-modified β-hemoglobin chains bound to inside-out membrane vesicles. Symbols: ■, ghosts with bound, digested β-hemoglobin; □, ghosts unbound, digested β-hemoglobin. For experimental details see Pontremoli *et al.* (1984b).

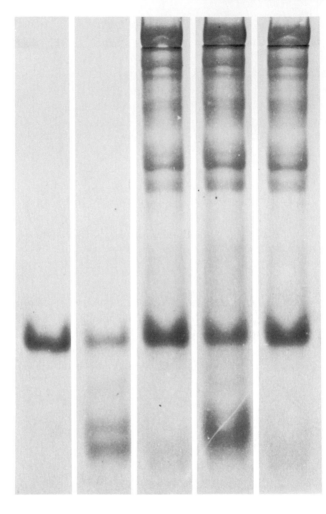

A   B   C   D   E

Fig. 18. Analysis by SDS–polyacrylamide gel electrophoresis of digestion products formed from calpain-modified β-hemoglobin chains bound to membrane vesicles. Lane A, calpain-modified β-hemoglobin chains; lane B, calpain modified β-hemoglobin chains digested with acidic endopeptidase purified from erythrocyte membranes; lane C, unbound calpain-modified β-hemoglobin chains incubated with inside-out membrane vesicles; lane D, membrane vesicles containing bound calpain-modified β-hemoglobin chains incubated as in lane C; lane E, native β-hemoglobin incubated with inside-out membrane vesicles. For experimental details see Pontremoli et al. (1984b).

al., 1984b). Our findings uphold a recent report indicating an extensive degradation of membrane proteins in erythrocytes loaded with calcium ions (Lorand *et al.*, 1983).

In their membrane-bound form, CaP-hemoglobin chains are further degraded by the intrinsic membrane endopeptidase. This proteolytic degradation produces acid-soluble products (Fig. 17) and peptide products which are identical, in gel electrophoretic analysis, to those formed by digestion of CaP-hemoglobin with purified acidic membrane endopeptidases (Fig. 18). These results support the hypothesis that three consecutive steps are required for the intracellular degradation of globin chains. They also revealed the existence of an additional requirement for this multistep proteolytic process. The inner surface of the membrane appears in fact to require a limited proteolytic digestion, catalyzed by calpain, in order to bind CaP-hemoglobin.

Hence the physiological role of human erythrocyte calpain is to initiate the digestion of "excess" globin chains through two distinct but equally important processes: (1) removal of an octapeptide from the NH_2 terminus of β-hemoglobin and (2) modification of the inner surface of the erythrocyte membrane. These proteolytic events promote binding of the partially degraded substrate to the modified membrane, where complete degradation of the bound globin chains occurs.

Elucidation of the proteolytic process shows that activation of procalpain by the substrate at micromolar concentrations of Ca^{2+} provides a mechanism for selective intracellular protein turnover. It will be interesting to identify the protein(s) that require subtle modifications in order to bind CaP-hemoglobin and to study the possible role of this proteolytic system in the removal of abnormal or denatured proteins. The cooperation between two distinct classes of proteinases with different location and properties may occur in other cell types and may be responsible for the selectivity of intracellular protein turnover.

ACKNOWLEDGMENTS

This work was supported in part by a grant from the Consiglio Nazionale delle Ricerche: Progetto Finalizzato Ingegneria Genetica e Basi-Molecolari delle Malattie Ereditarie.

REFERENCES

Clark, M. R., Mohandas, N., Feo, C., Jacobs, M. S., and Shohet, S. B. (1981). Separate mechanisms of deformability loss in ATP-depleted and Ca-loaded erythrocytes. *J. Clin Invest.* **67**, 531–539.

Eaton, J. W., Skelton, T. D., Swofford, H. S., Kolpin, C. E., and Jacob, H. S. (1973). Elevated erythrocyte calcium in sickle cell disease. *Nature (London)* **246**, 105–106.

Golovtchenko-Matsumoto, A. M., Matsumoto, I., and Osawa, T. (1982). Degradation of Band-3 glycoprotein in vitro by a protease isolated from human erythrocyte membranes. *Eur. J. Biochem.* **121**, 463–467.

Ishiura, S., Murofushi, H., Suzuki, K., and Imahori, K. (1978). Studies of a calcium-activated neutral proteinase from chicken skeletal muscle. *J. Biochem. (Tokyo)* **84**, 225–230.

Lorand, L., Bjerrum, O. J., Hawkins, M., Lowe-Krentz, L., and Siefring, G. E., Jr. (1983). Degradation of transmembrane proteins in Ca^{2+}-enriched human erythrocytes. *J. Biol. Chem.* **258**, 5300–5305.

MacGregor, J. S., Singh, V. N., Davoust, S., Melloni, E., Pontremoli, S., and Horecker, B. L. (1980). Evidence for formation of a rabbit liver aldolase: rabbit liver fructose 1,6-bisphosphatase complex. *Proc. Natl. Acad. Sci. U.S.A.* **77**, 3889–3892.

Melloni, E., Salamino, F., Sparatore, B., Michetti, M., and Pontremoli, S. (1982a). Cooperation between soluble and membrane-bound proteinases in the degradation of β-hemoglobin chains in intact human erythrocytes. *Arch. Biochem. Biophys.* **216**, 495–502.

Melloni, E., Sparatore, B., Salamino, F., Michetti, M., and Pontremoli, S. (1982b). Cytosolic calcium dependent proteinase of human erythrocytes: Formation of an enzyme-natural inhibitor complex induced by Ca^{2+} ions. *Biochem. Biphys. Res. Commun.* **106**, 731–740.

Melloni, E., Sparatore, B., Salamino, F., Michetti, M., and Pontremoli, S. (1982c). Cytosolic calcium dependent proteinase of human erythrocytes: the role of calcium ions on the molecular and catalytic properties of the enzyme. *Biochem. Biophys. Res. Commun.* **107**, 1053–1059.

Melloni, E., Pontremoli, S., Salamino, F., Sparatore, B., Michetti, M., and Horecker, B. L. (1984a). Two cytosolic, Ca^{2+}-dependent, neutral proteinases from rabbit liver: Purification and properties of the proenzymes. *Arch. Biochem. Biophys.* **232**, 505–512.

Melloni, E., Salamino, F., Sparatore, B., Michetti, M., and Pontremoli, S. (1984b). Ca^{2+}-dependent neutral proteinase from human erythrocytes: Activation by Ca^{2+} ions and substrate and regulation by the endogenous inhibitor. *Biochem. Int.* **8**, 477–489.

Melloni, E., Salamino, F., Sparatore, B., Michetti, M., and Pontremoli, S. (1984c). Characterization of the single peptide generated from the amino-terminus end of α- and β-hemoglobin chains by the Ca^{2+}-dependent neutral proteinase. *Biochim. Biophys. Acta* **788**, 11–16.

Michell, B. (1983). Ca^{2+} and protein kinase C: Two synergistic cellular signals. *Trends Biochem. Sci.* **8**, 262–265.

Murachi, T. (1983). Calpain and calpastatin. *Trends Biochem. Sci.* **8**, 167–169.

Murakami, T., Hatanaka, M., and Murachi, T. (1981). The cytosol of human erythrocytes contains a highly Ca^{2+}-sensitive thiol protease (calpain I) and its specific inhibitor protein (calpastatin). *J. Biochem. (Tokyo)* **90**, 1809–1816.

Palek, J. (1973). Calcium accumulation during sickling of hemoglobin S (HbSS) red cells. *Blood* **42**, 988 (abstr.).

Pontremoli, S., Salamino, F., Sparatore, B., Melloni, E., Morelli, A., Benatti, U., and De Flora, A. (1979). Isolation and partial characterization of three acidic proteinases in erythrocyte membranes. *Biochem. J.* **181**, 559–568.

Pontremoli, S., Melloni, E., Salamino, F., Sparatore, B., Michetti, M., Benatti, U., Morelli, A., and De Flora, A. (1980a). Identification of proteolytic activities in the cytosolic compartment of mature human erythrocytes. *Eur. J. Biochem.* **110**, 421–430.

Pontremoli, S., Sparatore, B., Melloni, E., Salamino, F., Michetti, M., Morelli, A., Benatti, U., and De Flora, A. (1980b). Differences and similarities among three acidic endopeptidases associated with human erythrocytes membranes. Molecular and functional studies. *Biochim. Biophys. Acta* **630**, 313–322.

Pontremoli, S., Sparatore, B., Melloni, E., Michetti, M., and Horecker, B. L. (1984a). Activation by hemoglobin of the Ca^{2+}-requiring neutral proteinase of human erythrocytes: Structural requirements. *Biochem. Biophys. Res. Commun.* **123,** 331–337.

Pontremoli, S., Melloni, E., Sparatore, B., Michetti, M., and Horecker, B. L. (1984b). A dual role for the Ca^{2+}-requiring proteinase in the degradation of hemoglobin by erythrocyte membrane proteinases. *Proc. Natl. Acad. Sci. U.S.A.* **81,** 6714–6717.

Pontremoli, S., Melloni, E., and Horecker, B. L. (1985a). The regulation of mammalian cytosolic Ca^{2+}-requiring neutral proteinases. *Curr. Top. Cell. Regul.* (in press).

Pontremoli, S., Sparatore, B., Salamino, F., Michetti, M., and Melloni, E. (1985b). The reversible activation by Mn^{2+} ions of the Ca^{2+}-requiring neutral proteinase of human erythrocytes. *Arch. Biochem. Biophys.* **239,** 517–522.

Wiley, J. S., McCulloch, K. E., and Bowden, D. (1982). Increased calcium permeability of cold-stored erythrocytes. *Blood* **60,** 92–98.

Yoshida, N., Weksler, B., and Nachman, R. (1983). Purification of human platelet calcium-activated protease. *J. Biol. Chem.* **258,** (11), 7168–7174.

Chapter 6

Toxicological Implications of Perturbation of Ca²⁺ Homeostasis in Hepatocytes

STEN ORRENIUS

Department of Toxicology
Karolinska Institute
Stockholm, Sweden

GIORGIO BELLOMO

Medical Clinic
University of Pavia
Pavia, Italy

185

CALCIUM AND CELL FUNCTION, VOL. VI

I. INTRODUCTION

Since its discovery in 1808 by Humphry Davy, the calcium ion has been found to be critically involved in an increasing number of vital cell functions, including growth and differentiation, motility and contractility, endocytosis, exocytosis and secretion, and regulation of intermediary metabolism (Campbell, 1983). More recent research has shown that most of the regulatory functions of Ca^{2+} are mediated by specific binding proteins, of which calmodulin is the most widely studied (Cheung, 1980).

Intracellular Ca^{2+} compartmentation is governed by several transport systems operating in a highly regulated fashion. Because of the many vital functions of the calcium ion, it is easily realized that the maintenance of Ca^{2+} homeostasis is critical for the cell, and that its perturbation may result in loss of cell functions and, eventually, cell death. The importance of normal Ca^{2+} homeostasis for various cell functions and for the survival of cells and tissues has been clearly established in a number of studies during the last century (Ringer, 1882; Locke, 1894; Berliner, 1933; Shooter and Grey, 1952; Eagle, 1956; Curtis, 1962; Willmer, 1965, 1966).

The aim of this chapter is to discuss the regulation of Ca^{2+} compartmentation in the hepatocyte and the toxicological implications of its perturbation. Based on recent studies, we have concluded that loss of normal Ca^{2+} homeostasis appears to be an early and critical event in the development of toxic cell injury. An increase in cytosolic free Ca^{2+} concentration, resulting from mobilization of intracellularly sequestered Ca^{2+} and/or inhibition of cellular Ca^{2+} extrusion, seems to be critically linked to the development of cytotoxicity. We propose mechanisms by which hepatotoxic agents may affect Ca^{2+} homeostasis and discuss, finally, alternative effects of enhanced cytosolic Ca^{2+} concentration that could trigger cytotoxicity.

II. CALCIUM AND TISSUE TOXICITY

There is now ample experimental evidence for a close association between tissue damage and calcium overload. Thus, myocardial injury caused by ischemia and overdoses of β-adrenergic agents or vitamin D_3 is associated with a severalfold increase in the calcium content of the injured tissue (Fleckenstein *et al.*, 1984). Similarly, the degeneration of the vascular wall during the atherosclerotic process has been shown to be closely related to calcium accumulation in the endothelial and subendothelial layers. Moreover, the development of the diabetic cataract is invariably associated with an increase in lenticular calcium content (Fleckenstein et al., 1981). The importance of calcium accumulation in the development of these lesions is emphasized by the observation that in all cases they were prevented by pretreatment of the animals with drugs that block calcium entry (Fleckenstein et al., 1981, 1984).

Liver damage has also been found to be associated with the accumulation of large amounts of calcium in the necrotic tissue. Although the generality of this phenomenon has not been extensively investigated, no situation has been described in which necrotic liver did not contain an increased amount of calcium (Farber, 1981, 1984). Results of studies in vitro have further emphasized the importance of calcium accumulation in the development of toxic liver injury. Thus, Schanne and associates (1979) found that 10 hepatotoxins known to act directly on the plasma membrane were less toxic to hepatocyte cultures when calcium was omitted from the medium. Moreover, removal of calcium from the medium resulted in complete reversibility of the plasma membrane damage induced by the toxins. These and other observations led Farber and his associates to propose that influx of extracellular Ca^{2+} represents a critical step in the development of irreversible cell injury (Schanne et al., 1979; Kane et al., 1980).

Although an increase in the permeability of the plasma membrane to Ca^{2+} leading to an equilibration of extra- and intracellular Ca^{2+} concentrations would certainly be expected to result in cell death, subsequent studies have shown that development of irreversible toxic injury in hepatocytes does not critically depend on influx of extracellular Ca^{2+}. In contrast to the findings with hepatocyte cultures (Schanne et al., 1979), lowering of the Ca^{2+} concentration in the incubation medium of isolated hepatocytes did not prevent the cytotoxicity of several hepatotoxins, including carbon tetrachloride and bromobenzene, which require metabolic activation to express toxicity (Smith et al., 1981). In fact, experimental conditions were reported under which removal of Ca^{2+} from the incubation medium potentiated, rather than prevented, cytotoxicity (Smith et al., 1981; Fariss et al., 1984). One has to be cautious when interpreting results obtained from incubations in Ca^{2+}-deficient media, since hepatocytes lose viability more rapidly during incubation at extreme Ca^{2+} concentrations. Nevertheless, it appears from these studies that influx of extracellular Ca^{2+} is not an obligatory step in development of irreversible toxic injury in hepatocytes.

However, the observation that toxic cell death may occur also in the absence of any apparent influx of extracellular Ca^{2+} does not exclude a critical role of Ca^{2+} in the development of toxicity. The calcium ionophore A23187 is toxic to hepatocytes in the absence of extracellular Ca^{2+} when present at concentrations high enough to permeate the intracellular Ca^{2+} compartments. Moreover, as discussed below, several of the other agents that produce toxicity when hepatocytes are incubated in Ca^{2+}-deficient media cause perturbations of intracellular Ca^{2+} homeostasis and a sustained increase in cytosolic free Ca^{2+} level. In fact, it now appears that loss of the ability of the cell to maintain the cytosolic free Ca^{2+} concentration at low, physiological levels may well be a critical factor in development of toxicity. If unaccompanied by compensatory extrusion of calcium ions from the cell, both influx of extracellular Ca^{2+} and release into the cytosol of Ca^{2+} sequestered in other intracellular compartments may cause a perturbation of Ca^{2+} homeostasis of toxicological significance.

III. CALCIUM COMPARTMENTATION IN THE HEPATOCYTE

The compartmentation of Ca^{2+} in the hepatocyte is illustrated schematically in Fig. 1. The mitochondria and the endoplasmic reticulum represent the predominant sites of intracellular Ca^{2+} sequestration. In contrast, the concentration of Ca^{2+} in the cytosol is very low ($\sim 10^{-7}\,M$), in spite of the high Ca^{2+} level in the extracellular fluid ($\sim 10^{-3}\,M$). This concentration gradient is maintained by the energy-dependent uptake of Ca^{2+} into the mitochondria and endoplasmic reticulum in conjunction with the active extrusion of calcium ions through the plasma membrane.

As briefly mentioned, many cell functions are regulated by fluctuations in the concentration of calcium ions. The activation of the phosphorylase system in liver by α_1-adrenergic agonists and vasoactive peptide hormones, mediated by an increase in cytosolic free Ca^{2+} concentration and subsequent activation of phosphorylase b kinase, is one of the most extensively investigated Ca^{2+}-mediated

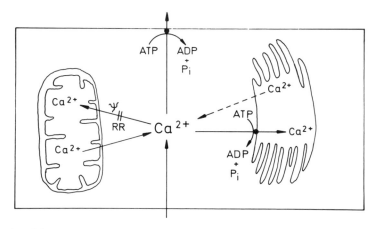

Fig. 1. Ca^{2+} compartmentation in the hepatocyte: intracellular compartments and transport systems. The figure illustrates the regulation of intracellular Ca^{2+} compartmentation by energy-dependent transport systems located in the mitochondrial, endoplasmic reticular, and plasma membranes. RR, Ruthenium Red.

responses (Exton, 1981; Malencik and Fischer, 1983). Although phosphorylase *b* kinase is activated at micromolar Ca^{2+} concentrations, the phosphorylase system is not fully active in "resting cells." This finding suggests that the cytosolic free Ca^{2+} concentration in hepatocytes is below 10^{-6} *M*. This assumption has received further experimental support in recent studies in which either a "null-point" titration method (Murphy *et al.*, 1980) or the fluorescent dye Quin 2 (Charest *et al.*, 1983) was used to measure cytosolic Ca^{2+} concentration in isolated hepatocytes. Based on these and other studies it now appears that the cytosolic free Ca^{2+} level in unstimulated hepatocytes is about 10^{-7} *M*, i.e., four orders of magnitude lower than the concentration in the extracellular fluid. However, in spite of this marked concentration gradient, promoting a constant influx of calcium ions from the extracellular space, the hepatocyte is able to maintain the cytosolic Ca^{2+} concentration within a very narrow range. This control is achieved by the operation of several transport systems located in the mitochondria, endoplasmic reticulum, and plasma membrane.

A. Mitochondria

Most of the calcium present in hepatocytes is sequestered in the mitochondria (Murphy *et al.*, 1980; Bellomo *et al.*, 1982b). The mitochondrial calcium content varies in response to the metabolic requirements of the hepatocyte, and it is also affected by extracellular stimuli.

Ca^{2+} movements into and out of the mitochondria are thought to occur via two different transport routes responsible for the uptake and release of Ca^{2+}, respectively. Ca^{2+} uptake is driven by the transmembrane potential generated across the mitochondrial inner membrane during coupled respiration (Bygrave, 1977; Carafoli and Crompton, 1978; Lehninger *et al.*, 1978b). If endogenous respiration is limited by oxygen availability, the driving force for Ca^{2+} uptake can be supplied by ATP hydrolysis, which also develops a stable transmembrane potential (Chance, 1965). The structural basis for Ca^{2+} uptake in mitochondria is represented by a uniport carrier in which each calcium ion carries two positive charges which are electrically compensated for by the extrusion of two H$^+$ from the matrix (Åkerman, 1978; Nicholls and Crompton, 1980). Kinetic parameters reported for Ca^{2+} uptake by liver mitochondria are summarized in Table I.

Release of Ca^{2+} from calcium-loaded mitochondria can be induced either by factors that cause a collapse of the transmembrane potential, or by selective inhibitors of the uniport carrier, e g , ruthenium red (Moore, 1971). When the transmembrane potential collapses, Ca^{2+} release occurs via reversal of the uniport carrier, but when ruthenium red is present, and this carrier is blocked, it occurs through a distinct release mechanism. The latter has been suggested to involve a H$^+$/Ca^{2+} exchange (Fiskum and Lehninger, 1979) and to be regulated by the oxidation–reduction state of intramitochondrial pyridine nucleotides

TABLE I

Kinetic Parameters of the Ca^{2+} Transport Systems in Liver Mitochondria, Endoplasmic Recticulum, and Plasma Membrane.

Ca^{2+} transport system	Kinetic parameters		References
	K_m	V_{max}	
Mitochondria	2–10 μM	0.5–1.0 μmol min^{-1} mg^{-1}	Bygrave (1977); Lehninger et al. (1978b); Carafoli and Crompton (1978)
Endoplasmic reticulum	0.2–20 μM	11–20 nmol min^{-1} mg^{-1}	More et al. (1975); Bygrave (1978); Dawson (1982)
Plasma membrane	10–17 nM	30 nmol min^{-1} mg^{-1}	Kraus-Friedman et al. (1982); Chan and Junger (1983)

(Lehninger et al., 1978a). However, there is still considerable controversy concerning both how Ca^{2+} release from liver mitochondria occurs under physiological conditions (Gunter et al., 1983), and how this release is regulated (Beatrice et al., 1982, 1984; Bernardi and Azzone, 1983; Vercesi, 1984).

B. Endoplasmic Reticulum

Most of the extramitochondrial Ca^{2+} present in hepatocytes is sequestered within the endoplasmic reticulum. Although the capacity for Ca^{2+} sequestration in this compartment is less in liver than in excitable tissues (Martonosi, 1983), up to 30% of the ionophore-releasable Ca^{2+} in isolated hepatocytes seems to be derived from the endoplasmic reticulum (Murphy et al., 1980; Bellomo et al., 1982b). In the absence of trapping agents such as oxalate, isolated liver microsomes incubated with ATP sequester 8–10 nmol Ca^{2+} per milligram protein (Dawson, 1982).

Ca^{2+} uptake by liver microsomes occurs via a specific translocase coupled to a Mg^{2+}-dependent, Ca^{2+}-activated ATPase (Moore et al., 1975; Dawson, 1982). Some kinetic parameters of this process are reported in Table I. The regulation of Ca^{2+} uptake by liver endoplasmic reticulum is not yet well characterized. Recent studies by Famulski and Carafoli (1984) have demonstrated, however, that calmodulin and Ca^{2+} stimulate the phosphorylation of a 20-kDa microsomal protein which in turn promotes Ca^{2+} uptake. Dephosphorylation of this protein, controlled by a Ca^{2+}-dependent phosphatase, can deactivate the uptake mechanism.

The mechanism by which Ca^{2+} is released from the liver endoplasmic reticulum is not yet well characterized, although there is evidence that it may be of considerable physiological importance. Several reports suggest that inositol 1,4,5-trisphosphate, a hydrolytic product of phosphatidylinositol 4,5-bisphosphate formed upon stimulation of hepatocytes by α_1-adrenergic agonists and vasoactive peptides (Creba et al., 1983), may function as a second messenger in mobilizing Ca^{2+} from nonmitochondrial, ATP-dependent, vesicular stores, i.e., presumably from the endoplasmic reticulum (Dawson and Irvine, 1984; Joseph et al., 1984; Burgess et al., 1984).

C. Plasma Membrane

Although liver mitochondria and microsomes, when incubated together in vitro, have been found capable of buffering the Ca^{2+} concentration of the incubation medium at levels similar to those found in the hepatocyte cytosol (Becker et al., 1980), there is little doubt that the plasma membrane Ca^{2+} translocase plays an important role in maintaining Ca^{2+} homeostasis in the hepatocyte in vivo. In view of the concentration gradient existing across the plasma membrane, promoting a continuous influx of Ca^{2+} from the extracellular fluid, it is obvious that long-term regulation of Ca^{2+} homeostasis in hepatocytes is critically dependent on the active extrusion of calcium ions through the plasma membrane.

Using a fraction containing inverted plasma membrane vesicles from rat liver, Carafoli and associates were able to demonstrate the existence of an active Ca^{2+} transport system associated with a Ca^{2+}-stimulated ATPase (Famulski and Carafoli, 1981; Kraus-Friedman et al., 1982). The Ca^{2+}-ATPase exhibits two distinct kinetic components with high and low affinity for Ca^{2+}, respectively (Iwasa et al., 1982). Kinetic parameters of the high-affinity component are given in Table I. In contrast to the Ca^{2+}-ATPases present in plasma membrane fractions isolated from several other tissues (Waisman et al., 1981; Guraj et al., 1983; Klaven et al., 1983), the liver enzyme appears to be insensitive to calmodulin (Lotersztajn et al., 1981; Kraus-Friedman et al., 1982).

The hepatic plasma membrane Ca^{2+} translocase has been found to be highly sensitive to vanadate, which has been shown to inhibit, each to a similar extent, the Ca^{2+}-dependent phosphorylation of a 110-kDa protein, Ca^{2+}-ATPase activity, and Ca^{2+} transport (Chan and Junger, 1983). Studies performed in our laboratories have further shown that the hepatic plasma membrane Ca^{2+} translocase is also sensitive to sulfhydryl reagents, suggesting that free sulfhydryl group(s) are critical for optimal activity (Bellomo et al., 1983).

IV. METHODS USED TO STUDY CALCIUM
COMPARTMENTATION IN HEPATOCYTES

Several methods have been proposed to study calcium compartmentation in hepatic tissue and isolated liver cells. An early approach was based on rapid cell disruption followed by isolation of the intracellular organelles and subsequent determination of the calcium content of the various organelle fractions (Tischler *et al.*, 1977). However, the difficulty of obtaining noncontaminated fractions and the possibility of a redistribution of Ca^{2+} during the isolation procedure are major disadvantages of this approach.

Based on studies of the uptake of $^{45}Ca^{2+}$ by hepatocytes, Claret-Berthon and associates (1977) identified three different calcium pools in liver cells: a dynamic cytoplasmic pool which can be divided into mitochondrial, microsomal, and cytosolic subfractions; a slowly exchangeable nuclear pool; and an apparently nonexchangeable pool. This technique has subsequently been improved by Barritt *et al.* (1981) and Parker *et al.* (1983), who have introduced a new computer model to analyze the data obtained by $^{45}Ca^{2+}$ distribution studies.

Using a nondisruptive method, Murphy *et al.* (1980) demonstrated that it was possible to release Ca^{2+} from different intracellular compartments of isolated hepatocytes and to measure the Ca^{2+} released by a spectrophotometric method using the metallochromic dye Arsenazo III (Kendrick, 1976). Modification of this procedure in our laboratories has led to the development of a method that allows quantitation of the two major pools of releasable Ca^{2+} present in isolated hepatocytes: a mitochondrial pool and an endoplasmic reticular pool (Bellomo *et al.*, 1982b). The method involves subsequent additions of uncoupler carbonyl cyanide *m*-chlorophenylhydrazone (CCCP) and ionophore A23187 to isolated hepatocytes suspended in Ca^{2+}- and Mg^{2+}-free Hank's medium in the presence of Arsenazo III. As shown in Fig. 2, the addition of CCCP causes the release of ~70% of the releasable Ca^{2+} from isolated hepatocytes, whereas the remainder is mobilized by the subsequent addition of ionophore A23187. The rationale for this method is based on the assumption that CCCP, by collapsing the mitochondrial transmembrane potential, causes a selective release of Ca^{2+} sequestered in the mitochondria and that ionophore A23187, when added after CCCP at a concentration high enough to permeabilize all intracellular compartments, will mobilize the extramitochondrial, i.e., endoplasmic reticular, Ca^{2+} pool. There is now ample experimental support for this assumption (Murphy *et al.*, 1980; Bellomo *et al.*, 1982b; Jones *et al.*, 1983).

Although the method just described can be employed to quantitate the mitochondrial and endoplasmic reticular Ca^{2+} pools in hepatocytes, it does not provide any information about the cytosolic free Ca^{2+} level. For this purpose, Murphy *et al.* (1980) developed a "null-point" titration technique which was used to measure fluctuations in cytosolic Ca^{2+} concentration in response to α-

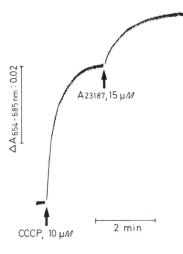

Fig. 2. Release of mitochondrial and extramitochondrial Ca^{2+} pools monitored by absorbance changes of Arsenazo III during sequential additions of carbonyl cyanide m-chlorophenylhydrazone (CCCP) and ionophore A23187 to isolated hepatocytes. Hepatocytes were suspended in Ca^{2+}- and Mg^{2+}-free Hanks' solution ($\sim 2 \times 10^6$ cells/ml). CCCP was added first to the cell suspension, and Ca^{2+} release was recorded until no further change in absorbance was observed. At this point, the Ca^{2+} ionophore A23187 was added and additional Ca^{2+} release was recorded. See Thor $et\ al.$ (1984) for experimental details.

adrenergic stimuli. However, this method is technically very difficult and has failed to yield reproducible results in our hands.

Tsien (1980) has developed a series of fluorescent calcium indicators for direct measurement of cytosolic Ca^{2+} concentration, and one of them, Quin 2, has also been used successfully in hepatocytes (Charest $et\ al.$, 1983). Unfortunately, the excitation–emission wavelength characteristics of Quin 2 are very similar to those of the reduced pyridine nucleotides NADH and NADPII, which are present at high concentrations in hepatocytes. Thus, Quin 2 cannot be used for calcium measurements in hepatocytes if there are concomitant changes in pyridine nucleotide autofluorescence, e.g., during oxidative stress.

In an attempt to overcome this problem, we have recently used phosphorylase a activity to monitor alterations in cytosolic free Ca^{2+} concentration in hepatocytes exposed to oxidative stress (Bellomo $et\ al.$, 1984a). Phosphorylase a activity has previously been demonstrated to be a valid indicator of fluctuations in cytosolic Ca^{2+} levels, if measured under conditions in which phosphorylase activation is strictly dependent on Ca^{2+}-requiring phosphorylase b kinase (Malencik and Fischer, 1983). As shown in Fig. 3, both stimulated influx of extracellular Ca^{2+} and mobilization of Ca^{2+} sequestered in the mitochondria result in enhanced phosphorylase a activity in isolated hepatocytes which probably is mediated by an increased cytosolic free Ca^{2+} concentration.

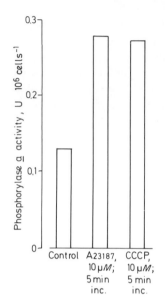

Fig. 3. Activation of phosphorylase by addition of carbonyl cyanide *m*-chlorophenylhydrazone (CCCP) and ionophore A23187 to isolated hepatocytes. Ionophore A23187 or CCCP was added to a suspension of hepatocytes (~10^6 cells/ml) in Krebs–Henseleit medium supplemented with 20 m*M* HEPES, pH 7.4. After 5 min incubation at 37°C, samples were taken for phosphorylase *a* assay. See Bellomo *et al.* (1984a) for experimental details.

V. PERTURBATION OF CALCIUM HOMEOSTASIS DURING OXIDATIVE CELL INJURY

For many years we have been particularly interested in biochemical mechanisms involved in the development of toxic cell injury. Although we have employed a variety of hepatotoxic agents in our studies, most of our recent investigations have been performed with redox-active compounds capable of inducing oxidative stress during their metabolism by hepatocytes. Menadione (2-methyl-1,4-naphthoquinone) and *t*-butylhydroperoxide belong to this group of compounds.

In hepatocytes, several flavoenzymes, including NADPH-cytochrome P-450 reductase, NADH-cytochrome b_5 reductase, and NADH-ubiquinone oxido-reductase, catalyze the one-electron reduction of menadione to the semiquinone free radical which, in the presence of dioxygen, is rapidly reoxidized to the parent quinone with the concomitant formation of a superoxide anion radical

($O_2^{\overline{\cdot}}$). Superoxide dismutases present in the cytosolic and mitochondrial compartments catalyze the dismutation of $O_2^{\overline{\cdot}}$ to form O_2 and H_2O_2. The selenoprotein glutathione peroxidase catalyzes the reduction of both H_2O_2 and organic hydroperoxides, including t-butylhydroperoxide. The glutathione disulfide (GSSG) formed in this reaction is rereduced to GSH by NADPH-linked glutathione reductase. If regeneration of NADPH becomes rate-limiting, there is an intracellular accumulation followed by active excretion of GSSG from the hepatocytes (Eklöw et al., 1984; Nicotera et al., 1984a) which may lead to depletion of cellular GSH under extreme conditions.

Thus, the exposure of hepatocytes to toxic levels of either menadione or t-butylhydroperoxide results in oxidative stress characterized by NADPH and GSH oxidation, which may lead to GSH depletion and cell death (Thor et al., 1982; Bellomo et al., 1982b). Our recent results suggest that GSH depletion is followed by depletion of protein thiols which, in conjunction with the effect of the oxidative stress on the mitochondrial Ca^{2+} pool, causes a perturbation of cellular Ca^{2+} homeostasis which is of critical importance in the development of toxic injury (Di Monte et al., 1984).

A. Studies with Intact Cells

As shown in Table II, the metabolism of toxic concentrations of menadione and t-butylhydroperoxide results in a decrease in the intracellular levels of exchangeable Ca^{2+} in hepatocytes (Thor et al., 1982; Bellomo et al., 1982b). The decrease in Ca^{2+} content is preceded by GSH and NADPH oxidation and appears to be intimately related to the decreased GSH/GSSG and NADPH/NADP$^+$ redox levels. In fact, inclusion of the thiol reducing agent dithiothreitol in the incubation prevents the alterations in intracellular Ca^{2+} content induced by t-butylhydroperoxide metabolism.

Glutathione reductase is the enzyme that links glutathione oxidation to NADPH oxidation in hepatocytes. Inhibition of glutathione reductase by pretreatment of the cells with N,N-bis(2-chloroethyl)-N-nitrosourea (BCNU) disrupts this link and allows glutathione oxidation to proceed without any appreciable change in the NADPH/NADP$^+$ redox level (Babson et al., 1981; Eklöw et al., 1984). When hepatocytes pretreated with BCNU are incubated with t-butylhydroperoxide, there is no apparent effect on the mitochondrial Ca^{2+} pool, whereas the loss of GSH and of the extramitochondrial Ca^{2+} pool is accelerated (Table II) (Bellomo et al., 1982b). These early findings suggested that pyridine nucleotide oxidation caused by t-butylhydroperoxide metabolism might be associated with release of intramitochondrially sequestered Ca^{2+} and that the cytosolic GSH level might be important in regulating the size of the extramitochondrial pool.

TABLE II

Effects of *t*-Butylhydroperoxide and Menadione Metabolism on Intracellular GSH and NADPH Levels and on the Intracellular Distribution of Ca^{2+} in Isolated Hepatocytes[a]

	Percentage of control level[b]			
Substrate	GSH	NADPH	Mitochondrial Ca^{2+}	Extramitochondrial Ca^{2+}
None	100	100	100	100
Menadione, 200 μ*M*; 30 min incubation	10	10	50	40
t-Butylhydroperoxide, 4 m*M*; 30 min incubation	35	10	55	25
t-Butylhydroperoxide, 4 m*M*; 5 min incubation (BCNU-treated cells)[c]	5	75	85	20

[a]Hepatocytes isolated from phenobarbital-treated rats suspended ($\sim 10^6$ cells/ml) in Krebs-Henseleit medium supplemented with 20 m*M* HEPES, pH 7.4, were incubated with or without menadione or *t*-butylhydroperoxide at 37°C. When indicated, samples were taken for assay of GSH and NADPH levels, and for spectrophotometric determination of the mitochondrial- and extramitochondrial-Ca^{2+} pools using Arsenazo III. See Bellomo *et al.* (1982b) and Thor *et al.* (1982) for experimental details.
[b]The results are expressed as a percentage of the values observed in absence of substrate addition. The absolute values were: GSH, 50 nmol $\times$ 10^6 cells^{-1}; NADPH, 2.8 nmol $\times$ 10^6 cells^{-1}; mitochondrial Ca^{2+}, 3.4 nmol $\times$ 10^6 cells^{-1}; extramitochondrial Ca^{2+}, 1.0 nmol $\times$ 10^6 cells^{-1}.
[c]To inhibit glutathione reductase, hepatocytes were preincubated with *N,N*-bis(2-chlorotheyl)-*N*-nitrosourea as described by Bellomo *et al.* (1982b).

B. Studies with Subcellular Fractions

To investigate the possible relationship between fluctuations in the glutathione and pyridine nucleotide redox levels and the ability of the endoplasmic reticulum and the mitochondria to sequester Ca^{2+}, experiments were designed to study the effects of menadione and *t*-butylhydroperoxide on the various Ca^{2+} translocases present in the mitochondria, microsomes, and plasma membrane under metabolically controlled conditions.

1. Mitochondria

Incubation of Ca^{2+}-lodaded liver mitochondria with either menadione or *t*-butylhydroperoxide results in a release of sequestered Ca^{2+}. This effect is dependent on both Ca^{2+} load and menadione or *t*-butylhydroperoxide concentration (Bellomo *et al.,* 1982a, 1984b; Moore *et al.,* 1983). Ca^{2+} release induced by *t*-butylhydroperoxide is preceded by GSH and NAD(P)H oxidation, and there is now ample experimental evidence that this release is initiated by NADPH oxidation resulting from the metabolism of the hydroperoxide by the glutathione

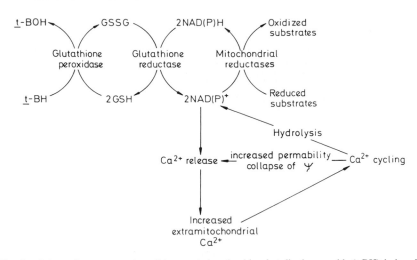

Fig. 4. Schematic representation of the events involved in t-butylhydroperoxide (t-BH)-induced Ca²⁺ release from liver mitochondria. The proposed mechanism implies that Ca²⁺ release is initiated by pyridine nucleotide oxidation resulting from metabolism of the hydroperoxide by the mitochondrial glutathione peroxidase–glutathione reductase system, and that subsequent mitochondrial damage is the result of influx–efflux cycling of the released Ca²⁺.

peroxidase–glutathione reductase enzyme system in the mitochondria (Lötscher *et al.*, 1979; Bellomo *et al.*, 1984b). The idea that pyridine nucleotide oxidation plays a critical role in initiating Ca²⁺ release during mitochondrial metabolism of menadione and *t*-butylhydroperoxide is supported by the finding that this release is prevented in the presence of reducing substrates such as β-hydroxybutyrate and isocitrate (Moore *et al.*, 1983).

Although the relationship between pyridine nucleotide oxidation and Ca²⁺ release by mitochondria incubated with *t*-butylhydroperoxide seems to be rather well established, less is known about the route by which Ca²⁺ release occurs. Work by Pfeiffer and associates (Beatrice *et al.*, 1982) revealed that Ca²⁺ release during hydroperoxide metabolism by liver mitochondria was associated with collapse of transmembrane potential and large-amplitude swelling. These findings suggested that Ca²⁺ release was due to permeability changes caused by damage to the mitochondrial inner membrane. However, more recent studies have shown that hydroperoxide-induced Ca²⁺ release can be dissociated from membrane damage by inclusion of either ruthenium red or EGTA in the incubation medium (Baumhüter and Richter, 1982; Moore *et al.*, 1983; Bellomo *et al.*, 1984b). It therefore appears that membrane damage is a consequence of Ca²⁺ cycling, rather than the cause of initial Ca²⁺ release by mitochondria incubated with *t*-butylhydroperoxide (Fig. 4). However, further work is required to characterize the route by which Ca²⁺ is released and its possible regulation by intramitochondrial pyridine nucleotides.

2. Microsomes

Although mitochondria appear to play the pivotal role in regulating cytosolic free Ca^{2+} concentrations, studies by Becker *et al.* (1980) have shown that mitochondrial Ca^{2+} influx–efflux cycling can be adjusted by energy-dependent Ca^{2+} sequestration in microsomes. The endoplasmic reticulum may therefore play an important regulatory role in maintaining intracellular Ca^{2+} homeostasis and low cytosolic free Ca^{2+} concentrations. If this is so, the microsomal Ca^{2+}-sequestering system, which utilizes ATP and is associated with Ca^{2+}-ATPase activity (Moore *et al.*, 1975), could be a primary target for damage by toxic metabolites of drugs and other foreign compounds metabolized by the microsomal monooxygenase system. In fact, Moore and his associates (1976; Moore, 1982) have shown that the liver microsomal Ca^{2+} pump is inhibited during the metabolism of both carbon tetrachloride and carbon disulfide. The inhibition of the endoplasmic reticular Ca^{2+} translocase and consequent disruption of intracellular Ca^{2+} homeostasis could therefore be a primary event in certain drug hepatotoxicities.

In order to investigate this further, we developed a relatively simple assay for measuring ATP-dependent Ca^{2+} sequestration in microsomes, involving the use of the metallochromic indicator Arsenazo III and ionophore A23187 (Jones *et al.*, 1983). The characteristics of the liver microsomal Ca^{2+}-sequestering system, as determined by this method, are essentially the same as those described previously (Moore *et al.*, 1975), and the uptake of Ca^{2+} is insensitive to both the mitochondrial Ca^{2+} translocase inhibitor ruthenium red and the protonophore CCCP. The microsomal Ca^{2+}-sequestering system is, however, very sensitive to oxidative damage. For example, relatively low concentrations of *t*-butylhydroperoxide rapidly inhibit Ca^{2+} uptake by this system (Fig. 5); and other forms of oxidative stress, such as the generation of active oxygen species by the redox cycling of menadione, also inhibit the microsomal Ca^{2+} translocase (Jones *et al.*, 1983). This inhibition is prevented when either GSH or the thiol reducing agent dithiothreitol (Fig. 5) is present in the incubation. Since previous studies by Moore *et al.* (1975) have shown that the microsomal Ca^{2+} pump is sensitive to thiol reagents, it seems probable that thiols, mainly GSH in liver cells, protect the Ca^{2+}-sequestering system of the endoplasmic reticulum by preventing the oxidation of thiol group(s) necessary for Ca^{2+}-ATPase activity.

3. Plasma Membrane

As discussed above, it is obvious that maintenance of a low cytosolic free Ca^{2+} concentration during any period of time requires that the entry of Ca^{2+} into the cell be balanced by the efflux of an equivalent amount of Ca^{2+} from the cell. This efflux is achieved by Ca^{2+} transport linked to an ATPase present in the plasma membrane (Chan and Junger, 1983). Inhibition of this activity will rapidly lead to enhanced intracellular Ca^{2+} levels and loss of Ca^{2+} homeostasis.

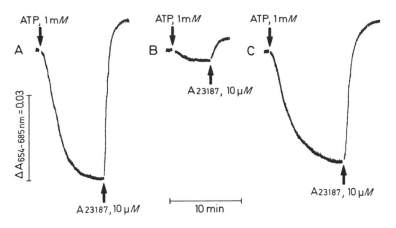

Fig. 5. Inhibition by *t*-butylhydroperoxide of Ca^{2+} sequestration by liver microsomes and its prevention by dithiothreitol. (A) ATP-induced Ca^{2+} uptake by liver microsomes; (B) ATP-induced Ca^{2+} uptake by liver microsomes preincubated for 20 min with 1 m*M* *t*-butylhydroperoxide; (C) ATP-induced Ca^{2+} uptake by liver microsomes preincubated for 20 min with 1 m*M* *t*-butylhydroperoxide and 1 m*M* dithiothreitol. Ca^{2+} release was induced by addition of ionophore A23187 to the microsomal suspension (~1 mg protein/ml), and Ca^{2+} fluxes were monitored by absorbance changes of Arsenazo III. See Jones *et al.* (1983) for experimental details.

We have developed a method for measuring ATP-dependent Ca^{2+} sequestration by an isolated liver plasma membrane fraction containing a high proportion of inside-out vesicles and have characterized the Ca^{2+} translocase in some detail (Bellomo *et al.*, 1983). By using this method we have found that preincubation of the plasma membrane fraction with *t*-butylhydroperoxide results in a decreased capacity to sequester Ca^{2+} (Table III). As in microsomes, GSH and dithiothreitol protect against the inhibition of Ca^{2+} sequestration by *t*-butylhydroperoxide. Moreover, GSH also has a stimulatory effect on Ca^{2+} sequestration by plasma membrane vesicles which have not been exposed to the hydroperoxide (Bellomo *et al.*, 1983). This observation provides further support for the assumption that thiol groups are important for Ca^{2+} sequestration, in plasma membrane vesicles as well as in microsomes, and that the effect of *t*-butylhydroperoxide on Ca^{2+} sequestration may be related to the interaction of the hydroperoxide, or, more probably, free-radical products formed from the hydroperoxide, with such critical thiol groups.

To investigate whether a similar inhibition of the plasma membrane Ca^{2+} translocase occurs in intact cells exposed to oxidative stress, we developed a method for isolating plasma membrane fragments from hepatocytes (Nicotera *et al.*, 1984a,b). Using this method we have found that the metabolism of menadione by isolated hepatocytes is associated with marked inhibition of the high-affinity component of the Ca^{2+}-stimulated ATPase, and that this inhibition is reversed when the preparation is treated with either GSH or dithiothreitol (Nic-

TABLE III

Inhibition of Plasma Membrane Ca^{2+} Transport and Ca^{2+}, Mg^{2+}-ATPase Activity by t-Butylhydroperoxide and Its Prevention by GSH and Dithiothreitol (DDT)[a]

Treatment	Ca^{2+} transport (nmol × mg^{-1})	Ca^{2+}, Mg^{2+}-ATPase activity (μmol P_i × h^{-1} × mg^{-1})
Control	9.9	25
t-Butylhydroperoxide, 1 mM; 10 min incubation	3.8	17
t-Butylhydroperoxide, 1 mM + GSH, 1 mM; 10 min incubation	8.5	24
t-Butylhydroperoxide, 1 mM + DTT, 1 mM; 10 min incubation	10	27

[a]Liver plasma membrane vesicles (~0.5 mg protein per milliliter) were incubated for 10 min at 37°C with or without 1 mM t-butylhydroperoxide in the absence or presence of 1 mM GSH or 1 mM DTT. ATP (1 mM) was then added and Ca^{2+} uptake and Ca^{2+}, Mg^{2+}-ATPase activity were measured by recording the absorbance change of Arsenazo III and the release of inorganic phosphate, respectively. See Bellomo *et al.* (1983) for experimental details.

otera *et al.*, 1984b). It therefore appears that oxidation of thiol groups critical to normal activity of the plasma membrane Ca^{2+} translocase may contribute to the perturbation of Ca^{2+} homeostasis in intact cells during oxidative stress.

VI. RELATIONSHIP BETWEEN PERTURBATION OF CALCIUM HOMEOSTASIS AND ALTERATIONS IN SURFACE MORPHOLOGY (HEPATOCYTE BLEBBING)

Freshly isolated rat hepatocytes by the collagenase perfusion technique (Moldéus *et al.*, 1978) appear to be perfectly spherical when incubated in suspension, and they are covered with surface microvilli (Fig. 6). Exposure of hepatocytes to a variety of toxic agents, including menadione and t-butylhydroperoxide, is associated with alterations in surface morphology, characterized by the disappearance of microvilli and the appearance of multiple surface blebs (Fig. 6). Often, these changes in surface morphology appear early during incubation, before there are any signs of perturbation of plasma membrane permeability; and they are reversible, i.e., surface morphology is normalized if the toxic agent is removed from the incubation. Surface blebbing appears not to be due to swelling caused by an alteration of the water permeability of the plasma membrane, since it is not affected by an increase in the osmolarity of the incubation medium (Orrenius *et al.*, 1983).

It is widely considered that cell surface morphology is determined by the organization of cortical microfilaments associated with the plasma membrane (Smith *et al.*, 1984, and references therein). This notion is supported by the fact that two classes of compounds, the cytochalasins and phalloidins, which are known to disrupt cortical microfilament structure, cause the formation of blebs on the surface of hepatocytes similar to those shown in Fig. 6 (Mesland *et al.*, 1981; Smith *et al.*, 1984). Thus, bleb formation appears to occur as a result of a disruption of cortical microfilament structure. We can find no evidence, however, that compounds which produce reactive intermediates interact directly with microfilament cytoskeletal components. It seems more likely that the key alterations in microfilament structure which produce membrane blebbing are brought about indirectly via alterations in levels of regulatory cofactors or ions. For example, alterations in the cytosolic concentrations of ATP, Ca^{2+}, Mg^{2+}, and/or H^+ would significantly affect the structure of the hepatocyte cytoskeleton. However, preliminary studies from our laboratories suggest that alterations in intracellular free Mg^{2+} and H^+ concentration are of little importance in bleb formation. For example, the transient application of 15 mM NH_4Cl, which loads cells with an excess of intracellular H^+, does not cause blebbing in isolated hepatocytes, and variation of extracellular Mg^{2+} concentration between 0 and 1 mM has no effect on bleb formation or toxicity induced by *t*-butylhydroperoxide (Smith *et al.*, 1984).

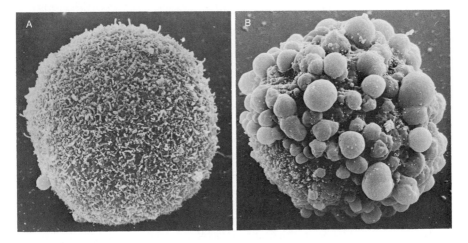

Fig. 6. Scanning electron micrographs of hepatocytes incubated in absence (A) or presence (B) of menadione. Hepatocytes isolated from phenobarbital-treated rats and suspended in Krebs–Henseleit medium supplemented with 20 mM HEPES, pH 7.4, were incubated for 60 min at 37°C in the absence or presence of 200 μM menadione, and samples were processed for electron microscopy in accordance with standard procedures involving glutaraldehyde and osmium fixation, followed by critical point drying and visualization in a JEOL scanning electron microscope (model JSM 35). Magnification: ×3000. See Thor *et al.* (1982) for experimental details.

The polymerization of G-actin (monomeric form) to F-actin (filamentous form) is dependent upon ATP; 1 mol of bound ATP is converted to ADP for every monomeric actin subunit polymerized. Thus, the microfilaments of the hepatocyte cytoskeleton require ATP for the maintenance of normal cell shape, and a lack of ATP could result in actin depolymerization, a breakdown of the actomyosin network and hence plasma membrane blebbing. We have previously reported (Smith et $al.$, 1984) that inhibition of ATP synthesis by treatment of isolated hepatocytes with antimycin A is in fact associated with extensive plasma membrane blebbing which precedes loss of cell viability. However, the alterations in surface structure occur well before ATP is depleted, and a detailed analysis has revealed that they correlate better to antimycin A–induced release of the mitochondrial Ca^{2+} pool than to ATP depletion (Smith et $al.$, 1984).

It is clear that a change in cytosolic free Ca^{2+} concentration resulting from either an increased influx of extracellular Ca^{2+} or a redistribution of intracellularly sequestered Ca^{2+} could alter the structure of the hepatocyte cytoskeleton, because Ca^{2+} and its Ca^{2+}-binding proteins play such a pivotal role in regulating cytoskeletal structure (Cheung, 1980). The observation that the calcium ionophore A23187 is able to produce plasma membrane blebbing in isolated hepatocytes supports the assumption that the alterations in surface morphology are in fact related to a perturbation of intracellular Ca^{2+} homeostasis (Jewell et $al.$, 1982). It therefore appears that the surface blebbing observed during exposure of hepatocytes to oxidative stress is caused by an increase in cytosolic free Ca^{2+} concentration. This increase results from impairment of Ca^{2+} sequestration by the mitochondria and the endoplasmic reticulum, in conjunction with inhibition of the Ca^{2+}-extruding system in the plasma membrane (Jewell et $al.$, 1982; Bellomo et $al.$, 1984a; Smith et $al.$, 1984; Di Monte et $al.$, 1984).

It is now well established that incubation of isolated hepatocytes in the presence of added ATP results in Ca^{2+} accumulation by the hepatocytes (Krell et $al.$, 1983; Thor et $al.$, 1984). This effect appears to be due to inhibition of the plasma membrane Ca^{2+}-extruding system by the extracellular ATP (Bellomo et $al.$, 1984c). Although most of the extra Ca^{2+} is sequestered by the mitochondria, phosphorylase activation during incubation of hepatocytes with ATP indicates that the cytosolic free Ca^{2+} concentration is increased (Thor et $al.$, 1984). Moreover, exposure of hepatocytes to extracellular ATP is also associated with extensive surface blebbing. When the extracellular ATP is removed, the hepatocytes resume normal surface morphology (Thor et $al.$, 1984). The relationship between phosphorylase activation and surface blebbing during incubation of hepatocytes with extracellular ATP is shown in Fig. 7.

Thus, it appears that hepatocyte blebbing, as studied under our experimental conditions, is caused by a redistribution of intracellular Ca^{2+} leading to an increase in the cytosolic concentration of free Ca^{2+}. Plasma membrane blebbing

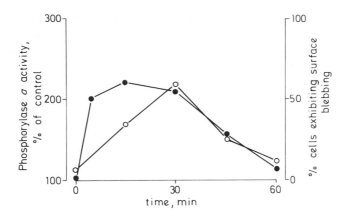

Fig. 7. Correlation between phosphorylase activation and surface blebbing during ATP-induced inhibition of Ca^{2+} efflux from isolated hepatocytes. Hepatocytes suspended in Krebs–Henseleit medium supplemented with 20 mM HEPES, pH 7.4, were incubated at 37°C in absence or presence of 1 mM ATP. Where indicated, samples were taken for phosphorylase a assay (●) and estimation of the percentage of cells exhibiting surface blebs (○). See Thor *et al.* (1984) for experimental details.

could be the result of F-actin depolymerization to G-actin in cortical microfilaments, although we have not been able to show this (Smith *et al.*, 1984), but it could also reflect other kinds of structural reorganization, such as actin cross-linking of the type induced by α-actinin or vinculin, or actin filament severing and bundling as mediated by gelsolin, villin, and other actin-binding proteins (Smith *et al.*, 1984). Villin is of special interest since it could well mediate a dispersion and fragmentation of actin microfilaments in response to a rise in cytosolic Ca^{2+} concentration. However, further studies are required to substantiate this hypothesis.

VII. CONCLUDING REMARKS

There is now substantial evidence that perturbation of Ca^{2+} homeostasis is an early and critical event in the development of irreversible hepatotoxicity. Impairment of the various Ca^{2+} translocases may lead to redistribution of intracellular Ca^{2+} and to Ca^{2+} accumulation. One of the earliest manifestations of this disturbed Ca^{2+} homeostasis appears to be an alteration in surface morphology known as plasma membrane blebbing. Such blebbing occurs well before cell death and seems to result from changes in hepatocyte cytoskeletal organization in response to an increase in cytosolic free Ca^{2+} concentration. However, plasma membrane blebbing is reversible and therefore does not per se represent irreversible toxicity. Thus, if a sustained rise in cytosolic free Ca^{2+} concentration is the common "point of no return" in toxic cell death, other effects of the increased Ca^{2+} concentration, such as protease and/or phospholipase activation, are probably more important for the development of irreversible toxicity. However, this can only be decided by further experimental work.

REFERENCES

Åkerman, K. E. O. (1978). Charge transfer during valinomycin-induced Ca^{2+} uptake in rat liver mitochondria. *FEBS Lett.* **93**, 293–296.

Babson, J. R., Abell, N. S., and Reed, D. J. (1981). Protective role of the glutathione redox cycle against adriamycin-mediated toxicity in isolated hepatocytes. *Biochem. Pharmacol.* **30**, 2299–2304.

Barritt, G. J., Parker, J. C., and Wadsworth, J. C. (1981). A kinetic analysis of the effects of adrenaline on calcium distribution in isolated rat liver parenchymal cells. *J. Physiol. (London)* **312**, 29–55.

Baumhüter, S., and Richter, C. (1982). The hydroperoxide-induced release of mitochondrial calcium occurs via a distinct pathway and leaves mitochondria intact. *FEBS Lett.* **148**, 271–275.

Beatrice, M. C., Stiers, D. L., and Pfeiffer, D. R. (1982). Increased permeability of mitochondria during Ca^{2+} release induced by t-butylhydroperoxide or oxaloacetate. The effect of ruthenium red. *J. Biol. Chem.* **257**, 7161–7171.

Beatrice, M. C., Stiers, D. L., and Pfeiffer, D. R. (1984). The role of glutathione in the retention of Ca^{2+} by liver mitochondria. *J. Biol. Chem.* **259**, 1279–1287.

Becker, G. L., Fiskum, G., and Lehninger, A. L. (1980). Regulation of free Ca^{2+} by liver mitochondria and endoplasmic reticulum. *J. Biol. Chem.* **255**, 9009–9012.

Bellomo, G., Jewell, S. A., and Orrenius, S. (1982a). The metabolism of menadione impairs the ability of rat liver mitochondria to take up and retain Ca^{2+}. *J. Biol. Chem.* **257**, 11558–11562.

Bellomo, G., Jewell, S. A., Thor, H., and Orrenius, S. (1982b). Regulation of intracellular calcium compartmentation: Studies with isolated hepatocytes and t-butylhydroperoxide. *Proc. Natl. Acad. Sci. U.S.A.* **79**, 6842–6846.

Bellomo, G., Mirabelli, F., Richelmi, P., and Orrenius, S. (1983). Critical role of sulfhydryl group(s) in ATP-dependent Ca^{2+} sequestration by the plasma membrane fraction from rat liver. *FEBS Lett.* **163**, 136–139.

Bellomo, G., Thor, H., and Orrenius, S. (1984a). Increase in cytosolic Ca^{2+} concentration during t-butylhydroperoxide metabolism by isolated hepatocytes involves NADPH oxidation and mobilization of intracellular Ca^{2+} stores. *FEBS Lett.* **168**, 38–40.

Bellomo, G., Martino, A., Richelmi, P., Moore, G. A., Jewell, S. A., and Orrenius, S. (1984b). Pyridine nucleotide oxidation, Ca^{2+} cycling and membrane damage during *tert*-butylhydroperoxide metabolism by rat liver mitochondria. *Eur. J. Biochem.* **140**, 1–6.

Bellomo, G., Nicotera, P. L., and Orrenius, S. (1984c). Alterations in intracellular calcium compartmentation following inhibition of calcium efflux from isolated hepatocytes. *Eur. J. Biochem.* **144**, 19–23.

Berliner, K. (1933). The action of calcium on the heart: A critical review. *Am. Heart J.* **8**, 548–562.

Bernardi, P., and Azzone, G. F. (1983). Regulation of calcium efflux in rat liver mitochondria. Role of membrane potential. *Eur. J. Biochem.* **134**, 377–383.

Burgess, G. M., Godfrey, P. P., McKinney, J. S., Berridge, M. J., Irvine, R. F., and Putney, J. W., Jr. (1984). The second messenger linking receptor activation to internal Ca release in liver. *Nature (London)* **309**, 63–66.

Bygrave, F. L. (1977). Mitochondrial calcium transport. *Curr. Top. Bioenerg.* **6**, 259–318.

Bygrave, F. L. (1978). Properties of energy-dependent calcium transport by rat liver microsomal fraction as revealed by initial-rate measurements. *Biochem. J.* **170**, 87–91.

Campbell, A. K. (1983). "Intracellular Calcium: Its Universal Role as Regulator." Wiley, London.

Carafoli, E., and Crompton, M. (1978). The regulation of intracellular calcium by mitochondria. *Ann. N.Y. Acad. Sci.* **307**, 269–284.

Chan, K. M., and Junger, K. D. (1983). Calcium transport and phosphorylated intermediate of $(Ca^{2+}\text{-}Mg^{2+})$-ATPase in plasma membranes of rat liver. *J. Biol. Chem.* **258**, 4404–4410.

Chance, B. (1965). The energy-linked reaction of calcium with mitochondria. *J. Biol. Chem.* **240**, 2729–2748.

Charest, R., Blackmore, P. F., Berthon, B., and Exton, J. H. (1983). Changes in free cytosolic Ca^{2+} in hepatocytes following α_1-adrenergic stimulation. Studies on Quin 2-loaded hepatocytes. *J. Biol. Chem.* **258**, 8769–8773.

Cheung, W. Y. (1980). Calmodulin plays a pivotal role in cellular regulation. *Science* **207**, 19–27.

Claret-Berthon, B., Claret, M., and Mazet, J. C. (1977). Fluxes and distribution of calcium in rat liver cells: Kinetic analysis and identification of pools. *J. Physiol. (London)* **272**, 529–555.

Creba, J. A., Downes, C. P., Hawkins, P. T., Brewster, G., Michell, R. H., and Kirk, C. J. (1983). Rapid break-down of phosphatidylinositol 4-phosphate and phosphatidylinositol 4,5 bisphosphate in rat hepatocytes stimulated by vasopressin and other Ca^{2+}-mobilizing hormones. *Biochem. J.* **212**, 733–747.

Curtis, A. S. G. (1962). Cell contact and cell adhesion. *Biol. Rev. Cambridge Philos. Soc.* **37**, 82–129.

Davy, H. (1808). On some new phenomena of chemical changes produced by electricity, particularly the decomposition of the fixed alkalies and the exhibition of the new substances which constitute their bases. *Philos. Trans. R. Soc. London* **98**, 1–44.

Dawson, A. P. (1982). Kinetic properties of the Ca^{2+} accumulation system of a rat liver microsomal fraction. *Biochem. J.* **206**, 73–79.

Dawson, A. P., and Irvine, R. F. (1984). Inositol(1,4,5)tris-phosphate-promoted Ca^{2+} release from microsomal fractions of rat liver. *Biochem. Biophys. Res. Commun.* **120**, 858–864.

Di Monte, D., Bellomo, G., Thor, H., Nicotera, P. L., and Orrenius, S. (1984). Menadione-induced cytotoxicity is associated with protein thiol oxidation and alterations in intracellular Ca^{2+} homeostasis. *Arch. Biochem. Biophys.* **235**, 343–350.

Eagle, H. (1956). Nutrition needs of mammalian cells in tissue culture. *Arch. Biochem. Biophys.* **61**, 356–366.

Eklöw, L., Moldéus, P., and Orrenius, S. (1984). Oxidation of glutathione during hydroperoxide metabolism. A study using isolated hepatocytes and the glutathione reductase inhibitor 1,3-bis(2-chloroethyl)-1-nitrosourea. *Eur. J. Biochem.* **138**, 459–463.

Exton, J. H. (1981). Molecular mechanisms involved in α-adrenergic responses. *Mol. Cell. Endocrinol.* **23**, 233–264.

Famulski, K. S., and Carafoli, E. (1981). Ca^{2+}-transporting activity of membrane fractions isolated from the postmitochondrial supernatant of rat liver. *Cell Calcium* **3**, 263–281.

Famulski, K. S., and Carafoli, E. (1984). Calmodulin-dependent protein phosphorylation and calcium uptake in rat-liver microsomes. *Eur. J. Biochem.* **141**, 15–20.

Farber, J. L. (1981). The role of calcium in cell death. *Life Sci.* **29**, 1289–1295.

Farber, J. L. (1984). The role of calcium in liver cell death. *In* "Mechanisms of Hepatocyte Injury and Death" (D. Keppler, H. Popper, L. Bianchi, and W. Reutter, eds.), pp. 111–118. MTP Press, Lancaster.

Fariss, M. W., Olofsdottir, K., and Reed, D. J. (1984). Extracellular calcium protects isolated rat hepatocytes from injury. *Biochem. Biophys. Res. Commun.* **121**, 102–110.

Fiskum, G., and Lehninger, A. L. (1979). Regulated release of Ca^{2+} from respiring mitochondria by $Ca^{2+}/2H^+$ antiport. *J. Biol. Chem.* **254,** 6236–6239.

Fleckenstein, A., Witzleben, H., Frey, M., and Millner, T. G. (1981). Prevention of cataracts of alloxan-diabetic rats by long-term treatment with verapamil. *Pfluegers Arch.* **391,** Suppl. R12.

Fleckenstein, A., Frey, M., and Fleckenstein-Grün, G. (1984). Cellular injury by cytosolic calcium overload and its prevention by calcium antagonists—a new principle of tissue protection. *In* "Mechanisms of Hepatocyte Injury and Death" (D. Keppler, H. Popper, L. Bianchi, and W. Reutter, eds.), pp. 321–335. MTP Press, Lancaster.

Gunter, T. E., Chace, J. H., Puskin, J. S., and Gunter, K. K. (1983). Mechanism of sodium independent calcium efflux from rat liver mitochondria. *Biochemistry* **22,** 6341–6351.

Guraj, P., Zurini, M., Murer, H., and Carafoli, E. (1983). A high affinity, calmodulin-dependent Ca^{2+} pump in the basal-lateral plasma membrane of kidney cortex. *Eur. J. Biochem.* **136,** 71–76.

Iwasa, Y., Iwasa, T., Higashi, K., Matsui, K., and Miyamoto, E. (1982). Demonstration of a high affinity Ca^{2+}-ATPase in rat liver plasma membranes. *Biochem. Biophys. Res. Commun.* **105,** 488–494.

Jewell, S. A., Bellomo, G., Thor, H., Orrenius, S., and Smith, M. T. (1982). Bleb formation in hepatocytes during drug metabolism is caused by disturbances in thiol and calcium ion homeostasis. *Science* **217,** 1257–1259.

Jones, D. P., Thor, H., Smith, M. T., Jewell, S. A., and Orrenius, S. (1983). Inhibition of ATP-dependent microsomal Ca^{2+} sequestration during oxidative stress and its prevention by glutathione. *J. Biol. Chem.* **258,** 6390–6393.

Joseph, S. K., Thomas, A. P., Williams, R. J., Irvine, R. F., and Williamson, J. R. (1984). *myo*-Inositol 1,4,5-trisphosphate. A second messenger for the hormonal mobilization of intracellular Ca^{2+} in liver. *J. Biol. Chem.* **259,** 3077–3081.

Kane, A. B., Young, E. E., Schanne, F. A. X., and Farber, J. L. (1980). Calcium dependence of phalloidin-induced liver cell death. *Proc. Natl. Acad. Sci. U.S.A.* **77,** 1177–1180.

Kendrick, N. C. (1976). Purification of arsenazo III, a Ca^{2+}-sensitive dye. *Anal. Biochem.* **76,** 487–501.

Klaven, N. B., Pershandsingh, A., Henius, G. V., Laris, P. C., Long, J. W., and McDonald, J. M. (1983). A high affinity, calmodulin sensitive (Ca^{2+}-Mg^{2+})-ATPase and associated calcium-transport pump in the Ehrlich ascites tumor cell plasma membrane. *Arch. Biochem. Biophys.* **226,** 618–628.

Kraus-Friedman, H., Biber, J., Murer, H., and Carafoli, E. (1982). Calcium uptake in isolated hepatic plasma-membrane vesicles. *Eur. J. Biochem.* **129,** 7–12.

Krell, H., Ernish, N., Kospereck, S., and Pfaff, E. (1983). On the mechanisms of ATP-induced and succinate-induced redistribution of cations in isolated rat liver cells. *Eur. J. Biochem.* **131,** 247–254.

Lehninger, A. L., Vercesi, A., and Bababunmi, E. (1978a). Regulation of Ca^{2+} release from mitochondria by the oxidation-reduction state of pyridine nucleotides. *Proc. Natl. Acad. Sci. U.S.A.* **75,** 1690–1694.

Lehninger, A. L., Reynafarje, B., Vercesi, A., and Tew, W. P. (1978b). Transport and accumulation of calcium in mitochondria. *Ann. N.Y. Acad. Sci.* **307,** 160–176.

Locke, F. S. (1894). Notiz über den Einfluss physiologischer Kochsalzlösung und die Erregbarkeit von Muskel und Nerve. *Zentralbl. Physiol.* **8,** 166–167.

Lotersztajn, S., Hanoune, J., and Pecker, F. (1981). A high affinity calcium-stimulated magnesium-dependent ATPase in rat liver plasma membranes. *J. Biol. Chem.* **256,** 11209–11215.

Lötscher, H. R., Winterhalter, K. H., Carafoli, E., and Richter, C. (1979). Hydroperoxides can modulate the redox state of pyridine nucleotides and the calcium balance in rat liver mitochondria. *Proc. Natl. Acad. Sci. U.S.A.* **76,** 4340–4344.

Malencik, D. A., and Fischer, E. H. (1983). Structure, function, and regulation of phophorylase kinase. *In* "Calcium and Cell Function" (W. Y. Cheung, ed.), Vol. 3, pp. 161–188. Academic Press, New York.

Martonosi, A. M. (1983). Transport of calcium by sarcoplasmic reticulum. *In* "Calcium and Cell Function" (W. Y. Cheung, ed.), Vol. 3, pp. 37–101. Academic Press, New York.

Mesland, D. A. M., Los, G., and Spiele, H. (1981). Cytochalasin B disrupts the association of filamentous web and plasma membrane in hepatocytes. *Exp. Cell Res.* **135**, 431–435.

Moldéus, P., Högberg, J., and Orrenius, S. (1978). Isolation and use of liver cells. *In* "Methods in Enzymology" (P. A. Hoffee and M. E. Jones, eds.), Vol. 51, pp. 60–71. Academic Press, New York.

Moore, C. L. (1971). Specific inhibition of mitochondrial Ca^{++} transport by ruthenium red. *Biochem. Biophys. Res. Commun.* **42**, 298–305.

Moore, G. A., Jewell, S. A., Bellomo, G., and Orrenius, S. (1983). On the relationship between Ca^{2+} efflux and membrane damage during t-butylhydroperoxide metabolism by liver mitochondria. *FEBS Lett.* **153**, 289–292.

Moore, L. (1982). Carbon disulfide hepatotoxicity and inhibition of liver microsome calcium pump. *Biochem. Pharmacol.* **31**, 1465–1467.

Moore, L., Chen, T., Knapp, H. R., Jr., and Landon, E. J. (1975). Energy-dependent calcium sequestration activity in rat liver microsomes. *J. Biol. Chem.* **250**, 4562–4568.

Moore, L., Davenport, G. R., and Landon, E. J. (1976). Calcium uptake of a rat liver microsomal subcellular fraction in response to *in vivo* administration of carbon tetrachloride. *J. Biol. Chem.* **251**, 1197–1201.

Murphy, E., Coll, K., Rich, T. L., and Williamson, J. R. (1980). Hormonal effects on calcium homeostasis in isolated hepatocytes. *J. Biol. Chem.* **255**, 6600–6608.

Nicholls, D. G., and Crompton, M. (1980). Mitochondrial calcium transport. *FEBS Lett.* **111**, 261–268.

Nicotera, P. L., Moore, M., Bellomo, G., Mirabelli, F., and Orrenius, S. (1984a). Demonstration and partial characterization of glutathione disulfide-stimulated ATPase activity in the plasma membrane fraction from rat hepatocytes. *J. Biol. Chem.* **260**, 1999–2002.

Nicotera, P. L., Moore, M., Mirabelli, F., Bellomo, G., and Orrenius, S. (1984b). Inhibition of hepatocyte plasma membrane Ca^{2+}-ATPase activity by menadione metabolism and its restoration by thiols. *FEBS Lett.* **181**, 149–153.

Orrenius, S., Jewell, S. A., Thor, H., Bellomo, G., Eklöw, L., and Smith, M. T. (1983). Drug-induced alterations in the surface morphology of isolated hepatocytes. *In* "Isolation, Characterization and Use of Hepatocytes" (R. A. Harris and N. W. Cornell, eds.), pp. 333–340. Am. Elsevier, New York.

Parker, J. C., Barritt, G. J., and Wadsworth, J. C. (1983). A kinetic investigation of the effects of adrenaline on ^{45}Ca^{2+} exchange in isolated hepatocytes at different Ca^{2+} concentrations at 20°C and in the presence of inhibitors of mitochondrial Ca^{2+} transport. *Biochem. J.* **216**, 51–62.

Ringer, S. (1882). Concerning the influence exerted by each of the constituents of the blood on the contraction of the ventricle. *J. Physiol. (London)* **3**, 380–393.

Schanne, F. A. X., Gilfor, D., and Farber, J. L. (1979). Calcium dependence of toxic cell death: A final common pathway. *Science* **206**, 700–702.

Shooter, R. A., and Grey, G. O. (1952). Studies of mineral requirements of mammalian cells. *Br. J. Exp. Pathol.* **36**, 341–350.

Smith, M. T., Thor, H., and Orrenius, S. (1981). Toxic injury to isolated hepatocytes is not dependent on extracellular calcium. *Science* **213**, 1257–1259.

Smith, M. T., Thor, H., Jewell, S. A., Bellomo, G., Sandy, M. S., and Orrenius, S. (1984). Free radical-induced changes in the surface morphology of isolated hepatocytes. *In* "Free

Radicals in Molecular Biology, Aging, and Disease'' (D. Armstrong, R. S. Sohal, R. G. Cutler, and T. F. Slater, eds.), pp. 103–118. Raven Press, New York.

Thor, H., Smith, M. T. Hartzell, P., Bellomo, G., Jewell, S. A., and Orrenius, S. (1982). The metabolism of menadione (2-methyl-1,4-naphthoquinone) by isolated hepatocytes. A study of the implications of oxidative stress in intact cells. *J. Biol. Chem.* **257,** 12419–12425.

Thor, H., Hartzell, P., and Orrenius, S. (1984). Potentiation of oxidative cell injury in hepatocytes which have accumulated Ca^{2+}. *J. Biol. Chem.* **259,** 6612–6615.

Tischler, M. E., Hecht, P., and Williamson, J. R. (1977). Determination of mitochondrial/cytosolic metabolite gradients in isolated rat liver cells by cell disruption. *Arch. Biochem. Biophys.* **181,** 278–292.

Tsien, R. Y. (1980). New calcium indicators and buffers with selectivity against magnesium and protons: Design, synthesis and properties of prototype structures. *Biochemistry* **19,** 2396–2404.

Vercesi, A. (1984). Dissociation of $NAD(P)^+$-stimulated mitochondrial Ca^{2+} efflux from swelling and membrane damage. *Arch. Biochem. Biophys.* **232,** 86–91.

Waisman, D. M., Gimble, J. M., Goodman, D. B. P., and Rasmussen, H. (1981). Studies of the Ca^{2+} transport mechanism of human erythrocyte inside-out plasma membrane vesicles. *J. Biol. Chem.* **256,** 409–414.

Willmer, E. N., ed. (1965). ''Cells and Tissues in Culture,'' Vols. 1 and 2. Academic Press, New York.

Willmer, E. N., ed. (1966). ''Cells and Tissues in Culture,'' Vol. 3. Academic Press, New York.

Chapter 7

The Role of Calcium
in Meiosis

GENE A. MORRILL
ADELE B. KOSTELLOW

Department of Physiology and Biophysics
Albert Einstein College of Medicine
Bronx, New York

209

CALCIUM AND CELL FUNCTION, VOL. VI

I. INTRODUCTION

Every animal embryo develops from a single cell, which is formed by the union of the egg and sperm haploid nuclei within the egg. This chapter will be concerned with the calcium-associated membrane and cytoplasmic events that lead to the formation of a haploid nucleus in the oocyte and that enable it to be fertilized. The emphasis will be on the oocytes of vertebrates and echinoderms, which begin their embryological development in a similar manner.

The fully grown oocytes of both starfish and vertebrates are four C cells blocked in prophase of the first meiotic division (the sea urchin egg completes meiosis as it completes growth). The hormone stimulus which induces spawning in starfish and ovulation in vertebrates leads to changes in membrane and cytoplasm which permit completion of two meiotic divisions in starfish, and one and one-half divisions in vertebrates, with the second division arrested at metaphase. This stimulus is said to "reinitiate" meiosis because the cells had already reached prophase.

In almost all vertebrates, the cytoplasmic changes following the stimulus of fertilization are necessary to induce anaphase of the last meiotic division and to obtain a haploid egg nucleus. In echinoderms, meiosis is complete before fertilization. Strictly speaking, the cytoplasmic response which follows their fertilization is relevant to syngamy (the union of the two haploid pronuclei) and not to meiosis. This response is, however, remarkably similar in both vertebrates and echinoderms, and certainly much of what we know about fertilization was first described in sea urchins.

Taken *in toto,* meiosis encompasses the period of complex chromosomal changes prior to prophase arrest followed by a period of oocyte growth as well as the completion of the meiotic or reductive divisions. Many investigators use the term "maturation" to describe the completion of meiosis as defined by Wilson (1925), who stated that maturation is accomplished in the animal oocyte by means of two successive meiotic divisions, also called "meiotic maturation" (Schuetz, 1969). The term "maturation" has also been used in the sense of being ripe for fertilization, which in vertebrates occurs at a stage before the completion of the second meiotic division. The changes normally caused by fertilization are designated "activation," and can be initiated by a variety of parthenogenetic stimuli. In general, the germ cell is referred to as an "oocyte" until ovulation, at which time it becomes an "egg."

Much of the research to date has centered about two transition points: (1) the hormone-mediated release of the prophase block, and (2) the release of second metaphase block at fertilization. It is generally recognized that the meiotic divisions involve a complex sequence or cascade of biochemical and biophysical events and that calcium may play a role in the initiation of many of the programmed changes.

II. OOCYTES AND EGGS FROM VARIOUS SPECIES AS MODEL SYSTEMS

Meiosis appears to be a very conservative process, using many of the same basic mechanisms for a stepwise activation of the oocyte and egg in species ranging from sea urchin through mammals. Development of techniques for obtaining gametes in large amounts from marine invertebrates (e.g., see Kanatani, 1973), amphibians (Masui, 1967; Wallace et al., 1973), and mammals (e.g., see Channing et al., 1982) have made it possible to carry out experiments on a year-round basis using a number of vertebrate and invertebrate oocytes and eggs as model systems.

Mammalian oocytes have been studied within the follicle, dissected out as cumulus cell-enclosed oocytes, or stripped of investing cumulus cells with the enzyme hyaluronidase. When fully grown oocytes are removed from antral follicles, they begin to mature spontaneously. They are relatively small (~ 100 μm in diameter) and are usually ovulated in quantities of not more than a dozen at a time. Superovulating drugs can induce ovulation of as many as 50 eggs at a time in mice. Techniques developed in recent years have made it possible to study free calcium levels even in single oocytes (e.g., Cutherbertson et al., 1981; Cobbold et al., 1983).

Amphibian oocytes are quite large (1.2–2 mm diameter and larger) and can be dissected from the ovary using fine-tipped forceps. Such dissected "oocytes" have a layer of follicle cells closely applied to the surface of the oocyte. The follicle cell layer can then be stripped off with jewelers' tweezers following brief exposure to Ca^{2+},Mg^{2+}-free Ringer's solution (Masui, 1967) or by treatment with proteolytic enzymes such as pronase or collagenase. The oocyte free of somatic cells is called a "denuded oocyte."

The South African clawed toad, *Xenopus laevis,* is the most widely used amphibian species, in part because of its availability in both Europe and North America and the relative ease of obtaining oocytes or eggs all year. Ovulation can be induced repeatedly in the same female since new populations of oocytes develop continually. Several mutant strains are available which are experimentally useful. For example, oocytes of the albino mutant of *X. laevis* can be microinjected with the calcium-specific photoprotein aequorin, and Ca^{2+}-induced luminescence monitored after addition of meiotic agonists (Wasserman *et al.,* 1980; Moreau *et al.,* 1980). This would not be possible with the highly pigmented oocytes of other amphibians.

Since breeding *Xenopus* contain oocytes at all stages of growth, overripe oocytes are often "cleared" by pretreatment of the female with gonadotropin. The use of different protocols for pretreating *Xenopus* females has introduced some uncertainty into comparison of data from various laboratories. Spontaneous maturation may occur when follicle cells are stripped from the oocytes (Vilain *et al.,* 1980). Gonadotropin-induced ovulation can yield several hundred eggs. The largest oocytes mature nearly synchronously; smaller oocytes that respond mature more slowly (Reynhout *et al.,* 1975). Up to 200 ovulated eggs accumulate in the oviduct, and 10–20 are deposited at a time. These can be fertilized and divide synchronously for the first few divisions. Ovulation continues for about 12 hr. Fully grown oocytes and eggs of *Xenopus* are 1.2–1.4 mm in diameter.

The North American leopard frog, *Rana pipiens,* is not as widely available and is more commonly used by investigators in eastern North America. *Rana pipiens* are collected before they enter hibernation in the fall and can be maintained without feeding in artificial hibernation at 4–5°C. Each female contains 1000–3000 large (1.5–2.0 mm diameter) prophase-arrested oocytes. The large oocytes are normally ovulated all at one time, in the spring. They can be hormonally induced to begin meiosis at any time between October and June, either *in vivo* or *in vitro,* and undergo nearly synchronous meiotic divisions. Following gonadotropin-stimulated ovulation, several thousand eggs accumulate in the oviducts before being deposited. They can be stripped from the female into sperm suspensions and develop synchronously (within seconds) for the first three divisions.

Large numbers of starfish oocytes can also be readily collected in prophase arrest and have been extensively used to study changes following resumption of the first meiotic division. Nuclear breakdown is not particularly synchronous (see Guerrier *et al.,* 1982).

In the sea urchin, oogenesis proceeds to up to second polar body extrusion and fertilization initiates the mitotic divisions. Thus, the sea urchin has not been used to study release of the first block, but has been a popular material for study of fertilization (see Epel, 1982). Fully grown oocytes and eggs from many invertebrate species are about 100 μm in diameter.

III. BIOCHEMICAL AND BIOPHYSICAL EVENTS DURING VERTEBRATE MEIOTIC DIVISIONS

The hormone stimulus that releases the oocyte from the first meiotic prophase initiates a cascade of membrane and cytoplasmic changes in the oocyte that culminate in ovulation and arrest at second meiotic metaphase. Fertilization releases the second block and initiates a second cascade of membrane and cytoplasmic events.

A. Resumption of the First Meiotic Division and Ovulation

The physiological stimulus which triggers meiosis in mammals is LH (reviewed in Channing *et al.*, 1982; Tsafriri *et al.*, 1982), whereas in amphibians it is probably progesterone (Schuetz, 1967; Masui, 1967). In starfish it is 1-methyl adenine (reviewed by Meijer and Guerrier, 1984b). Recent evidence indicates that progesterone may also have a limited but definite stimulatory effect on the resumption of meiosis in rabbit ova and that it mediates the meiosis-inducing action of human chorionic gonadotropin (hCG) (Mori *et al.*, 1983). In both starfish and vertebrates, the meiotic stimulus overcomes a "prophase block" which, in mammals, may have kept the oocyte in prophase for several decades.

The criterion often used for the successful reinitiation of the meiotic divisions is the subsequent breakdown of the large nucleus or "germinal vesicle." Nuclear breakdown is followed by the completion of the first meiotic division and the extrusion of one-half the chromosomes in the polar body. The temporal sequence of changes known to occur during the first and second meiotic divisions in *Rana* oocytes and eggs are outlined below and illustrated in Fig. 1.

1. Early responses—Initiation of processes leading to the completion of the first meiotic division. During the first 60–90 min:
   a. There is a release of bound membrane and intracellular calcium.
   b. Intracellular cAMP levels fall, if not already low, due to inhibition of adenylate cyclase.
   c. Intracellular pH rises.
   d. Creatine phosphate levels increase and inorganic phosphate levels fall.
2. Late responses—Events within the next 8–9 hr leading to nuclear breakdown:
   a. There is a rise in intracellular calcium and magnesium.
   b. Protein synthesis increases.
   c. A protein called "maturation promoting factor" (MPF) appears in oocyte cytoplasm.

  d. The electrogenic Na^+,K^+-ATPase disappears and the oocyte membrane undergoes partial depolarization.
  e. Protein and lipid phosphorylation increase.
  f. Ca^{2+}-ATPase activity increases.
  g. Ca^{2+} permeability decreases.
  h. The nucleus rises to lie just under the plasma membrane at the animal hemisphere and breaks down.
  i. Ovulation, the release from the ovarian follicle, occurs at about the time of nuclear breakdown.
3. Events leading to arrest at metaphase II (12–36 hr):
  a. K^+, Cl^-, and Ca^{2+} permeability disappear leading to membrane depolarization and the appearance of membrane excitability; the egg cytoplasm is nearly isopotential with the external environment.
  b. The first meiotic division is completed and one-half of the chromosomes are extruded into the polar body. The second meiotic division begins and continues until metaphase.
  c. A protein called "cytostatic factor" (CSF) appears in the egg cytoplasm.
  d. Endocytosis increases, accompanied by a rise in intracellular Na^+.
  e. Bound-calcium levels increase and calcium permeability virtually disappears.

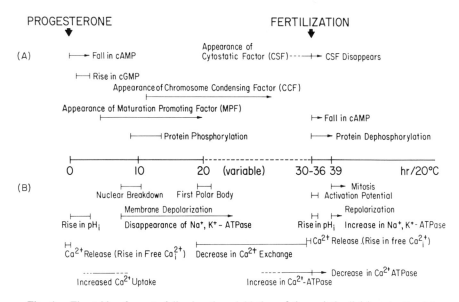

Fig. 1. Timetable of events following the reinitiation of the meiotic divisions in *R. pipiens* oocytes by progesterone. The events can be divided into molecular (A) and membrane (B) associated changes. The intervals shown are typical for development at 20–22°C. Zero time represents the prophase arrested oocyte. Adapted from Morrill *et al.* (1981).

The egg is now "mature" in the sense that it can be fertilized. Fertilization must generally occur within a fixed time period (average about 24 hr) or the egg will be overripe.

B. Fertilization and Completion of the Meiotic Divisions

Fertilization introduces the sperm haploid nucleus, and activates events which lead to the completion of the second meiotic division by the vertebrate egg nucleus. This is followed by DNA synthesis, syngamy of egg and sperm nucleus, and the first mitotic division.

1. Events associated with release of the metaphase II block:
   a. The sperm head fuses with the oocyte plasma membrane.
   b. Na^+ enters the egg; the egg membrane undergoes a transient hyperpolarization (activation potential); membrane changes defend against polyspermy.
   c. There is a cortical granule exocytosis accompanied by a release of bound calcium and rapid efflux from the egg; the CSF disappears or is inactivated.
   d. The sperm or sperm head is drawn into the egg cytoplasm, where it decondenses.
   e. Intracellular pH rises.
   f. Creatine phosphate levels increase and inorganic phosphate levels fall.
   g. Egg protein and lipid are dephosphorylated.
   h. Vertebrate egg chromosomes procede from metaphase II to completion of the second meiotic division (about 20 min in amphibians), and the second polar body is extruded, leaving a haploid egg nucleus. The egg nucleus has now completed the "maturation divisions."
   i. DNA synthesis is initiated (in mammals, the oocyte has not synthesized DNA since it was differentiated in the fetal ovary).
   j. The haploid egg and sperm nuclei fuse. Their chromosomes become aligned for the first mitotic division of the new organism.
2. Events associated with the transition between meiosis II and mitosis I:
   a. The egg plasma membrane repolarizes; the repolarization is associated with the reappearance of the membrane conductance to K^+ and the Na^+,K^+-ATPase.
   b. There is further protein and lipid dephosphorylation.

For a general review of meiosis see Masui and Clarke (1979). Specialized reviews of events following release of the prophase block include those covering calcium (Baulieu et al., 1978; Morrill et al., 1981; Guerrier et al., 1982; Meijer and Guerrier, 1984b), cyclic AMP (Maller and Krebs, 1980; Morrill et al., 1981; Maller, 1983; Schultz et al., 1983; Eppig and Downs, 1984), maturation inhibitor (Tsafriri et al., 1982), electrical changes (Schlichter, 1983; Morrill et al.,

1984b), cytoplasmic factors (Masui *et al.*, 1980), macromolecular changes (Maller, 1983; Morrill *et al.*, 1984b), and intracellular pH (Busa and Nuccitelli, 1984; Morrill *et al.*, 1984c).

IV. EXOGENOUS CALCIUM AND MAGNESIUM REQUIREMENTS DURING MEIOSIS

In all species studied, the hormonal reinitiation of the meiotic divisions causes an immediate elevation of intracellular free calcium, followed by a series of calcium-associated metabolic events. To what extent do these phenomena depend upon a supply of extracellular calcium?

Resumption and completion of meiosis I by oocytes from bovine follicles has been shown by Leibfried and First (1979) and by Jagiello *et al.* (1982) to be dependent on the inclusion of both Ca^{2+} and Mg^{2+} in the culture medium and to have a greater dependence on Mg^{2+} than Ca^{2+}. Mouse oocytes required both ions, but were less sensitive to a reduction in Mg^{2+} concentration (Jagiello *et al.*, 1982). Some maturation was observed in both species with media containing only Mg^{2+}. Ca^{2+} is an absolute requirement for meiosis in cumulus-enclosed swine oocytes (Bae, 1981). On the other hand, the meiosis of neither follicle-enclosed nor isolated rat oocytes was affected by incubation in Ca^{2+}-free medium containing EGTA or EDTA (Tsafriri and Bar-Ami, 1978).

Merriam (1971) and Ecker and Smith (1971) were the first to demonstrate that exogenous calcium and magnesium were necessary for the progesterone-induced resumption of meiosis in *Xenopus* follicles. Merriam found that the continuous presence of both 0.74 mM Ca^{2+} and 0.82 mM Mg^{2+} in the medium was essential for induction of nuclear breakdown. Either ion alone in the medium could support a partial but morphologically atypical response to progesterone, although Mg^{2+} was more efficient and both ions together gave the best result. Merriam (1971) suggested that the *Xenopus* oocyte contained diffusible Ca^{2+} and Mg^{2+} pools and that both were essential for completion of the first meiotic division.

In contrast to *Xenopus* oocytes, studies by Morrill *et al.* (1975) and by Kostellow and Morrill (1980) on *Rana* oocytes have shown that diffusible Ca^{2+} (but not Mg^{2+}) was essential only during the first 5–10 min following exposure to steroid. If oocytes were then transfered to medium free of both Ca^{2+} and Mg^{2+}, the time course of nuclear breakdown was identical to that in Ca^{2+}, Mg^{2+}-containing medium. In fact, nuclear membrane breakdown was inhibited by only 40% when oocytes were induced in a Ca^{2+},Mg^{2+}-free medium. If the intracellular diffusible Ca^{2+} pool was first diminished by preincubation in Ca^{2+}-free, EGTA-containing Ringer's solution, the induction of membrane breakdown was inhibited 70–80%. Preincubation at 4°C led to the loss of 55–60% of oocyte Ca^{2+} and the total inability to respond to progesterone induction (Kostellow and Morrill, 1980). These findings indicate that in *Rana*, a diffusible

Ca^{2+} component is essential for response to steroid, and that in the absence of extracellular Ca^{2+}, about 60% of the oocytes can release enough bound Ca^{2+} to support the initiation of meiosis.

The lack of an exogenous Mg^{2+} requirement for *Rana* oocytes may reflect the high Mg^{2+} levels (20–30 mM) present in the prophase-arrested oocyte (Morrill *et al.*, 1971). This independence of $[Mg^{2+}]_0$ does, however, make it possible to study Ca^{2+}-dependent events without the complication of a requirement for exogenous Mg^{2+}.

Endocytosis may make a major contribution to calcium uptake in prophase-arrested amphibian oocytes. The concept of membrane recycling was proposed some years ago to account for the enormous volumes of extracellular fluid internalized by cultured cells during pinocytosis and has been since studied in a wide variety of cells. The rate of membrane and fluid turnover in nonphagocytic cells can be appreciable. For example, mouse fibroblasts in culture internalize an amount of cell surface membrane equivalent to 50% of the cell surface every hour (reviewed in Steinman *et al.*, 1983). Endocytosis has been demonstrated in amphibian oocytes using the labeled yolk protein, vitellogen (e.g., Wallace, 1972; Schuetz *et al.*, 1974) and ^3H-labeled inulin (Morrill *et al.*, 1983, 1984a). Inulin, a 5000–5500 MW polysaccharide is neither actively transported nor metabolized by most cells and can be used as a marker for fluid phase uptake. ^3H-Labeled inulin uptake into the cytoplasm of the prophase-arrested oocyte is linear for at least 5 hr with no measurable ($<1\%$) efflux or metabolism. Based on the specific activity of the medium and ^3H uptake during the first hour, the fluid phase uptake was estimated to be about 24 nl per oocyte per hour. Since $[Ca^{2+}]_0$ is 1.1 mM, this fluid exchange would result in the internalization of about 26 pmol of Ca^{2+} per oocyte per hour. As measured by $^{45}Ca^{2+}$ uptake, Ca^{2+} uptake and exchange is also about 26 pmol per oocyte per hour (Morrill *et al.*, 1980), indicating that nearly all of the Ca^{2+} taken up by the prophase-arrested amphibian oocyte is due to endocytosis. Ca^{2+} efflux, on the other hand, is probably largely due to Ca^{2+}-ATPase activity and $Ca^{2+}:Na^+$ exchange systems.

The available evidence indicates that some oocytes require exogenous Ca^{2+} and/or Mg^{2+} to complete meiosis, and that the requirement depends in part upon the oocyte's endogenous sequestered supply of both ions. The significance of calcium entry into the oocyte at the time of the meiotic stimulus and during the following period is complex, and will be described in Section V.

V. CALCIUM REDISTRIBUTION ASSOCIATED WITH THE RESUMPTION OF MEIOSIS

Progesterone is probably the physiological stimulus which reinitiates amphibian meiosis (Schuetz, 1967; Masui, 1967). Most of the experimental evidence indicates that its effect is on the cell surface rather than at the classical DNA-binding steroid receptor. It is, in fact, one of the earliest examples of steroid action at the plasma membrane (reviewed in Baulieu et al., 1978; Masui and Clarke, 1979; Morrill et al., 1981), although cytosolic progestin receptors have been described in Rana oocytes (Kalimi et al., 1979). Two lines of evidence, however, indicate that it is the action of progesterone on the plasma membrane that is critical in meiotic maturation. First, an enucleated or actinomycin D–treated oocyte can be stimulated by progesterone to undergo the cytoplasmic changes which prepare it for fertilization. Second, a number of agents which act at the membrane will either substitute for progesterone as a meiotic stimulus, or else inhibit its action (see Table I).

TABLE I

Effectiveness of Various Agents on Resumption of the First Meiotic Division in Oocytes of Several Species

		Nuclear breakdown			
Agent[a]	Action	Starfish[b]	Xenopus[c]	Rana[d]	Mouse[e]
Progesterone	Physiological inducer	—	Induces	Induces	—
1-MeAde	Physiological inducer	Induces	—	—	—
dBcGMP		—	—	Induces	—
dBcAMP		Inactive	Inhibits	Inhibits	Inhibits
Insulin		—	Induces	Induces	—
Methylaxanthines	PDE inhibitors	Inhibits	Inhibits	Inhibits	Inhibits
Leupeptin	Protease inhibitor	Inhibits	Inhibits	Inhibits	—
Arachidonic acid		Induces	—	—	—
A23187	Ca-ionophore	Inactive	Induces	Inhibits	Inhibits
Tetracaine	Anesthetic	Inhibits	Induces	Inhibits	Inactive
Procaine	Anesthetic	Inhibits	—	Inhibits	—
Dibucaine	Anesthetic	Inhibits	Induces	—	—
Propranolol	β-blocker	Inhibits	Induces	Induces	—
Trifluoperazine	Ca^{2+}-CaM Antagonist	Inhibits	Inhibits	Inhibits	Inhibits
W-7	Antagonist	Inhibts	Inhibits	—	—
D-600, Verapamil	Ca^{2+}-channel blocker	Inhibits	Induces	Induces	Induces
La^{3+}	Ca-antagonist	Inhibits	Induces	Induces	Oocyte degenerates
Mn^{2+}		—	Induces	—	—

[a]Responses are for comparable drug concentrations, when possible.

[b]Doree et al., 1982; Meijer and Guerrier, 1984a,b.

[c]Schorderet-Slatkine et al., 1976, 1977; Hollinger and Alvarez, 1982; Maller and Krebs, 1977; Maller, 1983.

[d]Morrill et al., 1980; Kostellow et al., 1980; Kostellow and Morrill, 1980.

[e]Paleos and Powers, 1981; Jagiello et al., 1982; Powers and Paleos, 1982.

Many of the experiments with these agents which were carried out during the past 10 years were concerned with the role of cyclic AMP and/or calcium as potential second messengers for the meiotic stimulus. We will discuss the evidence that both are involved; cyclic AMP as an inhibitor and calcium as an effector of meiosis. Table I lists somes of the agents which induce and/or inhibit meiosis in echinoderms, amphibians, and mammals.

A. Induction of Meiosis by Agents Acting at the Plasma Membrane

In both *Rana* and *Xenopus,* the earliest response to progesterone is a transient release of $^{45}Ca^{2+}$ from the oocyte into the medium (O'Connor *et al.,* 1977; Morrill *et al.,* 1980; Kostellow *et al.,* 1980). As shown in Fig. 2, in *Rana* oocytes there is a biphasic release of $^{45}Ca^{2+}$ with an initial transient at 1–2 min followed by a second transient at 8–12 min. A number of nonhormonal inducers (La^{3+}, D-600, propranolol) also initiate a rapid release of Ca^{2+} within the first 1–2 min. The release is not additive with progesterone suggesting that these agents act through a common mechanism (Morrill *et al.,* 1980; Kostellow *et al.,* 1980). Agents that are competitive inhibitors of progesterone-induced nuclear breakdown block progesterone-induced Ca^{2+} release. These include the local anesthetic tetracaine (Morrill *et al.,* 1980) and trifluoperazine (Kostellow *et al.,* 1980), a compound that binds to the Ca^{2+}–calmodulin complex.

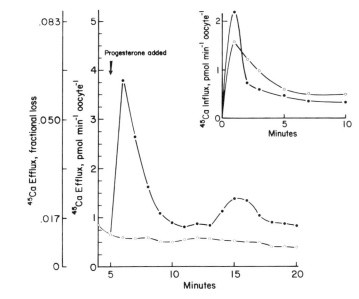

Fig. 2.   Rate of ^{45}Ca efflux and exchange (inset) in denuded *R. pipiens* oocytes. For efflux studies oocytes were preloaded with ^{45}Ca for 24 hr at 16°C and rinsed for 5 min, and progesterone was added at the time indicated. Symbols: Progesterone (●), Untreated (○). From Morrill *et al.* (1980).

The report that addition of 4.8 μM of the Ca^{2+},Mg^{2+}-ionophore A23187 to a medium containing 20 mM Ca^{2+} or Mg^{2+} will induce nuclear breakdown in *Xenopus* (Wasserman and Masui, 1975) has been frequently cited as evidence for calcium as a second messenger to progesterone. Additional evidence for a positive role of Ca^{2+} has been derived from induction by agents known to interact with Ca^{2+} binding sites (Schorderet-Slatkine *et al.*, 1977; Baulieu *et al.*, 1978). Schorderet-Slatkine and co-workers suggested that these agents acted by releasing calcium from the plasma membrane. Table I compares the effects of various agents on nuclear breakdown in oocytes of four species, starfish, *Xenopus, Rana,* and mouse. As can be seen, the ionophore A23187 induces nuclear breakdown in *Xenopus,* has no effect on starfish, and is inhibitory in both *Rana* and mouse oocytes. Only the methylxanthines and tripluoperizine (TFP) appear to be universal inhibitors. The inhibitory effect of TFP is questionable, since Hollinger and Alvarez (1982) report that 200 μM TFP will induce nuclear breakdown in *Xenopus* oocytes.

The responses summarized in Table I must be interpreted with caution, since many of these drugs have multiple actions. For example, local anesthetics (tetracaine, dibucaine, etc.) not only affect Ca^{2+} permeability, but also are reported to affect glucose or anion transport, exocytosis, and ligand-induced Ig receptor capping in other cells. Several enzymatic activities are also affected by local anesthetics, notably Ca^{2+},Mg^{2+}-ATPase in several tissues, calmodulin, phospholipase A_2, protein kinase C, and the response of adenylate cyclase to catecholamines (reviewed in Volpi *et al.*, 1981).

Increased calcium uptake is not, of itself, sufficient to induce the resumption of the meiotic divisions in *Rana*. For instance, the Ca^{2+},Mg^{2+}-ionophore A23187 increased $^{45}Ca^{2+}$ uptake about 5-fold over the first 10 min of treatment and over 10-fold in 2 hr, yet did not induce meiosis (Kostellow and Morrill, 1980), although A23187 induces in *Xenopus* (Wasserman and Masui, 1975). D-600, a Ca^{2+}-channel blocker, inhibited $^{45}Ca^{2+}$ uptake from the medium (Kostellow and Morrill, 1980), yet induced nuclear breakdown in both *Xenopus* (Schorderet-Slatkine *et al.*, 1977) and *Rana* (Kostellow and Morrill, 1980). Conversely, La^{3+}, which has been shown to be an effective inducer in both *Xenopus* (Schorderet-Slatkine *et al.*, 1977) and *Rana* (Morrill *et al.*, 1980), essentially blocks Ca^{2+} uptake. Furthermore, when denuded oocytes were treated with 10 mM La^{3+} in Ca^{2+},Mg^{2+}-free medium, we consistently observed 100% nuclear breakdown (Kostellow and Morrill, 1980). La^{3+} displaces membrane-bound Ca^{2+}, but does not itself enter the cell (see Section V,B).

Thus, the effectiveness of meiotic inducers is not correlated with the ability to induce calcium uptake from the medium. All inducers, however, release calcium from the oocyte, and competitive inhibitors of progesterone induction block the progesterone-induced calcium release (Morrill *et al.*, 1980).

We have found that meiosis can be induced in *Rana* oocytes by extremely low

$(10^{-10}\ M)$ concentrations of dBcGMP in the presence of elevated (10 m*M*) Ca^{2+} (Kostellow and Morrill, 1980). It may substitute for the rise in cyclic GMP which we found to occur about 2 hr after progesterone induction (Kostellow and Morrill, 1980). The elevation in intracellular calcium, described below, is known to stimulate guanylate cyclase although other factors, such as a release of arachidonic acid may play a role (see Section V,F).

B. Intracellular Free Calcium Transients

Masui *et al.* (1977) showed that when progesterone-treated *Xenopus* oocytes were injected with 4–8 m*M* EGTA (final concentration) and cultured in a Ca^{2+}-free medium containing Mg^{2+}, nuclear breakdown did not occur. They found

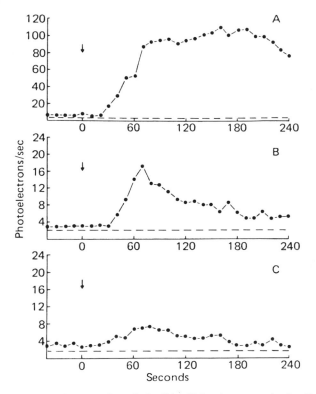

Fig. 3. Aequorin luminescence from single albino *X. laevis* oocytes stimulated by progesterone. Each oocyte was injected with 40 nl of 1% aequorin, allowed to equilibrate for 3 hr, and then monitored. Progesterone (1 or 5 μg/ml) was introduced at the time indicated by the arrow. (A) Frog II oocyte no. 3; (B) frog II oocyte no. 5; (C) frog I oocyte no. 3 (Table I). The dark noise (mean rate of thermoionic emission of the photomultiplier tube) is represented by the dashed line. From Wasserman *et al.* (1980).

that the later the oocytes were injected with EGTA, the greater the frequency of nuclear breakdown.

Early attempts to inject wild-type *Xenopus* oocytes with the photoprotein, aequorin, and to measure a progesterone-induced luminescence were without success (Belle *et al.*, 1977), probably because these oocytes were highly pigmented. Wasserman *et al.* (1980) subsequently injected the photoprotein aequorin into the pigmentless oocytes of an albino mutant of *Xenopus* and reported that about one-half of the progesterone-treated oocytes exhibited a transient increase in light emission within the first 5–6 min (Fig. 3), although all underwent nuclear breakdown. They concluded that the initial progesterone–oocyte interaction involved a transient increase in calcium located at or near the oocyte surface. They suggested that the failure to observe a Ca^{2+} transient in more than one-half of the oocytes might be because: (1) the Ca^{2+} activity that increased after progesterone is restricted to only a portion of the oocyte cytoplasm, and/or (2) the hormonal state of the oocytes may have been variable. They also note that, in *Xenopus,* "many of the so-called early events of maturation that occur in response to progesterone . . . are either by-passed or have already taken place" (see Section II).

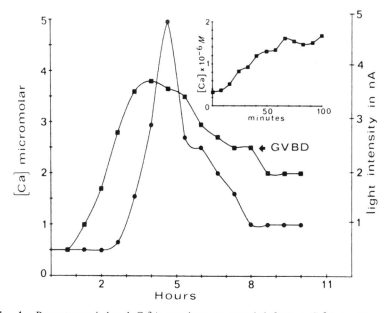

Fig. 4. Progesterone-induced Ca^{2+} transients as recorded from a Ca^{2+}-sensitive microelectrode (●) or 25 aequorin-injected oocytes taken from the same Ambystoma (■). Maturation ratio was 14:25. Inset illustrates initial changes in Ca^{2+} activity as recorded with the electrode. Right ordinate indicates light intensity in nA. Left ordinate refers to Ca^{2+} activity changes as measured with the Ca^{2+}-sensitive electrode. From Moreau *et al.* (1980).

Moreau *et al.* (1980) examined changes in free Ca^{2+} in the oocytes of three amphibians, *Xenopus, Ambystoma,* and *Pleurodeles,* using calcium-sensitive electrodes and the photoproteins aequorin and obelin. Resting Ca^{2+} levels in oocytes of *Xenopus* and *Ambystoma* ranged from 0.4 to $0.9 \times 10^{-6} M$. With the calcium electrode, a significant increase in free Ca^{2+} could be detected 10 min after progesterone addition, which was even earlier than the increase they observe with aequorin in the albino mutant (Fig. 4). The increases in free Ca^{2+} (which we estimate from the graphs to be less than twice the resting levels) are small compared with the 10-fold increases several hours later. These reached a maximum 2–3 hr before nuclear breakdown began, and declined to near resting levels as nuclear breakdown took place. By placing the Ca^{2+} electrode just outside the oocyte, they also made the important observation that Ca^{2+} efflux only occurred in the presence of external Ca^{2+}.

Both Mn^{2+} and *p*-hydroxymercuriphenylsulfonate, which induce nuclear breakdown in *Xenopus,* caused a transient flash from microinjected aequorin or obelin within seconds after exposure, which is comparable to the time of emission described by Wasserman *et al.,* above. Wasserman *et al.* (1980) did not, however, describe a continuing Ca^{2+} surge over the next few hours as reported by Moreau *et al.* (1980) and Vilain *et al.* (1980).

Important information has come from studies in which Ca^{2+} was introduced into the cytoplasm of the amphibian oocyte. Direct microinjection of Ca^{2+} into the *Xenopus* oocyte has no effect (Merriam, 1971; Moreau *et al.,* 1976). Moreau and his associates used a micropipette for Ca^{2+} iontophoresis into the cortex of the *Xenopus* oocyte, maintaining a constant current of 40 nA for 0.5–1 min. The iontophoresis of Ca^{2+}, but not of Mg^{2+} or K^+, induced nuclear breakdown. Iontophoresis of Ca^{2+}, into the endoplasm instead of the cortex did not induce breakdown (Moreau *et al.,* 1976). The outward-going iontophoretic current would have depolarized the *Xenopus* oocyte as well as opened voltage-dependent Ca^{2+} channels resulting in Ca^{2+} efflux. Studies on heart and skeletal muscle have been interpreted to suggest that the release of Ca^{2+} from the plasma membrane functions as a primer for release of larger amounts from internal sites of storage (cit. Langer, 1980). A similar relationship may explain the biphasic Ca^{2+} release seen in amphibian oocytes (Fig. 2).

We have suggested that the initial (1–2 min) Ca^{2+} release seen in *Rana* in response to progesterone and other meiotic agonists is from the oocyte surface, and that the second Ca^{2+} transient at 8–12 min reflects efflux of the "internal" release seen in *Xenopus* (Morrill *et al.,* 1981). The finding that La^{3+} induces a rapid Ca^{2+} release (Morrill *et al.,* 1980) also indicates that a primary action occurs at the oocyte plasma membrane, since La^{3+} has been shown to displace the exchangeable surface-bound Ca^{2+} in other systems and does not, itself, enter the cell (cit. Langer, 1980).

In *Rana* about 45% of the total prophase oocyte Ca^{2+} is associated with yolk platelets, very dense organelles which are rich in lipid and phosphoprotein. About 11% is associated with the pigment granules present in the cortex of the animal hemisphere (Morrill et al., 1971). Analysis of $^{45}Ca^{2+}$ exchange indicates that there are at least two Ca^{2+} components; a rapidly exchanging component (~5 μmol/liter cell water) with a $t_{1/2} = 5$ min, and a larger, slowly exchanging component (>2 mmol/liter) with an apparent $t_{1/2}$ of several days (Morrill et al., 1980). The earliest response to progesterone is the release of 5–10 μmol of the slowly exchanging component into the medium followed by a secondary release of 10–20 μmol from the oocyte 8–12 min later (Morrill et al., 1980). Thus, only a small fraction (<2%) of the total oocyte Ca^{2+} is apparently released from the oocyte; probably from ligands in the oocyte cortex. The identity of the Ca^{2+} binding site(s) is unknown. It seems unlikely, however, that this release could be from mitochondria for the following reasons. Mitochondrial calcium release is favored by a lowered pH and by cyclic AMP, but insulin inhibits it (Fiskum and Lehninger, 1982). The induction of oocyte meiosis is associated with an elevated pH_i and a fall in cyclic AMP (see Busa and Nuccitelli, 1984; Morrill et al., 1984c). Insulin is a good meiotic inducer (Baulieu et al., 1978). An alternate site of release is suggested by the affinity of isolated endoplasmic reticulum (ER) in several systems for Ca^{2+}, and by the proximity to the plasma membrane of ER subsurface cisterns (reviewed by Somlyo, 1984).

The rapid release of calcium seen for amphibian oocytes has parallels in invertebrate oocytes. In starfish, meiosis is reinitiated by a hormone produced by the follicle cells, which has been identified as 1-methyladenine (1-MeAde) (Kanatani, 1969; Hirai et al., 1973). Starfish oocytes injected with aequorin emit light within 2 sec following external application of 1-MeAde (Moreau et al., 1978). Plasma membrane-rich fractions prepared from isolated oocyte cortices also release Ca^{2+} less than 0.1 sec after hormone addition (Doree et al., 1978). Injection of EGTA into starfish oocytes supresses meiotic reinitiation if performed before or during the transient increase in free Ca^{2+} which follows 1-MeAde application, whereas meiosis can occur if EGTA is injected after the Ca^{2+}-transient (Moreau et al., 1978).

The results indicate that while there may be a hormone-induced calcium release within the oocyte cytosol, the initial release is localized, perhaps at the oocyte plasma membrane. In other words, the Ca^{2+} requirement for release of the prophase block seems to depend on the oocyte's ability to redistribute its Ca^{2+} rather than on the absolute level of intracellular free Ca^{2+}.

C. The Changes in Cyclic Nucleotides, Adenylate Cyclase, and Phosphodiesterase

Schatz and Morrill (1972) found that both caffeine and theophylline, which inhibit phosphodiesterase (PDE), and dibutryl cyclic AMP, which enters the oocyte, inhibited progesterone-induced meiosis in *Rana*. They observed later that exogenous Ca^{2+} was necessary to obtain this inhibitory effect (Morrill *et al.*, 1977). DBcAMP was also found to inhibit meiosis in mouse oocytes (Cho *et al.*, 1974), but no effect has been observed on meiosis in invertebrates (see review by Meijer and Guerrier, 1984b). Speaker and Butcher (1977) and Morrill *et al.* (1977) found a decrease of about 50% in endogenous cyclic AMP in progesterone-induced *Rana* oocytes. Maller and Krebs (1977) found that meiosis could be induced in *Xenopus* by injection of the regulatory subunit which inhibits cyclic AMP–dependent protein kinase. The magnitude of the cyclic AMP decrease in progesterone-treated *Xenopus* oocytes, first demonstrated by Maller *et al.* (1979), is very small. Mulner *et al.* (1983a) subsequently reported that if *Xenopus* cyclic AMP levels are increased by treatment of oocytes with cholera toxin or IBMX, it will be lowered significantly by progesterone.

Is the decrease in cyclic AMP due to an increase in PDE activity or to a decrease in adenylate cyclase activity? Mulner *et al.* (1979) found that progesterone inhibited cyclic AMP synthesis in *Xenopus* oocytes. Kostellow *et al.* (1980) measured a decrease in the rate of cyclic AMP synthesis following progesterone induction in *Rana,* but found no effect on its rate of hydrolysis, confirming that adenylate cyclase was the rate-limiting step in regulating oocyte cyclic AMP. Sadler and Maller (1981) and Finidori-Lepicard *et al.* (1981) reported that progesterone inhibited adenylate cyclase in a plasma membrane preparation from *Xenopus* oocytes. Thus, although PDE activity is essential for the induction of meiotic maturation, there is, as yet, no evidence that its activity increases following the meiotic stimulus.

Although PDE may not be activated during meiosis, calmodulin, first described by Cheung (1971) as an activator of PDE, is clearly involved.

D. The Role of Calmodulin

Even a transient, localized elevation in intracellular free Ca^{2+} levels to 10^{-6} M or more would cause calcium-binding proteins to form an active complex with calcium. Calmodulin is predominant in the homologous group of calcium-binding proteins which regulate critical cell processes (reviewed by Cheung, 1980), and it could not remain unaffected by the calcium transients associated with the reinitiation of meiosis. In a great variety of cell systems, calmodulin activates adenylate cyclase at submicromolar calcium concentrations, but cyclase activity is inhibited at higher micromolar concentrations (Bradham and Cheung, 1980; Mac Neil *et al.*, 1985). If oocyte adenylate cyclase resembles that of most other cell systems, the increased levels of free calcium resulting from the meiotic stimulus would be enough to inhibit its activity.

Kostellow *et al.* (1980) found that injected trifluoperazine, which binds to the Ca^{2+}–calmodulin complex (Levin and Weiss, 1977) was a competitive inhibitor for progesterone-induced nuclear breakdown, and blocked the progesterone-induced release of $^{45}Ca^{2+}$ from the *Rana* oocyte cortex during the first 1–2 min. However, trifluoperazine did not affect the *in situ* PDE activity, which is consistent with the findings of Wasserman and Smith (1981) concerning the effects of amphibian oocyte PDE.

Maller and Krebs (1980) found that the microinjection of purified bovine brain calmodulin initiated meiosis in some, but not all, batches of *Xenopus* oocytes. A low incidence of meiosis was reported for *Xenopus* oocytes which had been microinjected with calmodulin purified from ram testis (Huchon *et al.*, 1981). Mulner *et al.* (1983b) reported similar results with calmodulin from the same source, and found that anticalmodulin antibodies elevated cyclic AMP levels. Wasserman and Smith (1981) purified calmodulin from *Xenopus* oocytes. When their preparation was injected together with 10 μM Ca^{2+} (three times the measured K_d for oocyte calmodulin), up to 80% of the oocytes responded by undergoing nuclear breakdown. The exact amount of calmodulin injected was critical for obtaining a good response. They found that although the oocyte calmodulin stimulated bovine brain phosphodiesterase in a calcium-dependent manner, it had no apparent effect on ovarian PDE activity. This is consistent with the observations, cited above, that resumption of meiosis is associated with a decrease in adenylate cyclase activity rather than an increase in PDE activity. Mulner *et al.* (1983b) found that although microinjection of calcium–calmodulin could produce a decrease in cyclic AMP levels in *Xenopus* oocytes, injections of neither fluphenazine (which inhibits formation of the Ca^{2+}–calmodulin complex) nor anti-calmodulin antibodies could prevent the progesterone-induced fall in $[cAMP]_i$.

On the other hand, Allende's research group (cit. Orellana *et al.*, 1984) has described a cyclic nucleotide phosphodiesterase activity in *Xenopus* oocytes which could be stimulated by calcium–calmodulin. Miot and Erneux (1982) have confirmed their finding. This enzyme has been since purified about 2000-fold from *Xenopus* ovaries (Orellana *et al.*, 1984). It can be activated both by proteolysis, after which it is no longer calmodulin-dependent, and by certain phospholipids, in a Ca^{2+}-independent manner.

This work is clearly of major interest in determining the role of the early calcium transient in triggering the resumption of meiosis. At this time it seems impossible to propose a model for the role of calmodulin which would be consistent with all of the research reports.

E. The Role of Phosphatidylinositol and Protein Kinase C

A number of years ago Michell (1975) proposed that di- and triphosphoinositides may have a function in receptor systems in which the stimulus results either in an increase in intracellular free Ca^{2+} or in an increased permeability to Ca^{2+}

ions. An elevation of intracellular cyclic GMP is frequently associated with the rise in calcium (Michell, 1975). Phosphatidic acid derived from diacylglycerol has also been postulated to act as a Ca^{2+} ionophore (Tyson et al., 1976; Michell et al., 1977). Although the exact mechanisms have not been clearly established, they have led to several general hypotheses for calcium ion-linked receptors. According to one view, Ca^{2+} is replaced as a second messenger by inositol triphosphate and diacylglycerol, the two products released from membrane polyphosphoinositide as a consequence of receptor activation. The inositol triphosphate itself causes a release of intracellular Ca^{2+} (e.g., Streb et al., 1983), which modulates further cell reactions, but calcium is now a "third messenger". In the system described by Nishizuka, the diacylglycerol and Ca^{2+} in turn activate protein kinase C. In this model, activation of the C-kinase is a prerequisite for modulating cell function, and it acts synergistically with mobilized Ca^{2+} to elicit the full physiological cellular responses (see review by Nishizuka et al., 1984).

Several observations support the hypothesis that inositol phospholipid turnover is concerned in the hormonal stimulation of meiosis. Schorderet-Slatkine et al. (1977) reported some years ago that Gammexane, an agent known to block phosphatidyl inositol synthesis, inhibits progesterone-induced nuclear breakdown in Xenopus oocytes in a dose dependent manner. Phospholipid-interacting drugs such as trifluoperazine, chlorpromazine, dibucaine, and fluphenazine, have been found to be meiotic inhibitors. Their effect has been attributed to their known inhibition of calmodulin. They also strongly inhibit protein kinase C (Nishizuka et al., 1984). A specific inhibitor for C kinase has never been described; at present the data on calmodulin inhibitors in oocytes might just as well be applied to C kinase. More recently Picard and Doree (1983) have shown that microinjected LiCl reversibly inhibits 1-MeAde-induced meiotic maturation of starfish oocytes. Li^+ has been shown to inhibit phosphoinositide turnover, probably by inhibition of the enzyme that catalyzes the removal of the phosphate from inositol 1-phosphate. In the same species, Meijer and Guerrier (1984b) report that arachidonic acid (which can be derived from diacylglycerol) mimics 1-MeAde in inducing nuclear breakdown, and find that the presence of Ca^{2+} in the external medium facilitates induction by arachidonic acid.

Reinitiation of meiosis by the ionophore A23187 in Xenopus oocytes (Wasserman and Masui, 1975) might be explained in terms of an activation of protein kinase C. In Xenopus oocytes, addition of 5 μM A23187 required elevated levels (10–40 mM) of either calcium or magnesium for induction. Nishizuka (1984) has reported that at concentrations greater than 0.5 μM A23187 will itself cause the phosphorylation of a 40 kDa protein due to the nonspecific activation of phospholipases and phosphokinase C by a large increase in Ca^{2+} concentration.

Although much of the evidence linking meiotic events to C kinase and/or phosphoinositide turnover is circumstantial, the possibilities are intriguing enough to justify further studies.

F. The Role of Arachidonic Acid

Stimulation of the receptors that are related to inositol phospholipid turnover often releases arachidonic acid, and often increases cyclic GMP but not cyclic AMP (for reviews, see Michell, 1975; Michell et al., 1977). Arachidonic acid peroxide and prostaglandin endoperoxide may serve as activators for guanylate cyclase (Hidaka and Asano, 1977; Graff et al., 1978). The report that arachidonic acid and Ca^{2+} induce nuclear breakdown in star fish oocytes (Meijer and Guerrier, 1984b), whereas cyclic GMP and Ca^{2+} induce nuclear breakdown in *Rana* oocytes (Kostellow and Morrill, 1980) suggests that Ca^{2+} mobilization, arachidonic acid release, and cyclic GMP increase may be integrated into a single receptor cascade system.

Meijer and Guerrier (1984b) have suggested that in the starfish, 1-MeAde induces the activation of a plasma membrane phospholipase responsible for arachidonic acid release from the oocyte plasma membrane. In support of this hypothesis, they found that inhibitors of phospholipase A_2 (quinacrine, bromophenacyl bromide) inhibit 1-MeAde-induced nuclear breakdown, and that phospholipase A_2 from bee venom will induce nuclear breakdown.

Epel *et al.* (1982) have also proposed a role for arachidonic acid in fertilization of the sea urchin egg. In their model, a phospholipase in sperm degrades phospholipid and releases arachidonic acid. This is oxidized to peroxy compounds that induce Ca^{2+} release from membranes. They suggest that lipid peroxides act as calcium ionophores; alternatively, arachidonic acid peroxide may serve as an activator of guanylate cyclase.

G. The Role of Calcium in Regulating Intracellular pH

The hormonal stimulus at prophase and the subsequent stimulus of fertilization are each followed by a 0.15–0.3 unit rise in intracellular pH. Cytoplasmic alkalinization following progesterone induction was first demonstrated in *Xenopus* oocytes with pH-sensitive microelectrodes by Lee and Steinhardt (1981). This has been confirmed by Cicirelli et al. (1983) and by Houle and Wasserman (1983) using [^{14}C]dimethyloxozolidine dione (DMO) in *Xenopus* and by ^{31}P-NMR by Morrill et al. (1983, 1984c) in *Rana*. In general, cytoplasmic alkalinization began shortly after exposure to progesterone and reached a maximal value prior to nuclear breakdown.

Several observations indicate that the alkalinization of the oocyte cytoplasm following hormone stimulus may be more directly related to $Ca^{2+}:H^+$ exchange than to $Na^+:H^+$ exchange. Agents that block Ca^{2+} release following the meiotic stimulus (procaine, phenylmethylsulfonyl fluoride) also block progesterone-induced alkalinization (Morrill et al., 1983, 1984c). As shown in Fig. 5, removal of extracellular Ca^{2+} produces a transient rise in pH_i, whereas the subsequent addition of progesterone leads to a sustained elevation of pH_i. The progesterone-induced rise in pH_i coincides with the release and efflux of bound Ca^{2+} (see Figs. 2–4) and can be mimicked by the Ca^{2+},Mg^{2+}-ionophore A23187. Furthermore, if Ca^{2+} permeability of the plasma membrane is increased by A23187 in the presence of elevated (10 mM) external Ca^{2+}, pH_i rises and remains elevated (Fig. 5). Carboxylic acid antibiotics like A23187 reportedly catalyze an electroneutral exchange of endogenous alkali metal ions for extramitochondrial protons (e.g., Pressman, 1969). The A23187-induced rise in pH_i could therefore be due to $Ca^{2+} : H^+$ exchange, with Ca^{2+} moving down its chemical gradient into the cell in exchange for protons. However, if the ionophore is added to Ca^{2+}-free medium and the Ca^{2+} leaving the cell is trapped by EGTA, pH_i rises as Ca^{2+} is released from intracellular stores (Morrill et al., 1980; Kostellow et al., 1980), suggesting a $Ca^{2+} : H^+$ cotransport.

An A23187-dependent $Ca^{2+} : H^+$ exchange may play a role in the reinitiation of meiosis under some circumstances. For example, Xenopus oocytes can be induced to resume meiosis by treatment with Barth's medium at high pH (Kofoid et al., 1979; Robinson, 1979). A23187-induced alkalinization may explain why Xenopus oocytes will resume the first meiotic division in the presence of A23187 and high Mg^{2+} concentrations (Wasserman and Masui, 1975). In each case, a

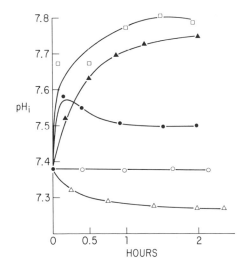

Fig. 5. Changes in intracellular pH in *Rana* follicles as a function of time after exposure to 1.6 μM progesterone (▲), 5 μM A23187 in Ringer's solution containing 10 mM Ca^{2+} (□), Ca^{2+}-free Ringer's solution containing 1.0 mM EGTA (●), and 1.6 μM progesterone treatment of follicles pretreated with 1.0 mM phenylmethylsulfonyl fluoride (PMSF) (△). The control is represented by (○). pH_i was measured by ^{31}PNMR. Sampling time was 4–8 min. Adapted from Morrill et al. (1984c).

rise in pH_i appears to be associated with an increase in intracellular free Ca^{2+}, although the mechanism for a Ca^{2+}-regulated proton exchange is not obvious.

It is important to note that the pH_i continues to rise for 2–3 hr in both *Rana* and *Xenopus*. Lee and Steinhardt (1981) have shown that microinjection of cytoplasm containing maturation promoting factor will also produce cytoplasmic alkalinization. Thus, different mechanisms may be involved in the short-term (0–15 min) alkalinization of the cytoplasm and its long-term (0.25–3 hr) maintenance. After the first 10–15 min, H^+ efflux (alkalinization) may be linked to the progesterone-stimulated Ca^{2+} uptake (see Section VI,A). Mechanisms for a net Ca^{2+} increase and H^+ efflux over a period of hours might include: (1) the calmodulin-activated Ca^{2+}-ATPase which catalyzes an electroneutral $Ca^{2+} : H^+$ exchange in reconstituted erythrocyte membrane vesicles (Carafoli *et al.*, 1982) and/or (2) the $Ca^{2+} : H^+$ antiporter which has been proposed as a mechanism for calcium entry into liver cells (Racker, 1980).

H. The Role of Serine Proteases

A direct activation of C-kinase by the Ca^{2+}-dependent serine proteases may also play a role in the resumption of meiosis (see section V, E and Addendum). Protein kinase C was first found as a proteolytically activated protein kinase, and later shown to be a Ca^{2+}-activated, phospholipid-dependent enzyme (reviewed in Nishizuka *et al.*, 1984). It was first shown by Peaucellier (1977) that serine proteases induce the initiation of meiosis in oocytes of the annelid *Sabellaria*. Three serine proteases were isolated from *Sabellaria* body fluids and two of them were active in concentrations as low as 0.2 nM. Commercial serine proteases were also active, but at doses at least three orders of magnitude greater than those isolated from *Sabellaria* fluids. Subsequent studies demonstrated that serine protease inhibitors blocked nuclear breakdown both in *Xenopus* (Guerrier *et al.*, 1978) and in *Rana* (Morrill *et al.*, 1983) and prevented the progesterone-induced rise in intracellular pH in *Rana* (Morrill *et al.*, 1983). As shown in Fig. 6, the serine protease inhibitor phenylmethylsulfonyl fluoride (PMSF) prevented progesterone-induced release of oocyte $^{45}Ca^{2+}$. Inhibition of both membrane and nuclear events occured only if protease inhibitors (PMSF, leupeptin, etc.) were added prior to the inducing agent. No effect was seen if the inhibitor was added 5 min later, and the inhibitors alone had minimal effect on membrane potential, phosphocreatine levels, and intracellular pH (Morrill *et al.*, 1983). These results suggest that (1) stimulation of protease activity occurs within the first few minutes after exposure to steroid, and (2) these protease inhibitors are not themselves toxic to the oocyte. Figure 7 shows the activity of a trypsin-like protease present in isolated plasma membranes prepared from prophase-arrested *Rana* oocytes. It is stimulated by either progesterone or insulin. This trypsin-like activity requires Ca^{2+} and has a pH optimum of about 7.8. Thus, it is tempting to speculate that an initial release of free Ca^{2+} by the meiotic agonists results in an alkalinization

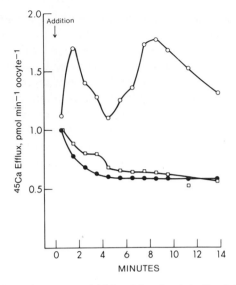

Fig. 6. Effect of a serine protease inhibitor (phenylmethylsulfonyl fluoride, PMSF) on progesterone-induced $^{45}Ca^{2+}$ efflux from denuded *R. pipiens* oocytes. For efflux studies oocytes were preloaded with ^{45}Ca for 24 hr at 16°C and rinsed for 5 min, and 1.6 μM progesterone (○) or progesterone plus 100 μM PMSF (□) was added at the time indicated. The control is represented by (●).

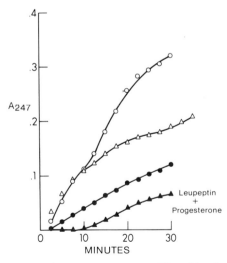

Fig. 7. Effect of two meiotic inducers, progesterone (○) and insulin (△), on serine protease activity in isolated plasma membranes from prophase arrested *Rana* oocytes. *p*-Tosyl-1-arginine methyl ester (TAME) was used as a trypsin-specific substrate and the assay (Hummel, 1959) was carried out in 0.2 M Tris buffer at pH 8.1 containing 10 mM $CaCl_2$. Addition of 0.1 mM leupeptin (a protease inhibitor) 10 min prior to the hormone (▲) blocked protease activation. The control is represented by (●).

of the oocyte cortex and activation of membrane protease, with a further release of oocyte Ca^{2+} and activation of C-kinase.

Proteases may also be involved in egg activation at fertilization. In sea urchin eggs, a trypsin-like protease is present in the cortical granules and is released when the eggs are fertilized (cit. Brachet, 1976). Wittenberg *et al.* (1978) subsequently reported the existence of a powerful trypsin inhibitor in the cytoplasm and yolk of *Rana* eggs and embryos and found that cytoplasmic levels of this inhibitor increased between the hormonal stimulus and metaphase II arrest, and then decreased abruptly following fertilization. They propose that fluctuations in free Ca^{2+} levels control the distribution of inhibitor between cytoplasm and yolk platelets as "a device for controlling limited proteolysis in the cytoplasm." Wittenberg *et al.* (1978) suggested that this limited proteolysis was involved in the unmasking of mRNA, but more recent findings suggest that the egg protease may equally well activate C-kinase.

VI. CALCIUM REDISTRIBUTION ASSOCIATED WITH NUCLEAR BREAKDOWN AND ARREST AT METAPHASE II

A. The Effect of Free Calcium on Maturation Promoting Factor

An adequate maturation-inducing stimulus initiates a series of metabolic changes in the oocyte (outlined for the amphibian *Rana* above, in Fig. 1, and Section III). In *Rana,* cycloheximide inhibition of protein synthesis at any time during the first 4–5 hr after exposure to the hormone will block maturation. Masui and Markert (1971) found that a "maturation promoting factor" (MPF) appeared in the cytoplasm of *Rana* oocytes about 4 hr after the meiotic stimulus. The transfer of cytoplasm containing MPF could initiate meiosis in a prophase-arrested oocyte, without an obligatory period of protein synthesis. (Serial transfers indicate that, once formed, MPF is autocatalytic.) MPF, a protein with a molecular weight of about 100,000 has since been found to be associated with mitotic activity in cells from species ranging from yeast to mammals.

Stimulated, then, by MPF, the oocyte completes the first meiotic division, discards half of its chromosomes into a small polar body, and immediately begins the second meiotic division, which continues until metaphase. There it becomes arrested, and can proceed no further until the oocyte is fertilized or "activated" by experimental means. Masui and Markert (1971) found that following nuclear membrane breakdown in *Rana,* the cytoplasm developed a second factor, called "cytostatic factor" (CSF) capable of arresting mitosis at metaphase when injected into two-cell embryos (reviewed in Masui *et al.*, 1980). Masui proposed that CSF was responsible for the arrest of meiosis at the second metaphase.

Changes in intracellular calcium are clearly critical in controlling the activity of both MPF and CSF. Both of these factors are extremely sensitive to calcium and are rapidly inactivated by micromolar concentrations of $CaCl_2$. ($MgCl_2$ stabilizes both factors.) Fertilization or egg activation by various experimental techniques is, in all species, accompanied by the egg's release of sequestered calcium (see Section VII). Masui has proposed that the consequent inactivation of CSF permits meiosis to proceed to completion after fertilization (reviewed in Masui et al., 1980). Newport and Kirschner (1984), in studies on the cell cycle in early Xenopus embryos, present evidence that CSF acts through stabilizing MPF, which, acting alone, can bring the cell cycle to metaphase and arrest it there. They agree that a rise in intracellular free calcium inactivates CSF and leads to the disappearance of MPF activity.

B. The Decrease in Free Calcium

Are the changes in oocyte and egg calcium distribution before fertilization consistent with this hypothesis? On the whole, they seem to be. Calcium electrode determinations of free calcium in amphibian oocytes by Moreau et al. (1980) (Fig. 4) showed that free calcium began to decline sharply several hours before nuclear breakdown (the aequorin-measured decline is less sharp, because it represents an average of oocytes which do not undergo nuclear breakdown quite synchronously). By this time, maturation in Xenopus oocytes can no longer be inhibited by injected EGTA (Masui et al., 1977). Morrill and his associates (1971, 1974) found that the oubain-sensitive, Na^+,K^+-activated ATPase, which was the principle ATPase of the uninduced prophase-blocked oocyte, began to disappear after hormone induction. Before the first meiotic division occurred, it was replaced by a Ca^{2+},Mg^{2+}-activated ATPase, an enzyme known to be stimulated by elevated free intracellular calcium and calmodulin. At the same time, there was an increase in calcium binding by a cortical granule fraction (Morrill et al., 1971; Morrill and Murphy, 1972). Gardiner and Grey (1983) have described subsurface cisternae in Xenopus oocytes which may also be a location for sequestered calcium. These cisternae increase 10-fold by second metaphase. Both an increase in calcium sequestration and the active calcium pump would serve to lower the level of free calcium in the oocyte and to stabilize MPF and CSF.

C. The Effect of Free Calcium on Microtubules

Formation of the spindle apparatus for the meiotic division must be sensitive to free Ca^{2+} levels. Microtubules in vitro can be induced to depolymerize with micromolar amounts of calcium (Weisenberg, 1972). In experiments with the isolated mitotic apparatus from sea urchin, Silver and his associates (1980) presented data to establish that the vesicles associated with the mitotic apparatus sequester calcium (as originally postulated by Harris, 1975). Their preparations used ATP as an energy source, and the activity seemed similar to that of the Ca^{2+},Mg^{2+}-ATPase of the sarcoplasmic reticulum in muscle. These vesicles would provide an additional storage site for calcium. Deery *et al.* (1984) have described the nature of calmodulin association with both the cytoplasmic microtubular complex and with centrosomes in 3T3 and SV3T3 cells. They found that low levels of free calcium allowed calmodulin to associate with microtubules without inducing disassembly.

D. The Change in Transmembrane Potential

Mammalian and amphibian oocytes undergo a progressive decrease in transmembrane potential which begins prior to breakdown of the nuclear membrane (Moreau *et al.*, 1976; Ziegler and Morrill, 1977; Wallace and Steinhardt, 1977; Kado *et al.*, 1981; Powers, 1982). We have found that in *Rana*, a decrease in Ca^{2+} permeability occurs at the same time as the decrease in transmembrane potential (Morrill *et al.*, 1984a). This decrease in permeability may be causally related to the previous increase in intracellular free calcium in a manner analogous to that proposed for inactivation of voltage-dependent Ca^{2+} channels in a variety of cell types (reviewed in Tsien, 1983). Brehm *et al.* (1980) have proposed that intracellular Ca^{2+} accumulates as a result of Ca^{2+} entry and produces a specific inhibitory effect on Ca^{2+} channel conductance that appears as inactivation. The hypothesis states that free intracellular Ca^{2+} is a necessary as well as sufficient cause of channel inactivation; a more restricted version states only that free intracellular Ca^{2+} plays an important role, but does not exclude the participation of membrane depolarization.

E. The Effect of Calcium on Oocyte Fluid Phase Turnover and Membrane Recycling

We have measured endocytotic activity in the *Rana* oocyte using [³H]inulin as an index of membrane and fluid uptake and found that an area equivalent to the entire surface of the prophase oocyte is recycled several times an hour (Morrill *et al.*, 1983, 1984a). About the time of nuclear breakdown in *Rana*, membrane capacitance measurements indicated that there was a net loss of plasma membrane surface area (Weinstein *et al.*, 1982) which coincided with the disappearance of membrane Ca^{2+} permeability, with the disappearance of the Na^+-pump ATPase, and with ouabain binding to the oocyte plasma membrane (Weinstein *et al.*, 1982; Morrill *et al.*, 1984b). Ovulation occurs shortly after nuclear breakdown and is associated with the disappearance of oocyte microvilli and separation of the oocyte's Ca^{2+},Mg^{2+}-dependent contact with follicle cells. Receptor-mediated endocytosis of vitellogenin seemed to disappear at about the same time in *Xenopus* oocytes (Schuetz *et al.*, 1974) as did insulin-induced endocytosis in *Rana* oocytes (Morrill *et al.*, 1984a). These changes in membrane area, membrane permeability, and response to external ligands must alter Ca^{2+} uptake and exchange and reflect a programmed internalization of plasma membrane components.

F. The Increase in Total Calcium

Total calcium continues to rise up to the time of nuclear breakdown in the *Rana* oocyte, increasing 2- to 3-fold between hormone stimulus and second metaphase (Morrill *et al.*, 1971). In *Xenopus*, calcium accumulation nearly doubles before nuclear breakdown (O'Connor *et al.*, 1977). Batta and Knudson (1980) showed that the total calcium in cumulus-enclosed rat oocytes increases about 3-fold by nuclear breakdown.

In summary, the decrease in free Ca^{2+} which occurs before the first meiotic division would serve to promote microtubule stability and to stabilize CSF, the cytoplasmic factor essential for metaphase II arrest. The increase in total calcium preloads the oocyte for the extensive calcium release which occurs at fertilization.

VII. CALCIUM REDISTRIBUTION FOLLOWING FERTILIZATION

From the time that the egg membrane reacts to contact with the sperm, the cortex of most species of vertebrates and echinoderm eggs show signs that the egg has been "activated". (The term "activation" is often used to describe the experimental initiation of these changes by artificial membrane-perturbing stimuli, such as pricking with a needle or the application of chemical agents.) Eggs from most animal species respond to activation by visible changes in the cortex. Cortical granules or alveoli disappear, material is secreted into the space between the plasma membrane and its envelopes, and cortical pigments migrate. In some species the egg membrane suddenly depolarizes in a manner similar to an action potential. Free calcium rises dramatically.

A. Exogenous Calcium Requirements

There is no clear evidence that the egg requires exogenous Ca^{2+} and/or Mg^{2+} to become activated. Vertebrate sperm, however, require calcium to undergo the acrosome reaction which permits fusion of the egg and sperm membranes (Dan, 1954). If sea urchin eggs are fertilized in and remain in calcium-free sea water, they will progress normally to the first cell division (reviewed in Epel, 1982). When starfish oocytes are pretreated with calcium-free sea water, the fertilization membrane rises without having been exposed to the normal inducer, 1-methyladenine. It should be pointed out that, in starfish, calcium-free sea water partially mimics one step (hormone stimulus) necessary for developing a fertilizable egg (reviewed in Meijer and Guerrier, 1984b). In contrast to the salinity of the milieu of marine fish and invertebrates, fertilization of many fish and amphibian eggs takes place in the nearly ion-free water of streams and lakes, although small amounts of monovalent and divalent cations may be present both in the semen and in the jelly coat surrounding the freshly laid egg.

Probably the best evidence that activation, per se, does not require external divalent cations is the report by Steinhardt and Epel (1974) that the ionophore A23187 will activate both amphibian and hamster eggs in a Ca^{2+},Mg^{2+}-free medium. The rate of cortical granule breakdown and pronuclear development in Ca^{2+},Mg^{2+}-free media was, in fact, comparable to the rate of sperm-activated hamster eggs. However, Wolf (1974) found that exogenous calcium was necessary to elicit the cortical response (cortical granule breakdown) in *Xenopus* eggs when pricked with forceps. The effectiveness of this stimulus may depend on a response to Ca^{2+} entering the site of injury.

B. The Change in Membrane Potential and Calcium Release

The earliest measurable response to sperm–egg interaction is a transient change in the egg transmembrane potential, known as the fertilization potential. This electrical change is often seen as a very rapid depolarization almost coinciding (<1 min) with initial sperm–egg contact and was first described in starfish eggs by Tyler *et al.* (1956), and subsequently for fish (Hori, 1958), frog (Morrill and Watson, 1966), sea urchin (Steinhardt *et al.*, 1971; Ito and Yoshiaka, 1972), and only recently for mammalian eggs (Miyazaki and Igusa, 1981). This fertilization potential is similar to the ''activation potential'' seen upon pricking the egg with a needle (e.g., Maeno, 1959) or treatment with agents such as benzyl alcohol (Eusebi and Siracusa, 1983).

In studies with *Rana,* Morrill *et al.* (1971) found that fertilization triggers a major calcium efflux within the first few minutes. The changes in membrane potential and intracellular Ca^{2+} and Mg^{2+} are illustrated in Fig. 8. Sperm penetration caused a positive-going fertilization potential and a rapid efflux of both Ca^{2+} and Mg^{2+} during the rising phase of the fertilization potential. Intracellular Ca^{2+} levels fell below that present in the prophase-arrested oocyte.

Some of the same experimental stimuli which reinitiate meiosis from its arrest in prophase I will activate the ovulated vertebrate egg which is arrested in metaphase II. Fulton and Whittingham (1978), using mouse oocytes, have shown that iontophoresis of Ca^{2+}, but not of Na^+ or Mg^{2+}, induces the normal sequence of events known to follow fertilization: cortical granule breakdown, release of the oocyte from the block at metaphase II, and subsequent cell divi-

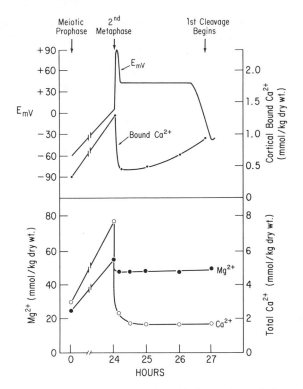

Fig. 8. Changes in membrane potential, calcium binding to the cortical granule fraction, and total calcium and magnesium levels between the terminal stage of first meiotic prophase (zero time on the abscissa) and first cleavage in *R. pipiens*. Fertilization occurred after 24 hr. Membrane potentials and Ca^{2+} and Mg^{2+} levels are average values for five or more frogs. Adapted from Morrill *et al.* (1971; Morrill and Murphy, 1972).

sion. Fulton and Whittingham used a total iontophoretic current in the range 100–150 nC, which would have, as noted by the authors, produced a large depolarization of the oocyte plasma membrane potential (from -14 mV to 170–350 mV, inside positive). Similarly, Cross (1981) found that iontophoretic injection of Ca^{2+} activates *Rana* eggs to develop and initiates a transient, positive-going shift in the membrane potential (the activation potential) which is like the sperm-induced fertilization potential in amplitude, duration, and Cl^- dependence. Injection of K^+, Na^+, Cl^-, or Mg^{2+} does not result in activation potentials, but the Ca^{2+} analogs, Sr^{2+} and Ba^{2+}, do substitute for Ca^{2+}. Cross (1981) concluded that the sperm induces an increase in intracellular Ca^{2+} which either "directly or indirectly opens Cl^- channels which are already present in the egg." Young *et al.* (1984) have incorporated plasma membrane fractions from *Xenopus* eggs and oocytes into lipid bilayers and demonstrated anion-selective channels that are activated by micromolar calcium concentrations. They found similar channels in membranes from both prophase-arrested oocytes and ovulated, unfertilized eggs and suggest that an increase in free Ca^{2+} directly activates these channels, and that the resultant Cl^- efflux (Morrill *et al.*, 1971) forms the ionic basis of the fertilization potential.

C. Intracellular Free Calcium Transients and Calcium Release

Early studies with marine invertebrates led to the concept that fertilization caused "a release of calcium from organic combinations in the cortex of the cell into the main body of protoplasm" (Mazia, 1937). Mazia (1937) did, in fact, demonstrate a 15% decrease in "bound" calcium following fertilization of sea urchin eggs.

A transient increase in intracellular free calcium following fertilization has been demonstrated in eggs of the mouse (Cuthbertson *et al.*, 1981), *Xenopus* (Busa and Nuccitelli, 1985), medaka fish (Ridgeway *et al.*, 1977; Gilkey *et al.*, 1978), and sea urchin (Steinhardt *et al.*, 1977) in studies with microelectrodes and with the Ca^{2+}-sensitive photoprotein, aequorin.

Sea urchin eggs were injected with aequorin, then the pooled eggs were fertilized simultaneously (Steinhardt *et al.*, 1977). Although the eggs are extremely small, this technique allowed the investigators to monitor luminescence, which increased shortly after fertilization at about the time the cortical reaction was observed. Luminescence remained elevated for several minutes, and decreased by the time the cortical reaction was complete.

The most spectacular demonstration of calcium release has been shown in the large egg of the medaka fish. Ridgeway *et al.* (1977) described an "explosion" of free calcium following fertilization of aequorin-injected eggs. These investigators later used image intensifier techniques and were able to observe that the luminescence began at the point of sperm entry and moved across the egg as a narrow band of light (Gilkey *et al.*, 1978). The wave of free calcium disappeared within minutes, indicating that the calcium had either left the cell or that it had been resequestered.

Cuthbertson *et al.* (1981) reported that in aequorin-injected mouse eggs the free Ca^{2+} rose exponentially from a resting level below 0.1 μM to nearly 5 μM within 2–3 min after exposure to a chemical activating stimulus (ethyl or benzyl alcohol). Free calcium remained elevated for 10–20 min. In contrast, activation with sperm produced a series of Ca^{2+} transients followed by a dramatic Ca^{2+} transient after 1.5–2.5 hr. Using intracellular microelectrodes to measure the membrane potential, Miyazaki and Igusa (1981) found that a single sperm produced a series of recurring membrane hyperpolarizations in the hamster egg, which they attributed to increased K^+ permeability produced by an increase in intracellular free Ca^{2+}. The first hyperpolarization appeared at the time that the sperm first made contact with the egg, presumably as fusion began. The recurring membrane transients for both membrane potential and free calcium changes indicate that there may be a series of coupled electrical-calcium events in mammalian eggs at fertilization and that calcium release in mammals differs from the single explosive event seen in the eggs of lower species.

The aequorin technique can, at best, only yield estimates for peak levels and usually cannot address localized Ca^{2+} changes (i.e., cortical ones). Busa and Nuccitelli have applied Ca^{2+}-selective microelectrodes to monitor free-Ca^{2+} changes in fertilized *Xenopus* eggs (Fig. 9). An electrode in the animal hemisphere showed that free Ca^{2+} increased from 0.4 to 1.2 μM over the course of 2 min following fertilization, and returned to its original value during the next 10 min. No further changes in Ca^{2+} were detected through the first cleavage division. In eggs impaled with two Ca^{2+} electrodes, the Ca^{2+} pulse was observed to travel as a wave from the animal to the vegetal hemisphere. The apparent delay between the start of the membrane potential change associated with fertilization and initiation of the Ca^{2+} wave at the sperm entry site was about 1 min. As described for both sea urchin and medaka, the timing and magnitude of the Ca^{2+} pulse coincided with exocytosis of the cortical granules.

The high levels of intracellular free calcium would be expected to inactivate MPF and CSF (as discussed in Section VI,A), and in fact MPF activity has been shown to decline immediately after fertilization (Masui and Clarke, 1979). This permits the vertebrate egg to complete its second meiotic division.

In addition to Ca^{2+} release both from and within the egg, several investigators have shown that fertilization is also accompanied by large changes in Ca^{2+} influx. In sea urchin eggs, isotope studies show two phases of Ca^{2+} uptake (Azarnia and Chambers, 1976; Johnson and Paul, 1977; Paul and Johnson,

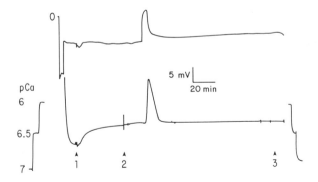

Fig. 9. Membrane potential (top) and animal hemisphere $[Ca^{2+}]_i$ in a fertilizing *Xenopus* egg. At 1, chlorobutanol was removed from the bathing medium. At 2, minced testis were added. At 3, cleavage furrow formation began. Traces at the far left and right (bottom) are Ca^{2+} electrode calibrations at the indicated pCa levels. From Busa and Nuccitelli (1985).

1978). The first begins almost as soon as sperm are added and ends 30–60 sec after insemination (Paul and Johnson, 1978). The second phase begins about the time of formation of an extracellular coat referred to as the hyaline layer. Epel (1982) has suggested that the "uptake" may be actually a binding of extracellular Ca^{2+} to this layer.

D. Calcium Efflux and Intracellular Sites of Calcium Binding

Increased calcium efflux at fertilization has been demonstrated with sea urchin eggs prelabeled with $^{45}Ca^{2+}$. For example, fertilization or activation with the ionophore A23187 resulted in a large efflux of calcium from the egg which lasted for 30 min (Steinhardt and Epel, 1974). This efflux can be of considerable magnitude; if one corrects for loosely associated calcium, there is actually a net decrease in total egg calcium (Azarnia and Chambers, 1976). As shown by Morrill *et al.* (1971), about 65% of the Ca^{2+} bound to the cortical pigment granules leaves the *Rana* egg in the first few minutes following fertilization; overall there is a net loss in egg Ca^{2+} during fertilization. This Ca^{2+} release is probably from the egg cortex and from other internal sources in the egg (Morrill *et al.*, 1971). Jaffe (1983) has proposed that these internal sources are "specialized regions of the egg's endoplasmic reticulum analogous to those in muscle cells." The main basis of this suggestion was the correlation between the observations that (1) in some (and perhaps most) muscle cells, natural contraction is mediated by "regenerative" or calcium-stimulated calcium release from subsurface cisternae, and (2) similar subsurface cisternae have been described in *Xenopus* eggs (Gardiner and Grey, 1983). These cisterane are 2–3 times more concentrated in the cortex of the animal hemisphere than in the cortex of the vegetal hemisphere; this distribution might contribute to the greater depolarization wave speed and excitability of the animal hemisphere. They become about 10-fold more abundant prior to metaphase arrest, disappear locally within a minute after the exocytotic wave at fertilization reaches their vicinity, and show some continuities with deeper portions of the endoplasmic reticulum.

The experimental results described above are consistent with the assumption that at least three early events—Ca^{2+} release from the egg cortex, reversal of the membrane potential, and opening of Cl^- channels—are responsible for release of the metaphase block. The fact that iontophoresis of Ca^{2+}, but not Mg^{2+}, into the egg cortex elicits the activation potential (Cross, 1981) indicates that a localized change in Ca^{2+} is the primary event. That observation, combined with the report that Cl^- channels in the metaphase egg are voltage-dependent (Schlichter, 1983), suggests that in the simplest model the membrane perturbing stimulus (sperm) displaces Ca^{2+} from membrane sites which in turn produces a partial membrane depolarization, the opening of Cl^- channels, further depolarization, and release and/or efflux of Ca^{2+}. As suggested by Jaffe (1983), activation could be a regenerative process similar to that seen in muscle (e.g., Fabiato, 1982).

E. The Role of Calcium in Regulating Intracellular pH

The large transient increase in intracellular free Ca^{2+} that occurs within the first few minutes after fertilization is associated with an increase in intracellular pH in both invertebrate (Johnson et al., 1976; Johnson and Epel, 1981, 1982; Shen and Steinhardt, 1978, 1979; Payan et al., 1983) and vertebrate eggs (Webb and Nuccitelli, 1981, 1982; Morrill et al., 1984c). In sea urchin eggs, this pH rise requires a small amount of external Na^+ (Johnson et al., 1976; Shen and Steinhardt, 1979), whereas Xenopus eggs can be perfused for at least 2 hr in Na^+-free solution with 1.0 mM amiloride and still exhibit a pH_i rise at fertilization (Webb and Nuccitelli, 1981).

Activation of the Xenopus egg can be induced by A23187 in a Ca^{2+},Mg^{2+}-containing medium. Activation by ionophore is associated with increased pH_i (Webb and Nuccitelli, 1981, 1982). As discussed above, A23187 may be a $Ca^{2+}:H^+$ ionophore, exchanging external Ca^{2+} and/or Mg^{2+} for intracellular protons.

Several workers have tried to determine whether the pH_i change plays a role in egg activation. Gilkey (1983) injected unfertilized medaka (fish) eggs, which are very large, with pH buffers to produce short-term, localized increases and decreases in cytoplasmic pH. The eggs were then fertilized at various times after injection. Unfertilized and fertilized eggs were also exposed to media containing either NH_4Cl or CO_2 to produce long-term, more extensive changes in cytoplasmic pH. These treatments neither activated the eggs nor interfered with the normal development of fertilized eggs, indicating that even though a natural change in cytoplasmic pH is induced by activation, it has no apparent role in medaka egg activation.

Lee and Steinhardt (1981) proposed that alkalinization of the oocyte cytoplasm is "a form of metabolic insurance and that there may be both pH-dependent and pH-independent pathways" associated with the meiotic divisions. More recently, Busa and Nuccitelli (1984) have suggested that pH_i and Ca^{2+} changes can be interdependent, as in the pronounced pH-dependence of Ca^{2+} binding to calmodulin. They propose that "pH_i functions as a synergistic messenger, providing a metabolic context within and through which the actions of other effectors are integrated."

As we have shown in Rana, a second fall in intracellular cyclic AMP coincides with the rising phase of intracellular pH. When unfertilized and fertilized eggs from the same female were compared, cyclic AMP decreased by about one-third within the first 30 min. following fertilization (Morrill et al., 1977, 1984b). The significance of changes in cyclic nucleotide metabolism is discussed in Section V,C.

VIII. SUMMARY

Meiosis is actually equivalent to two cell divisions, in each of which all of the cytoplasm remains with one daughter cell and the nucleus of the other cell is discarded. Although the oocyte appears to represent an extreme case of terminal differentiation, its final product is a totipotent cell. The prophase oocyte is, in a sense, produced at leisure. It has synthesized a tetraploid supply of DNA, stored up ribosomes complete with messages, and been packed with energy reserves in the form of yolk platelets. It represents a kind of model of a cell poised for division. When the signal arrives, it needs to carry out only the last essential steps. It is remarkable that almost all of these steps are either calcium dependent or are associated with a change in free intracellular calcium.

Free and bound calcium undergo a sequence of changes during the meiotic divisions. The stimulus which reinitiates meiosis triggers a rapid release of bound calcium from the oocyte membrane and internal sites. Some calcium is lost, but both bound and free calcium in the oocyte increase during the next few hours. The level of free intracellular calcium decreases just before nuclear breakdown, but total calcium remains high. Fertilization triggers an even greater calcium release; in sea urchins and amphibians there is a net loss compared with the prophase oocyte. The calcium-associated events within the oocyte which accompany the completion of meiosis and fertilization are surprisingly similar in echinoderms, amphibians, and mammals.

The stimulus that releases the prophase block is normally hormone receptor-mediated, while the fertilizing or activating stimulus that completes meiosis appears to be essentially a kind of membrane perturbation. Both stimuli, however, result in a fast rise in intracellular free calcium, accompanied by a rise in pH_i, and a fall in $[cAMP]_i$. In both events, $Ca^{2+}:H^+$ exchange may contribute to the rise in intracellular pH.

Michell (1975) and Takai et $al.$ (1982) have pointed out that cyclic AMP decreases in a number of different cell systems following stimuli that induce calcium-dependent phosphorylation and the turnover of inositol phospholipids. Since intracellular calcium levels increase to >1 μM following both the hormonal stimulus and fertilization, a significant fraction of the oocyte calmodulin should be activated. Its exact role in the reinitiation of meiosis is still unclear, as is the physiological significance of the Ca^{2+}-activated serine proteases. Calcium efflux is essential to the hormone stimulus and to fertilization. It is interesting that, depending on the species, both calcium ionophore (A23187) and calcium iontophoresis can trigger either event, suggesting some type of calcium-mediated calcium release.

The decrease in free intracellular calcium at the time of the first meiotic division would stabilize the maturation promoting factor (MPF) and microtubule assembly. In amphibians and mammals, we know that the oocyte plasma membrane is changed profoundly during and immediately following nuclear membrane breakdown. There is a net loss of membrane as the microvilli disappear, Na^+,K^+-ATPase activity disappears, and the plasma membrane undergoes a stepwise depolarization. By completion of the depolarization phase, the calmodulin-dependent Ca^{2+}-ATPase predominates. In vertebrates, the arrest at second metaphase is probably maintained by the highly unstable, Ca^{2+}-sensitive cytostatic factor (CSF) in the oocyte cytoplasm.

The burst of intracellular free calcium following fertilization inactivates MPF and CSF, permitting the second meiotic division to go to completion. It might be expected to stimulate another episode of calcium-dependent phosphorylation and to activate calmodulin-dependent enzymes. This calcium transient is probably of very short duration. Calcium floods out of the egg during the rising phase of the fertilization potential (at least in *Rana* and sea urchins). In many species the cortical reaction includes an exocytosis, probably calmodulin-dependent, and the insertion of new membrane. As sperm chromatin decondenses and DNA synthesis begins, the fertilized egg becomes a new organism. Studies with *Xenopus* embryos demonstrate that their mitotic cycle (MPF-driven) can be restarted with injected Ca^{2+} (Newport and Kirschner, 1984). Clearly, this is an area in which even more interesting discoveries are still to come.

ADDENDUM

A number of recent findings further support a role for inositol 1,4,5-trisphosphate (IP$_3$) in the release of bound calcium at fertilization. Whitaker has reported (1984) that microinjection of 10 pM (final concentration) of IP$_3$ into the sea urchin egg will induce at least three responses characteristic of fertilization (rise in intracellular pH, elevation of the surface envelope, and exocytosis of cortical granules). Similarly, Picard *et al.* (1985) have reported that microinjection of IP$_3$ brought about cortical granule exocytosis and elevation of a fertilization membrane in metaphase II arrested *Xenopus* oocytes and in starfish eggs arrested either at first meiotic prophase or after completion of meiosis. Using Ca^{2+}-selective microelectrodes, Busa *et al.* (1985) have shown that iontophoresis of IP$_3$ into *Xenopus* eggs at metaphase II produced a transient increase in free intracellular Ca^{2+} essentially indistinguishable from that seen at fertilization as well as other events typical of fertilization (membrane depolarization, cortical contraction, cortical granule exocytosis, and abortive cleavage furrow formation). They propose that sperm–egg interaction activates a local (at the sperm entry site) hydrolysis of phosphatidylinositol 4,5-bisphosphate (PIP_2), releasing sufficient IP_3 into the cortical cytosol to elicit a local increase in $[Ca^{2+}]_i$ via release from the IP_3-sensitive pool. This local response, in turn, may then trigger a more extensive Ca^{2+}-induced Ca^{2+} release from the second (presumably IP_3-insensitive) pool, which would propagate across the egg via further Ca^{2+}-induced release (as previously described by Gilkey *et al.*, 1978) giving rise to the Ca^{2+} wave. The principal role of this secondary release may be to protect against polyspermy by triggering the propagated exocytosis of the cortical granules.

We have recently found that early responses to meiotic agonists in *Rana* oocytes includes a transient five-fold increase in intracellular IP_3 levels and a rise in protein kinase C activity. Diacylglycerol, a product of PIP_2 hydrolysis by phospholipase C, will induce nuclear breakdown in *Rana* oocytes when added to the medium (A. B. Kostellow, E. J. Chein, and G. A. Morrill, unpublished). This suggests that progesterone acts via the phospholipid–calcium–protein kinase C system and that release of IP_3 may be responsible for the transient rise in free Ca^{2+} seen during the first 10–15 min (see Section V, above). Picard *et al.* (1985) reported that injection of IP_3 into prophase-arrested *Xenopus* oocytes did not induce nuclear breakdown. This suggests that either IP_3 was not delivered to the proper site or that additional events (e.g., release of diacylglycerol and activation of protein kinase C) must occur for release of the prophase block.

ACKNOWLEDGEMENTS

This work was supported in part by a research grant from the National Institute of Child Health and Human Development (HD 10463) and by funds from the NIH Biomedical Research Support Program (SO7RR).

REFERENCES

Azarnia, R., and Chambers, E. L. (1976). The role of divalent cations in activation of the sea urchin egg. I. Effect of fertilization on divalent cation content. *J. Exp. Zool.* **198,** 65–78.

Bae, I.-H. (1981). Role of calcium in resumption of meiosis of cultured porcine cumulus-enclosed oocytes. *Biol. Reprod.* **24,** Suppl. 1, 92A.

Batta, S. K., and Knudson, J. F. (1980). Calcium concentration in cumulus enclosed oocytes of rats after treatment with pregnant mares serum. *Biol. Reprod.* **22,** 243–246.

Baulieu, E. E., Godeau, F., Schorderet, M., and Schorderet-Slatkine, S. (1978). Steroid-induced meiotic maturation in *Xenopus laevis* oocytes: Surface and calcium. *Nature (London)* **275,** 593–598.

Belle, R., Ozon, R., and Stinnakre, J. (1977). Free calcium in full grown Xenopus laevis oocyte following treatment with ionophore A23187 or progesterone. *Mol. Cell. Endocrinol.* **8**, 65–72.

Brachet, J. (1976). Effects of two protease inhibitors on sea urchin and amphibian egg development. *Dev., Growth Differ.* **18**, 235–243.

Bradham, L. S., and Cheung, W. Y. (1980). Calmodulin-dependent adenylate cyclase. *In* "Calcium and Cell Function," Vol. I (W. Y. Cheung, ed.) Academic Press Inc., New York, pp. 109–125.

Brehm, P., Eckert, R., and Tillotson, D. (1980). Calcium-mediated inactivation of calcium current in Paramecium. *J. Physiol. (London)* **306**, 193–218.

Busa, W. B., and Nuccitelli, R. (1984). Metabolic regulation via intracellular pH. *Am. J. Physiol.* **246**, R409–R438.

Busa, W. B., and Nuccitelli, R. (1985). An elevated free cytosolic Ca^{2+} wave follows fertilization in eggs of the frog, *Xenopus laevis. J. Cell Biol.* **100**, 1325–1329.

Busa, W. B., Ferguson, J. E., Joseph, S. K., Williamson, J. R., and Nuccitelli, R. (1985). Activation of frog (Xenopus *laevis*) eggs by inositol trisphosphate. I. Characterization of Ca^{2+} Release from Intracellular Stores. *J. Cell. Biol.* **101**, 677–682.

Carafoli, E., Zurini, M., Niggli, V., and Krebs, J. (1982). The calcium-transporting ATPase of erythrocytes. *Ann. N.Y. Acad. Sci.* **402**, 304–328.

Channing, C. P., Pomerantz, S. H., Bae, I.-H., Evans, V. W., and Atlas, S. J. (1982). Action of hormones and other factors upon oocyte maturation. *Adv. Exp. Med. Biol.* **147**, 189–210.

Cheung, W. Y. (1971). Cyclic 3′,5′-nucleotide phosphodiesterase. Evidence for and properties of a protein activator. *J. Biol. Chem.* **246**, 2859–2869.

Cheung, W. Y. (1980). Calmodulin plays a pivotal role in cellular regulation. *Science* **207**, 19–27.

Cho, W. K., Stern, S., and Biggers, J. D. (1974). Inhibititory effect of dibutyryl cAMP on mouse maturation in vitro. *J. Exp. Zool.* **187**, 383–386.

Cicirelli, M. F., Robinson, K. R., and Smith, L. D. (1983). Internal pH of Xenopus laevis oocytes: A study of the mechanism and role of pH changes during meiotic maturation. *Dev. Biol.* **100**, 133–146.

Cobbold, P. H., Cutherbertson, K. S. R., Goyns, M. H., and Rice, V. (1983). Aequorin measurements of free calcium in single mammalian cells. *J. Cell Sci.* **61**, 123–136.

Cross, N. L. (1981). Initiation of the activation potential by an increase in intracellular calcium in eggs of the frog, Rana pipiens. *Dev. Biol.* **85**, 380–384.

Cuthbertson, K. S. R., Whittingham, D. G., and Cobbold, P. H. (1981). Free Ca^{2+} increases in exponential phases during mouse oocyte activation. *Nature (London)* **294**, 754–757.

Dan, J. C. (1954). Studies on the acrosome. III. Effect of calcium deficiency. *Biol. Bull. (Woods Hole, Mass.)* **107**, 335–349.

Deery, W. J., Means, A. R., and Brinkley, B. R. (1984). Calmodulin-microtubule association in cultured cells. *J. Cell Biol.* **98**, 904–910.

Dorée, M., Moreau, M., and Guerrer, P. (1978). Hormonal control of meiosis. In vitro induced release of calcium ions from the plasma membrane in starfish oocytes. *Exp. Cell Res.* **115**, 251–260.

Dorée, M., Picard, A., Cavadore, J. C., Le Peuch, C., and Demaille, J. G. (1982). Calmodulin antagonists and hormonal control of meiosis in starfish oocytes. *Exp. Cell Res.* **139**, 135–144.

Ecker, R. E., and Smith, L. D. (1971). Influence of exogenous ions on the events of maturation in Rana pipiens oocytes. *J. Cell Physiol.* **77**, 61–70.

Epel, D. (1982). The physiology and chemistry of calcium during the fertilization of eggs. *In* "Calcium and Cell Function" (W. Y. Cheung, ed.), Vol. 2, pp. 355–383. Academic Press, New York.

Epel, D., Perry, G., and Schmidt, T. (1982). Intracellular calcium and fertilization: Role of the cation and regulation of intracellular calcium levels. *In* "Membranes in Growth and Development," pp. 171–183. A. R. Liss, Inc., New York.

Eppig, J. J., and Downs, S. M. (1984). Chemical signals that regulate mammalian oocyte maturation. *Biol. Reprod.* **30,** 1–11.

Eusebi, F., and Siracusa, G. (1983). An electrophysiological study of parthenogenic activation in mammalian oocytes. *Dev. Biol.* **96,** 386–395.

Fabiato, A. (1982). Calcium release in skinned cardiac cells: Variations with species, tissues, and development. *Fed. Proc., Fed. Am. Soc. Exp. Biol.* **41,** 2238–2244.

Finidori-Lepicard, J., Hanoune, J., and Baulieu, E. E. (1981). Adenylate cyclase in Xenopus laevis oocytes: Characterization of the progesterone-sensitive membrane-bound form. *Mol. Cell. Endocrinol.* **28,** 211–227.

Fiskum, G., and Lehninger, A. L. (1982). Mitochondrial regulation of intracellular calcium. *In* "Calcium and Cell Function" (W. Y. Cheung, ed.), Vol. 2, pp. 39–71. Academic Press Inc., New York.

Fulton, B. P., and Whittingham, D. G. (1978). Activation of mammalian oocytes by intracellular injection of calcium. *Nature (London)* **273,** 149–151.

Gardiner, D. M., and Grey, R. D. (1983). Membrane junctions in Xenopus eggs: Their distribution suggests a role in calcium regulation. *J. Cell Biol.* **96,** 1159–1163.

Gilkey, J. C. (1983). Roles of calcium and pH in activation of eggs of the medaka fish, Oryzias latipes. *J. Cell Biol.* **97,** 669–678.

Gilkey, J. C., Jaffee, L. F., Ridgeway, E. B., and Reynolds, G. T. (1978). A free calcium wave traverses the activating egg of the medaka, Oryzias latipes. *J. Cell Biol.* **76,** 448–466.

Graff, G., Stephenson, J. H., Glass, D. B., Haddox, M. K., and Goldberg, N. D. (1978). Activation of soluble splenic cell guanylate cyclase by prostaglandin endoperoxides and fatty acid hydroperoxides. *J. Biol. Chem.* **253,** 7662–7676.

Guerrier, P., Moreau, M., and Doree, M. (1978). Control of meiosis reinitiation in starfish. Calcium ion as primary effective trigger. *Ann. Biol. Anim., Biochim., Biophys.* **18,** 441–452.

Guerrier, P., Moreau, M., Meijer, L., Mazzei, G., Vilain, J. P., and Dube, F. (1982). The role of calcium in meiosis reinitiation. *In* "Membranes in Growth and Development," pp. 139–155. A. R. Liss, Inc., New York.

Harris, P. (1975). The role of membranes in the organization of the mitotic apparatus. *Exp. Cell Res.* **94,** 409–425.

Hidaka, H., and Asano, T. (1977). Stimulation of human platelet guanylate cyclase by unsaturated fatty acid peroxides. *Proc. Natl. Acad. Sci. U.S.A.* **74,** 3657–3661.

Hirai, S., Chida, K., and Kanatani, H. (1973). Role of follicle cells in maturation of starfish oocytes. *Dev., Growth Differ.* **15,** 21–31.

Hollinger, T. G., and Alvarez, I. M. (1982). Trifluoperizine-induced meiotic maturation in Xenopus laevis. *J. Exp. Zool.* **224,** 461–464.

Hori, R. (1958). On the membrane potential of the unfertilized egg of the medaka Oryzias Latipes and changes accompanying activation. *Embryologia* **4,** 79–91.

Houle, J. G., and Wasserman, W. J. (1983). Intracellular pH plays a role in regulating protein synthesis in Xenopus oocytes. *Dev. Biol.* **97,** 302–312.

Huchon, D., Ozon, R., Fischer, E. H., and Demaille, J. G. (1981). The pure inhibitor of cAMP-dependent protein kinase initiates Xenopus laevis meiotic maturation. A 4-step scheme for meiotic maturation. *Mol. Cell. Endocrinol.* **22,** 211–222.

Hummel, B. C. W. (1959). A modified spectrophotometric determination of chymotrypsin, trypsin, and thrombin. *Can. J. Biochem. Physiol.* **37,** 1393–1399.

Ito, S., and Yoshiaki, K. (1972). Real activation potential observed in sea urchin egg during fertilization. *Exp. Cell Res.* **72,** 547–551.

Jaffe, L. F. (1983). Sources of calcium in egg activation: A review and hypothesis. *Dev. Biol.* **99,** 265–276.

Jagiello, G., Ducayen, M. B., Downey, R., and Jonassen, A. (1982). Alterations of mammalian oocyte meiosis I with divalent cations and calmodulin. *Cell Calcium* **3,** 153–162.

Johnson, J. D., Epel, D., and Paul, M. (1976). Intracellular pH and activation of sea urchin eggs after fertilization. *Nature (London)* **262,** 661–664.

Johnson, R. N., and Epel, D. (1981). Intracellular pH of sea urchin eggs measured by the dimethyloxazolidinedione (DMO) method. *J. Cell Biol.* **89,** 284–291.

Johnson, R. N., and Epel, D. (1982). Starfish oocyte maturation and fertilization: Intracellular pH is not involved in activation. *Dev. Biol.* **92,** 461–469.

Johnson, R. N., and Paul, M. (1977). Calcium influx following fertilization of Urechis caupo eggs. *Dev. Biol.* **57,** 364–374.

Kado, R. T., Marcher, K., and Ozon, R. (1981). Electrical membrane properties of the Xenopus laevis oocyte during progesterone-induced meiotic maturation. *Dev. Biol.* **84,** 471–476.

Kalimi, M., Ziegler, D. H., and Morrill, G. A. (1979). Characterization of a progestin-binding macromolecule in the amphibian oocyte cytosol. *Biochem. Biophys. Res. Commun.* **86,** 560–567.

Kanatani, H. (1969). Induction of spawning and oocyte maturation by L-methyl-adenine in starfishes. *Exp. Cell Res.* **57,** 333–337.

Kanatani, H. (1973). Maturation-inducing substance in starfishes. *Int. Rev. Cytol.* **35,** 253–298.

Kofoid, E. C., Knauber, D. C., and Allende, J. E. (1979). Induction of amphibian oocyte maturation by polyvalent cations and alkaline pH in the absence of potassium ions. *Dev. Biol.* **72,** 374–380.

Kostellow, A. B., and Morrill, G. A. (1980). Calcium dependence of steroid and guanine-3′,5′-monophosphate induction of germinal vesicle breakdown in *Rana pipiens* oocytes. *Endocrinology (Baltimore)* **106,** 1012–1019.

Kostellow, A. B., Ziegler, D. H., and Morrill, G. A. (1980). Regulation of Ca^{2+} and cyclic AMP during the first meiotic division in amphibian oocytes by progesterone. *J. Cyclic Nucleotide Res.* **6,** 347–358.

Langer, G. A. (1980). The role of calcium in the control of myocardial contractility: An update. *J. Mol. Cell. Cardiol.* **12,** 231–239.

Lee, S. C., and Steinhardt, R. A. (1981). pH changes associated with meiotic maturation in oocytes of Xenopus laevis. *Dev. Biol.* **85,** 358–369.

Leibfried, L., and First, N. L. (1979). Effects of divalent cations on in vitro maturation of bovine oocytes. *J. Exp. Zool.* **210,** 575–580.

Levin, R. M., and Weiss, B. (1977). Binding of trifluoperizine to the calcium-dependent activator of cyclic nucleotide phosphodiesterase. *Mol. Pharmacol.* **13,** 690–697.

Mac Neil, S., Lakey, T., and Tomlinson, S. (1985). Calmodulin regulation of adenylate cyclase activity. *Cell Calcium* **6,** 213–226.

Maeno, T. (1959). Electrical characteristics and activation potential of Bufo eggs. *J. Gen. Physiol.* **43,** 139–157.

Maller, J. L. (1983). Interaction of steroids with the cyclic nucleotide system in amphibian oocytes. *Adv. Cyclic Nucleotide Res.* **15,** 295–335.

Maller, J. L., and Krebs, E. G. (1977). Progesterone-stimulated cell division in Xenopus oocytes. Induction by regulatory subunit and inhibition by catalytic subunit of adenosine-3′,5′-monophosphate dependent protein kinase. *J. Biol. Chem.* **252,** 1712–1718.

Maller, J. L., and Krebs, E. G. (1980). Regulation of oocyte maturation. *Curr. Top. Cell. Regul.* **16,** 271–311.

Maller, J. L., Butcher, F. R., and Krebs, E. G. (1979). Early effect of progesterone on cyclic adenosine-3′,5′-monophosphate in Xenopus oocytes. *J. Biol. Chem.* **254,** 579–582.

Masui, Y. (1967). Relative roles of the pituitary, follicle cells and progesterone in the induction of oocyte maturation in Rana pipiens. *J. Exp. Zool.* **166,** 365–376.

Masui, Y., and Clarke, H. J. (1979). Oocyte maturation. *Int. Rev. Cytol.* **57,** 185–282.

Masui, Y., and Markert, C. L. (1971). Cytoplasmic control of nuclear behavior during meiotic maturation of frog oocytes. *J. Exp. Zool.* **177,** 129–146.

Masui, Y., Meyerhof, P. G., Miller, M. A., and Wasserman, W. J. (1977). Roles of divalent cations in maturation and activation of vertebrate oocytes. *Differentiation* **9,** 49–57.

Masui, Y., Meyerhof, P. G., and Miller, M. A. (1980). Cytostatic factor and chromosome behavior in early development. *In* "The Cell Surface: Mediator of Developmental Processes." (S. Subtelny and N. K. Wessells, eds.), pp. 235–256. Academic Press, New York.

Mazia, D. (1937). The release of calcium in Arbacia eggs on fertilization. *J. Cell. Comp. Physiol.* **10,** 291–304.

Meijer, L., and Guerrier, P. (1984a). Calmodulin in starfish oocytes. II. Trypsin treatment suppresses the trifluoperazine step. *Dev. Biol.* **101,** 257–262.

Meijer, L., and Guerrier, P. (1984b). Maturation and fertilization in starfish oocytes. *Int. Rev. Cytol.* **86,** 129–196.

Merriam, R. W. (1971). Progesterone-induced maturational events in oocytes of Xenopus laevis. I. Continuous necessity for diffusible calcium and magnesium. *Exp. Cell Res.* **68,** 75–80.

Michell, R. H. (1975). Inositol phospholipids and cell surface receptor function. *Biochem. Biophys. Acta* **415,** 81–147.

Michell, R. H., Jafferji, S. S., and Jones, J. M. (1977). The possible involvement of phosphatidylinositol breakdown in the mechanism of stimulus-response coupling at receptors which control cell surface calcium agtes. *Adv. Exp. Med. Biol.* **147,** 447–464.

Miot, F., and Erneux, C. (1982). Characterization of the soluble cyclic nucleotide phosphodiesterases in Xenopus laevis oocytes. Evidence for a calmodulin-dependent enzyme. *Biochim. Biophys. Acta* **701,** 253–259.

Miyazaki, S., and Igusa, Y. (1981). Fertilization potential in golden hamster eggs consists of recurring hyperpolarizations. *Nature (London)* **290,** 702–704.

Moreau, M., Doree, M., and Guerrier, P. (1976). Electrophoretic introduction of calcium ions into the cortex of Xenopus laevis oocytes triggers meiosis reinitiation. *J. Exp. Zool.* **197,** 443–449.

Moreau, M., Guerrier, P., and Doree, M. (1978). Hormonal control of meiosis reinitiation in starfish oocytes. *Exp. Cell Res.* **115,** 245–249.

Moreau, M., Vilain, J. P., and Guerrier, P. (1980). Free calcium changes associated with hormone action in amphibian oocytes. *Dev. Biol.* **78,** 201–214.

Mori, T., Morimoto, N., Kohda, H., Nishimura, T., and Kambegawa, A. (1983). Meiosis-inducing effects in vivo of antiserum to progesterone on follicular ova in immature rats treated with gonadotropins. *Endocrinol. Jpn.* **30,** 593–599.

Morrill, G. A., and Murphy, J. B. (1972). Role for protein phosphorylation in meiosis and in the early cleavage phase of amphibian embryonic development. *Nature (London)* **238,** 282–284.

Morrill, G. A., and Watson, D. E. (1966). Transmembrane electropotential changes in amphibian eggs at ovulation, activation and first cleavage. *J. Cell. Physiol.* **67,** 85–92.

Morrill, G. A., Kostellow, A. B., and Murphy, J. B. (1971). Sequential forms of ATPase activity correlated with changes in cation binding and membrane potential from meiosis to first cleavage in R. pipiens. *Exp. Cell Res.* **66,** 289–298.

Morrill, G. A., Kostellow, A. B., and Murphy, J. B. (1974). Role of Na$^+$, K$^+$-ATPase in early embryonic development. *Ann. N.Y. Acad. Sci.* **242,** 543–559.

Morrill, G. A., Schatz, F., and Zabrenetzky, V. S. (1975). RNA and protein synthesis during progesterone-induced germinal vesicle breakdown in R. pipiens ovarian tissue. *Differentiation* **4,** 143–152.

Morrill, G. A., Schatz, F., Kostellow, A. B., and Poupko, J. M. (1977). Changes in cyclic AMP levels in the amphibian ovarian follicle following progesterone induction of meiotic maturation. *Differentiation* **8**, 97–104.

Morrill, G. A., Ziegler, D. H., and Kostellow, A. B. (1980). Kinetics of calcium efflux and exchange from Rana pipiens oocytes immediately following reinitiation of the first meiotic division: Comparison of various meiotic agonists and antagonists. *Cell Calcium* **1**, 359–370.

Morrill, G. A., Ziegler, D. H., and Kostellow, A. B. (1981). Minireview: The role of Ca^{2+} and cyclic nucleotides in progesteroneinitiation of the meiotic divisions in amphibian oocytes. *Life Sci.* **29**, 1821–1835.

Morrill, G. A., Kostellow, A. B., Weinstein, S. P., and Gupta, R. K. (1983). NMR and electrophysiological studies of insulin action on cation regulation and endocytosis in the amphibian oocyte: Possible role of membrane recycling in the meiotic divisions. *Physiol. Chem. Phys. Med. NMR* **15**, 357–362.

Morrill, G. A., Kostellow, A. B., and Weinstein, S. P. (1984a). Endocytosis in the amphibian oocyte: Effect of insulin and progesterone on membrane and fluid internalization during the meiotic divisions. *Biochim. Biophys. Acta* **803**, 71–77.

Morrill, G. A., Ziegler, D. H., Kunar, J., Weinstein, S. P., and Kostellow, A. B. (1984b). Biochemical correlates of progesterone-induced plasma membrane depolarization during the first meiotic division in Rana oocytes. *J. Membr. Biol.* **77**, 201–212.

Morrill, G. A., Kostellow, A. B., Mahajan, S., and Gupta, R. K. (1984c). Role of calcium in regulating intracellular pH following the stepwise release of the metabolic blocks at first-meiotic prophase and second-meiotic metaphase in amphibian oocytes. *Biochim. Biophys. Acta* **804**, 107–117.

Mulner, O., Huchon, D., Thibier, C., and Ozon, R. (1979). Cyclic AMP synthesis in Xenopus laevis oocytes Inhibition by progesterone. *Biochim. Biophys. Acta* **582**, 179–184.

Mulner, O., Belle, R., and Ozon, R. (1983a). cAMP-dependent protein kinase regulates in ovo cAMP level of the Xenopus oocyte: Evidence for an intracellular feedback mechanism. *Mol. Cell. Endocrinol.* **31**, 151–160.

Mulner, O., Tso, J., Huchon, D., and Ozon, R. (1983b). Calmodulin modulates the cyclic AMP level in Xenopus oocyte. *Cell Differ.* **12**, 211–218.

Newport, J. W., and Kirschner, M. (1984). Regulation of the cell cycle during early Xenopus development. *Cell* **37**, 731–742.

Nishizuka, Y., Takai, Y., Kishimoto, A., Kikkawa, U., and Kaibuchi, K. (1984). Phospholipid turnover and hormone action. *Recent Prog. Horm. Res.* **40**, 301–345.

O'Connor, C. M., Robinson, K. R., and Smith, L. D. (1977). Calcium, potassium, and sodium exchange by full-grown and maturing Xenopus laevis oocytes. *Dev. Biol.* **61**, 28–40.

Orellana, O., Jedlicki, E., Allende, C. C., and Allende, J. E. (1984). Properties of a cyclic nucleotide phosphodiesterase of amphibian oocytes that is activated by calmodulin and calcium, by tryptic proteolysis, and by phospholipids. *Arch. Biochem. Biophys.* **231**, 345–354.

Paleos, G. A., and Powers, R. D. (1981). The effect of calcium on the first meiotic division of the mammalian oocyte. *J. Expl. Zool.* **217**, 409–416.

Paul, M., and Johnson, R. N. (1978). Uptake of Ca^{2+} is one of the earliest responses to fertilization in sea urchin eggs. *J. Exp. Zool.* **203**, 143–149.

Payan, P., Girard, J. P., and Ciapa, B. (1983). Mechanisms regulating intracellular pH in sea urchin eggs. *Dev. Biol.* **100**, 29–38.

Peaucellier, G. (1977). Initiation of meiotic maturation by specific proteases in oocytes of the polychaete annelid Sabellaria Alveolata. *Exp. Cell Res.* **106**, 1–14.

Picard, A., and Doree, M. (1983). Lithium inhibits amplification or action of the maturation-promoting factor (MPF) in meiotic maturation of starfish oocytes. *Exp. Cell Res.* **147**, 41–50.

Picard, A., Giraud, F., LeBouffant, F., Sladeczek, F., LePeuch, C., and Doree, M. (1985). Inositol 1,4,5-triphosphate microinjection triggers activation, but not meiotic maturation in amphibian and starfish oocytes. *FEBS Lett.* **182**, 446–450.

Powers, R. D. (1982). Changes in mouse oocyte membrane potential and permeability during meiotic maturation. *J. Exp. Zool.* **221**, 365–371.80.

Powers, R. D., and Paleos, G. A. (1982). Combined effects of calcium and dibutyryl cyclic AMP on germinal vesicle breakdown in the mouse oocyte. *J. Reprod. Fert.* **66**, 1–8.

Pressman, B. C. (1969). Mechanism of action of transport-mediating antibiotics. *Ann. N.Y. Acad. Sci.* **147**, 829–841.

Racker, E. (1980). Fluxes of Ca^{2+} and concepts. *Fed. Proc., Fed. Am. Soc. Exp. Biol.* **39**, 2422–2426.

Reynhout, J. K., Taddei, C., Smith, L. D., and La Marca, M. J. (1975). Response of large oocyte of Xenopus laevis to progesterone in Vitro in relation to oocyte size and time after previous HCG-induced ovulation. *Dev. Biol.* **44**, 375–379.

Ridgeway, E. B., Gilkey, J. C., and Jaffe, L. F. (1977). Free calcium increases explosively in activating medaka eggs. *Proc. Natl. Acad. Sci. U.S.A.* **74**, 623–627.

Robinson, K. R. (1979). Electrical current through full grown and maturing Xenopus laevis oocytes. *Proc. Natl. Acad. Sci. U.S.A.* **76**, 837–841.

Sadler, S. E., and Maller, J. L. (1981). Progesterone inhibits adenylate cyclase in Xenopus oocytes. Action on the guanine nucleotide regulatory protein. *J. Biol. Chem.* **256**, 6368–6373.

Schatz, F., and Morrill, G. A. (1972). Effects of dibutryl cyclic 3′, 5′-AMP, caffeine, and theophylline on meiotic maturation and ovulation in R. pipiens. *Pharmacol. Future Man. Proc. Int. Congr. Pharmacol., 5th, 1972*, Abstracts, p. 203.

Schlichter, L. C. (1983). Spontaneous action potentials produced by Na and Cl channels in maturing Rana pipiens oocyte. *Dev. Biol.* **98**, 47–59.

Schorderet-Slatkine, S., Schorderet, M., and Baulieu, E.–E. (1976). Initiation of meiotic maturation in *Xenopus laevis* oocytes by Lanthanum. *Nature (London)* **262**, 289–290.

Schorderet-Slatkine, S., Schorderet, M., and Baulieu, E. E. (1977). Progesterone-induced meiotic reinitiation in vitro in Xenopus laevis oocytes. *Differentiation* **9**, 67–76.

Schuetz, A. W. (1967). Action of hormones on germinal vesicle breakdown in frog (Rana pipiens) oocytes. *J. Exp. Zool.* **166**, 347–354.

Schuetz, A. W. (1969). Oogenesis: Processes and their regulation. *Adv. Reprod. Physiol.* **4**, 99–148.

Schuetz, A. W., Wallace, R., and Dumont, J. (1974). Steroid inhibition of protein incorporation by isolated amphibian oocytes. *J. Cell Biol.* **61**, 26–35.

Schultz, R. M., Montgomery, R. R., and Belanoff, J. R. (1983). Regulation of mouse oocyte meiotic maturation: Implication of a decrease in oocyte cAMP and protein dephosphorylation in commitment to resume meiosis. *Dev. Biol.* **97**, 264–273.

Shen, S. S., and Steinhardt, R. A. (1978). Direct measurement of intracellular pH during metabolic derepression of the sea urchin egg. *Nature (London)* **272**, 253–254.

Shen, S. S., and Steinhardt, R. A. (1979). Intracellular pH and the sodium requirement at fertilization. *Nature (London)* **282**, 87–89.

Silver, R. B., Cole, R. D., and Cande, W. Z. (1980). Isolation of mitotic apparatus containing vesicles with calcium sequestration activity. *Cell* **19**, 505–516.

Somlyo, A. P. (1984). Cellular site of calcium regulation. *Nature (London)* **309**, 516–517.

Speaker, M. G., and Butcher, F. R. (1977). Cyclic nucleotide fluctuations during steroid induced meiotic maturation of frog oocyte. *Nature (London)* **267**, 848–849.

Steinhardt, R. A., and Epel, D. (1974). Activation of sea urchin eggs by Ca^{2+}-ionophore. *Proc. Natl. Acad. Sci. U.S.A.* **71**, 1915–1919.

Steinhardt, R. A., Lundin, L., and Mazia, D. (1971). Bioelectric responses of the echinoderm egg to fertilization. *Proc. Natl. Acad. Sci. U.S.A.* **68**, 2426–2430.

Steinhardt, R. A., Zucker, R., and Schatten, G. (1977). Intracellular calcium release at fertilization in the sea urchin egg. *Dev. Biol.* **58**, 185–196.

Steinman, R. M., Mellaman, I. S., Muller, W. A., and Cohn, Z. A. (1983). Endocytosis and recycling of the plasma membrane. *J. Cell Biol.* **96**, 1–27.

Streb, H., Irvine, R. F., Berridge, M. J., and Schulz, I. (1983). Release of Ca^{2+} from a non-mitochondrial intracellular store in pancreatic acinar cells by inositol-1,4,5-trisphosphate. *Nature (London)* **306**, 67–68.

Takai, Y., Kishimoto, A., and Nishizuka, Y. (1982). Calcium and phospholipid turnover as trans-membrane signaling for protein phosphorylation. *In* "Calcium and Cell Function" (W. Y. Cheung, ed.), Vol. 2, pp. 385–412. Academic Press, New York.

Tsafriri, A., and Bar-Ami, S. (1978). Role of divalent cations in the resumption of meiosis of rat oocytes. *J. Exp. Zool.* **205**, 293–300.

Tsafriri, A., Dekel, N., and Bar-Ami, S. (1982). The role of oocyte maturation inhibitor in follicular regulation of oocyte maturation. *J. Reprod. Fertil.* **64**, 541–551.

Tsien, R. W. (1983). Calcium channels in excitable cell membranes. *Annu. Rev. Physiol.* **45**, 341–358.

Tyler, A., Monroy, A., Kao, C. Y., and Grundfest, H. (1956). Membrane potential and resistance of the starfish egg before and after fertilization. *Biol. Bull. (Woods Hole, Mass.)* **111**, 153–177.

Tyson, C. A., Van de Zande, H., and Green, D. E. (1976). Phospholipids as ionophores. *J. Biol. Chem.* **251**, 1326–1332.

Vilain, J. P., Moreau, M., and Guerrier, P. (1980). Propionate and cycloheximide reversibly block progesterone induced calcium surge in Ambystoma Mexicanum oocytes. *Dev., Growth Differ.* **22**, 25–31.

Volpi, M., Sha'afi, R. I., Epstein, P. M., Andrenyak, D. M., and Feinstein, M. B. (1981). Local anesthetics, mepacrine, and propranolol are antagonists of calmodulin. *Proc. Natl. Acad. Sci. U.S.A.* **78**, 795–799.

Wallace, R. A., and Steinhardt, R. A. (1977). Maturation of Xenopus oocytes. II. Observations on membrane potential. *Dev. Biol.* **57**, 305–316.

Wallace, R. A., Jared, D. W., Dumont, J. N., and Sega, M. W. (1973). Protein incorporation by iso-lated amphibian oocytes. III. Optimum incubation conditions. *J. Exp. Zool.* **184**, 321–334.

Wasserman, W. J., and Masui, Y. (1975). Initiation of meiotic maturation in Xenopus lacvis oocytes by the combination of divalent cations and ionophore A23187. *J. Exp. Zool.* **193**, 369–375.

Wasserman, W. J., and Smith, L. D. (1981). Calmodulin triggers the resumption of meiosis in amphibian oocytes. *J. Cell Biol.* **89**, 389–394.

Wasserman, W. J., Pinto, L. H., O'Connor, C. M., and Smith, L. H. (1980). Progesterone induces a rapid increase in Ca of Xenopus laevis oocytes. *Proc. Natl. Acad. Sci. U.S.A.* **77**, 1534–1536.

Webb, D. J., and Nuccitelli, R. (1981). Direct measurement of intracellular pH changes in Xenopus eggs at fertilization and cleavage. *J. Cell Biol.* **91**, 562–567.

Webb, D. J., and Nuccitelli, R. (1982). Intracellular pH changes accompanying the activation of development in frog eggs: Comparison of pH microelectrodes and ^{31}P-NMR measure-ments. *In* "Intracellular pH: Its Measurement, Regulation, and Utilization in Cellular Functions," pp. 293–324. A. L. Liss, Inc., New York.

Weinstein, S. P., Kostellow, A. B., Ziegler, D. H., and Morrill, G. A. (1982). Progesterone-induced down-regulation of an electrogenic Na^+, K^+-ATPase during the first meiotic division in amphibian oocytes. *J. Membr. Biol.* **69**, 41–48.

Weisenberg, R. C. (1972). Microtubule formation in vitro in solutions containing low calcium concentrations. *Science* **177**, 1104–1105.

Whitaker, M. (1984). Inositol 1,4,5-trisphosphate microinjection activates sea urchin eggs. *Nature (London)* **312**, 636–639.

Wilson, E. B. (1925). "The Cell in Development and Heredity." Macmillan, New York.

Wittenberg, C., Kohl, D. M., and Triplett, E. L. (1978). Amphibian embryo protease inhibitor V. Effect of calcium on the distribution of amphibian trypsin inhibitor during fertilization and subsequent development of Rana pipiens. *Cell Differ.* **7**, 11–20.

Wolf, D. P. (1974). The cortical response in Xenopus laevis ova. *Dev. Biol.* **40**, 102–115.

Young, G. P. H., Ding-E Young, J., Deshpande, A. K., Goldstein, M., Koide, S. S., and Cohn, Z. A. (1984). A Ca^{2+}-activated channel from Xenopus laevis oocyte membranes reconstituted into planar bilayers. *Proc. Natl. Acad. Sci. U.S.A.* **81**, 5155–5159.

Ziegler, D. H., and Morrill, G. A. (1977). Regulation of the amphibian oocyte membrane ion permeability by cytoplasmic factors during the first meiotic division. *Dev. Biol.* **60**, 318–325.

Chapter 8

Calcium and the Control of Insulin Secretion

BO HELLMAN
ERIK GYLFE

Department of Medical Cell Biology
University of Uppsala
Uppsala, Sweden

253

CALCIUM AND CELL FUNCTION, VOL. VI
Copyright © 1986 by Academic Press, Inc.
All rights of reproduction in any form reserved.

I. INTRODUCTION

It is essential for glucose homeostasis that the secretory machinery of the pancreatic β-cells respond rapidly and with high sensitivity to variations in the extracellular glucose concentration. When the sensitivity of the insulin secretory mechanism is reduced, the consequently impaired glucose tolerance may result in clinically overt diabetes mellitus. Since the original discovery that extracellular Ca^{2+} is a prerequisite for glucose stimulation of insulin release (Grodsky and Bennett, 1966; Milner and Hales, 1967), numerous studies have emphasized the fundamental role of Ca^{2+} in the insulin secretory process. Reference to this work is given in reviews from principal laboratories involved (Malaisse *et al.*, 1975, 1978b; Hellman, 1976b; Hellman *et al.*, 1979a,b, 1980a; Herchuelz and Malaisse, 1981; Wollheim and Sharp, 1981).

Major uncertainties still exist both with regard to the pools of Ca^{2+} involved in insulin secretion and the actions of Ca^{2+} at the molecular level. However, there has been a considerable extension in our knowledge during the last 3 years as a consequence of methodological innovations such as measurements of net fluxes of Ca^{2+} and its activity in the cytoplasm, as well as the establishment of continuous lines of insulin-producing cells. The main purpose of the present contribution is to provide an assessment, in the light of recent data, of the role of Ca^{2+} in the regulation of insulin release. We pay particular attention to how the β-cell handling of Ca^{2+} is affected by glucose, which is the major physiological stimulator of insulin secretion.

II. INSULIN RELEASE AND EXTRACELLULAR CALCIUM

The presence of extracellular Ca^{2+} is a prerequisite not only for glucose stimulation of insulin release, but also for most other initiators of the secretory process (Hedeskov, 1980). The fact that omission of extracellular Ca^{2+} results in a prompt decrease of the secretion down to a basal rate has been taken to reflect the rapid turnover of Ca^{2+} in an intracellular pool controlling insulin release (Malaisse *et al.*, 1978c; Hellman *et al.*, 1979b). Although it is well established that both the first and second phases of glucose-stimulated insulin release depend on extracellular Ca^{2+}, different opinions have been expressed as to which of the phases is the most sensitive to removal of extracellular Ca^{2+}. Henquin (1978b) observed that theophylline selectively restored the late phase when the secretory activity was suppressed by preventing the entry of Ca^{2+} either by omission of the ion or addition of D-600. However, to regard the first phase as particularly sensitive to the influx of Ca^{2+}, we need more precise knowledge of how cyclic AMP potentiates insulin secretion (see Section VIII,B). Other studies of the dynamics of the glucose-stimulated insulin release under conditions of restricted entry of Ca^{2+} have instead provided evidence that the first phase of insulin release is essentially due to mobilization of intracellular calcium (Wollheim *et al.*, 1978b, 1980, 1981; Siegel *et al.*, 1980c; Wollheim and Sharp, 1981).

The concentration of extracellular Ca^{2+} necessary for glucose stimulation of insulin release increases with the medium concentration of Mg^{2+} (Malaisse *et al.*, 1976a; Curry *et al.*, 1977; Somers *et al.*, 1979b; Frankel *et al.*, 1981; Berggren *et al.*, 1983b) and is inversely related to the cyclic AMP content of the β-cells (unpublished data). The significance of cyclic AMP in the action of Ca^{2+} on glucose-stimulated insulin release is illustrated in Fig. 1. With the rise of

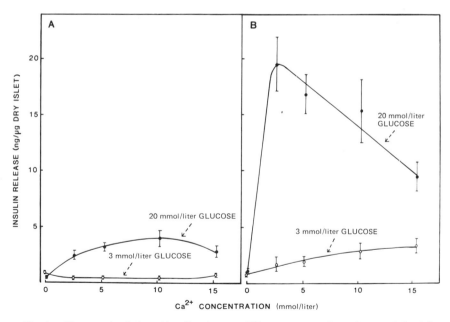

Fig. 1. Glucose stimulation of insulin release at different concentrations of extracellular Ca^{2+} in the absence and presence of 3-isobutyl-1-methylxanthine (IBMX). After 40 min of preliminary incubation at different concentrations of Ca^{2+} in a HEPES-buffered medium containing 3 m*M* glucose and 1 mg/ml albumin, *ob/ob*-mouse islets were incubated without (A), and with (B) 1 m*M* IBMX in the same type of medium containing 3 or 20 m*M* D-glucose. Mean values ±SEM for 8 experiments.

cyclic AMP obtained by inhibiting the phosphodiesterase activity with 3-iso-butyl-1-methylxanthine (IBMX), there is a potentiation of glucose-stimulated insulin release. Moreover, the maximal effect is reached at lower concentrations of extracellular Ca^{2+}. It is evident that an increase of extracellular Ca^{2+} above a certain concentration results in a reduction of insulin release from ob/ob-mouse islets. Suppression of glucose-stimulated insulin release by excessive concentrations of extracellular Ca^{2+} has also been observed with pieces of rat pancreas (Hales and Milner, 1968) and in monolayer cultures of the newborn rat pancreas (Wollheim et $al.$, 1978a).

The requirement for extracellular Ca^{2+} may indicate that a certain concentration of the cation is essential for the proper function of the secretory machinery. However, supporting a direct regulatory role for Ca^{2+} is the fact that merely increasing the extracellular concentration was sufficient for stimulating insulin release in the absence of glucose and other initiators of secretory activity. Devis et $al.$ (1975) observed that addition of 2–10 mM Ca^{2+} to a glucose-free medium resulted in a short-lived burst of insulin release when the rat pancreas had first

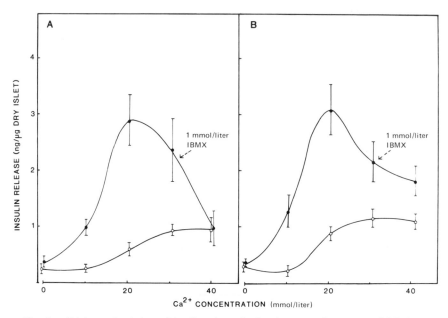

Fig. 2. Calcium stimulation of insulin release in the absence and presence of 3-isobutyl-1-methylxanthine (IBMX). Islets from ob/ob-mice were preliminarily incubated for 40 min at different concentrations of Ca^{2+} in HEPES-buffered medium containing 3 mM glucose and 1 mg/ml albumin. Insulin release was measured during 60 min of further incubation with and without 1 mM IBMX in the same type of medium lacking glucose. The increase in osmotic pressure obtained by raising the Ca^{2+} concentration was either compensated for (B) or not (A) by reduction of Na^+. Mean values ±SEM for 8 experiments.

been perfused with a Ca^{2+}-deficient medium containing EGTA. Since it was assumed that this pretreatment increased the Ca^{2+} permeability of the β-cell membrane, the results were interpreted as indicating that the divalent cation initiates the secretory activity. Subsequent studies provided more conclusive evidence for a directly stimulatory pool of Ca^{2+} which rapidly equilibrates with the exterior of the β-cells, by demonstrating that Ca^{2+} also initiates insulin release when added to a medium containing physiological concentrations of cations (Hellman, 1976a). This phenomenon is illustrated in Fig. 2, indicating the amounts of insulin released during 60 min from *ob/ob*-mouse islets exposed to different Ca^{2+} concentrations in the presence and absence of 1 m*M* IBMX. The experiments were performed with (A) and without (B) osmotic compensation by reduction of Na^+. We found that the stimulation of insulin release obtained with increasing concentrations of extracellular Ca^{2+} is more pronounced in the presence of IBMX. However, in the latter case, there was no further stimulation but actually a decreased secretory activity when we raised the Ca^{2+} concentration above 20 m*M*. The stimulation of insulin release obtained by raising extracellular Ca^{2+} has been found to meet the criteria of a physiological

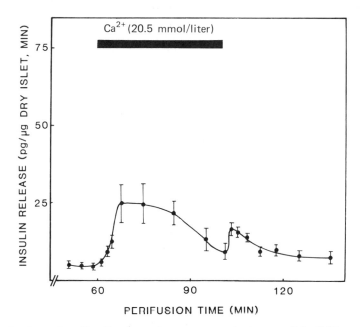

Fig. 3. Dynamics of insulin release after a temporary rise in extracellular Ca^{2+} concentration. Islets from *ob/ob*-mice were perifused with HEPES-buffered medium containing 2.56 m*M* Ca^{2+} and 5 m*M* theophylline. Between 60 and 100 min, the Ca^{2+} concentration was raised to 20.5 m*M*; the osmotic pressure at the lower Ca^{2+} concentration was equalized with choline chloride. Mean values ±SEM for 6 experiments. From Hellman *et al.* (1979b).

stimulus in being inhibited by L-epinephrine and 2,4-dinitrophenol (Hellman, 1976a).

In studying the dynamics of the stimulatory action of extracellular Ca^{2+} in perifusion experiments with isolated islets, we found that the effect is rapidly established and declines with time (Hellman, 1976b; Hellman et al., 1979b). Figure 3 shows alterations of the secretory rate obtained with a rise of the Ca^{2+} concentration from 2.6 to 20.5 mM. The gradual decrease of the stimulatory action of high extracellular concentrations of Ca^{2+} may be explained in terms of a progressive suppression of the Ca^{2+} permeability of the plasma membrane. Nevertheless, the action of Ca^{2+} certainly involves an inhibitory component, as demonstrated by the appearance of a new phase of stimulated secretion following the reduction of the Ca^{2+} to the original concentration.

III. INSULIN RELEASE AND INTRACELLULAR CALCIUM

Although stimulation of insulin release is mediated predominantly through increased Ca^{2+} permeability of the plasma membrane, alterations of the intracellular distribution of the ion may also be important. The latter alternative may explain the action of substances like pentobarbital, which stimulates insulin release in a Ca^{2+}-deficient medium (Hellman, 1977). It is difficult to demonstrate secretory activity evoked by intracellular redistribution of Ca^{2+} under conditions in which the entry of the cation is prevented. Since such an experimental design results in depletion of the pool of Ca^{2+} initiating the release (Malaisse et al., 1978b; Hellman et al., 1979b), it is possible that mobilization of Ca^{2+} from intracellular stores has often been overlooked as a mechanism contributing to the effect of various insulin secretagogues.

In the case of cyclic AMP, evidence has been provided that there is a redistribution of intracellular Ca^{2+} in favor of the cytoplasm in addition to the more important sensitization of the secretory machinery to Ca^{2+} (see Section VIII,B). In light of the rapid glucose action on phosphoinositide metabolism (see Section VIII,A), it is possible to envisage release of Ca^{2+} both from the inner part of the plasma membrane and from the endoplasmic reticulum as mechanisms contributing to the first phase of glucose-stimulated insulin release. Also, the long term action of glucose as a stimulator of insulin release has been considered to involve mobilization of intracellular calcium stores (Wollheim et al., 1978b, 1980, 1981; Kikuchi et al., 1979; Siegel et al., 1980b; Wollheim and Sharp, 1981; Janjic et al., 1982). However, recent experimental data suggest a reconsideration of this view. In fact, the overall effect of glucose on the intracellular distribution of calcium is instead to counteract the release of insulin by stimulating organelle sequestration of calcium at the expense of its activity in the cytoplasm (Hellman and Gylfe, 1984a,b, 1985; Hellman et al., 1985).

IV. INSULIN RELEASE AND BINDING OF CALCIUM TO THE β-CELL SURFACE

It is widely accepted that Ca^{2+} has a stabilizing effect on biological membranes, making cells less permeable to cations, water, and macromolecules (Rubin, 1982). The calcium responsible for this stabilization may well represent a superficial pool in the β-cell membrane with inhibitory effects on insulin secretion (Hellman, 1976b). Perifusion studies (Fig. 3) show that removal of excessive Ca^{2+} results in a prompt increase in the secretory rate. Moreover, the release of insulin from isolated islets is inhibited by exposure to analogs of Ca^{2+} which do not penetrate the β-cells (Hellman et al., 1979a,b; Flatt et al., 1980a,b). Figure 4 illustrates this type of action obtained during perifusion with La^{3+}. The effect of La^{3+} removal is similar to that obtained when reducing a high concentration of extracellular Ca^{2+}, in that it produces an immediate stimulation of insulin release.

A pertinent aspect of inhibitory membrane calcium is its possible regulation by modifiers of insulin secretion. Recordings of the intensity and polarization of the chlorotetracycline fluorescence in suspensions of islet cells have provided evi-

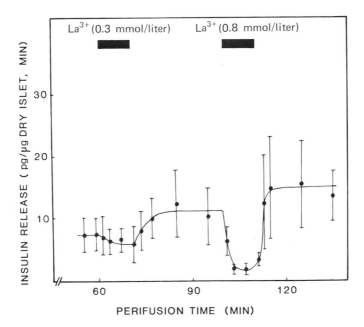

Fig. 4. Dynamics of insulin release in the presence of different concentrations of La^{3+}. Islets from *ob/ob*-mice were perifused with HEPES-buffered medium containing 2.56 mM Ca^{2+} and 1 mM 3-isobutyl-1-methylxanthine. La^{3+} was added to the medium between 60–70 min and 100–110 min at the concentrations indicated. Mean values ±SEM for 4 experiments.

dence for a glucose effect on calcium interactions with the β-cell membrane (Täljedal, 1978, 1979). Although the chlorotetracycline data strongly suggest that glucose increases the calcium mobility in the β-cell membrane, they do not distinguish between loss of Ca^{2+} and diminished membrane hydrophobicity. In a plasma membrane fraction of rat islets, Naber et al. (1980) observed two types of binding sites with high and low affinities for Ca^{2+}. The absence of a glucose effect on the steady state binding of Ca^{2+} to this membrane fraction may indicate that the β-cells must be intact for glucose to act on membrane calcium. Exposure to glucose increases the amount of superficial ^{45}Ca displaced by La^{3+}-washing of islets microdissected from ob/ob-mice. A prerequisite for this effect is that the radioactive loading of the islets be performed in the presence of trace amounts of Ca^{2+} (Hellman and Gylfe, 1978), or else that the loading be followed by brief rinsing in a nonradioactive medium (Hellman et al., 1976b). After introduction of an improved technique for discriminating between superficial and intracellular calcium by washing with a cold La^{3+} solution, this effect of glucose could not be reproduced using microdissected islets (Hellman, 1978) or islets with the surrounding capsule removed by extensive collagenase treatment (Flatt and Swanston-Flatt, 1983). It is therefore likely that the limited superficial pool of glucose-stimulated ^{45}Ca recovered by rinsing and the following washing with La^{3+} at 37°C originates from the functionally important calcium supposed to exist in the immediate submembrane space (Hellman, 1978).

Among the elements in the lanthanide series, thulium has proved to be particularly useful as a nonpenetrating analog for studying Ca^{2+} interactions with the exterior of the β-cells. Although it displaces ^{45}Ca from the cell periphery and inhibits basal and glucose-stimulated insulin release, Tm^{3+} differs from La^{3+} in not affecting transmembrane Ca^{2+} fluxes (Flatt et al., 1979, 1980a). These observations may indicate the existence of cationic binding sites on the surface of the β-cells, the occupancy of which inhibits insulin secretion by a mechanism different from general membrane stabilization. Using ^{171}Tm as a membrane probe, Flatt et al. (1979, 1981) demonstrated the existence of high- and low-affinity binding sites in β-cell-rich pancreatic islets isolated from ob/ob-mice. In contradistinction to sugars without stimulatory actions on insulin release, D-glucose reduced the high-affinity binding of ^{171}Tm. The latter effect appeared specific for the islets in not being possible to reproduce with artificial membranes or exocrine pancreatic cells.

Evidently the calcium associated with the exterior of the β-cells represents a rapidly exchangeable pool with significance for the insulin secretory activity. In addition to suppressing the insulin release following membrane stabilization, Ca^{2+} may also inhibit it by binding to high-affinity binding sites subject to glucose regulation. Electron microscopic studies have demonstrated reduced binding of cationic ferritin to the β-cell membrane at the sites of the granule extrusion (Howell and Tyhurst, 1977). The question remains whether the glucose interaction with the binding of cations is a primary event or one secondary to the exocytotic process. Glucose mobilization of calcium may occur not only at the exterior of the plasma membrane but also at the internal surface facing the cytoplasm. The evidence supporting a mobilizing effect of glucose on calcium with the latter location will be discussed separately (see Section VIII,A).

V. CALCIUM CONTENT OF THE β-CELL

Most of the information about the total calcium content of the β-cells has been obtained from studies of *ob/ob*-mouse islets, which consist of up to 90% β-cells (Hellman, 1965). Graphite furnace atomic absorption spectroscopy revealed that islets with intact capsules contain 58 mmol calcium per kilogram dry weight and specimens of β-cells microdissected from freeze-dried sections of the pancreas contain 31 mmol calcium per kilogram dry weight (Berggren *et al.*, 1978). This difference may reflect the calcium which can be displaced from the capsule with La^{3+} (Hellman *et al.*, 1979b; Flatt and Swanston-Flatt, 1983). β-Cell specimens microdissected from freeze-dried sections of the *ob/ob*-mouse pancreas have recently been analyzed also with a proton microprobe, indicating a calcium content of about 20 mmol per kilogram dry weight (Lindh *et al.*, 1985). According to a previously established relationship between islet dry weight and intracellular water (Hellman *et al.*, 1971a), the average concentration of calcium in the β-cells equals 16–25 mmol per liter. Thus we find a more than tenfold accumulation of calcium in the β-cells as compared with that in the extracellular medium. Studies of rat islets support the view that the pancreatic β-cells are rich in calcium. After electron microprobe analysis of samples obtained from deparaffinized sections, the islet content of calcium has been estimated to be 28 mmol per kilogram dry weight (Wolters *et al.*, 1979). Other measurements in collagenase-isolated islets using a fluorometric microtechnique or flame atomic absorption spectroscopy provided values of 30 (Wolters *et al.*, 1982) and 13–28 (Malaisse *et al.*, 1978b) mmol calcium per kilogram dry weight, respectively. Also islets from the guinea pig pancreas have been analyzed for total calcium (Berggren *et al.*, 1979). A procedure for eliminating the majority of the β-cells by streptozotocin injections allowed a comparison of collagenase-isolated islets, rich in α_2-cells, with the β-cell-rich islets from untreated guinea pigs. The calcium content of the β-cells was found to be 2- to 3-times higher than that in the α_2-cells and the exocrine pancreatic cells. Indeed, pancreatic β-cells have a calcium content which exceeds that in most other cells (Berggren *et al.*, 1978, 1979; Borle, 1981; Wolters *et al.*, 1982).

Figure 5 illustrates the proportions of superficial and intracellular calcium in β-cells from *ob/ob*-mice as determined in La^{3+}-wash experiments. Previous analyses of isolated islets have resulted in an overestimation of the superficial

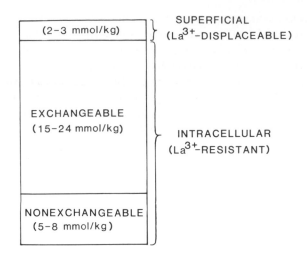

Fig. 5. Amounts of superficial and intracellular calcium in pancreatic β-cells from *ob/ob*-mice. About 25% of the intracellular calcium remains nonexchangeable during culture of islets for 7 days.

pool due to calcium binding to the surrounding capsule (Hellman *et al.*, 1979b; Flatt and Swanston-Flatt, 1983). Not only the superficial calcium, but also a major part of that originating intracellularly, is readily exchangeable. Removal of extracellular Ca^{2+} results in a considerable depletion of the islet calcium (Berggren *et al.*, 1978; Wolters *et al.*, 1984). After 120 min of exposure to a physiological concentration of saline, there was a 40% reduction in the calcium content of *ob/ob*-mouse islets. Studies with radioactive ^{45}Ca have considerably increased our knowledge about exchangeable β-cell calcium, emphasizing the existence of intracellular pools with a high turnover rate (see Sections VI,B and VI,C). Rat islets maintained in tissue culture with ^{45}Ca at different glucose concentrations have been reported to reach isotopic equilibrium after 24 hr (Ribes *et al.*, 1981). The comparison of the amounts of ^{45}Ca incorporated with the total calcium measured by graphite furnace atomic absorption spectroscopy has permitted estimates of the exchangeable fraction of the intracellular calcium. When *ob/ob*-mouse islets were cultured for 7 days at different glucose concentrations, the percentage of such calcium was found to be 66–80% (Gylfe *et al.*, 1978; Bergsten and Hellman, 1984). This value is lower than that obtained when including the superficial calcium bound to the microdissected *ob/ob*-mouse islets (Hellman *et al.*, 1979b).

An important aspect of the β-cell content of intracellular calcium is its regulation by glucose. After the original observations that the sugar stimulates ^{45}Ca incorporation into isolated islets from rats (Malaisse-Lagae and Malaisse, 1971) and *ob/ob*-mice (Hellman *et al.*, 1971b), the La^{3+}-wash technique established that the glucose effect involves an increased net uptake of ^{45}Ca into intracellular stores (Hellman *et al.*, 1976a). Since other studies indicated that glucose also has pronounced effects in stimulating intracellular calcium–calcium exchange (Hellman *et al.*, 1979b; Abrahamsson *et al.*, 1981), the sugar may well increase the ^{45}Ca uptake without much affecting the net uptake of calcium. Although it was claimed from a study of three single measurements with atomic absorption spectroscopy that 90 min of exposure to glucose more than doubles the calcium content of rat islets (Malaisse *et al.*, 1978b), this observation was not confirmed with a fluorometric technique (Wolters *et al.*, 1982) or by more extensive analyses of islets from *ob/ob*-mice with graphite furnace atomic absorption spectroscopy (Andersson *et al.*, 1982a).

In light of the difficulty of demonstrating a glucose-mediated increase of β-cell calcium, it is somewhat surprising that the ^{45}Ca content of rat islets loaded to isotopic equilibrium has been reported to be markedly affected by acute changes in the glucose concentration (Ribes *et al.*, 1981). However, these observations were not confirmed in a study compensating for changes of the medium associated with culture of the islets (Bergsten and Hellman, 1984). Even if the major effect of glucose is to stimulate intracellular calcium–calcium exchange, it seems likely that exposure to the sugar also results in some net uptake of Ca^{2+} into the β-cells. A modest (14%) increase in the amount of intracellular ^{45}Ca was found 30 min after raising the glucose concentration from 5.5 to 20 mmol/liter in *ob/ob*-mouse islets maintained at isotopic equilibrium (Bergsten and Hellman, 1984). Moreover, measurements with the Ca^{2+}-indicator Arsenazo III have indicated that glucose stimulates the removal of Ca^{2+} from the incubation medium when *ob/ob*-mouse islets are exposed to micromolar concentrations of this cation (Gylfe, 1982).

VI. CALCIUM TRANSPORT ACROSS THE PLASMA MEMBRANE

A. Electrophysiological Considerations

In the absence or at very low concentrations of glucose, the resting membrane potential (V_0) of the pancreatic β-cell is about -60 mV (Matthews and Sakamoto, 1975; Atwater and Beigelman, 1976; Meissner, 1976; Atwater *et al.*, 1978, 1980; Ribalet and Beigelman, 1979; Meissner *et al.*, 1979). This resting potential depends essentially on the K^+ permeability (Atwater *et al.*, 1978; Meissner *et al.*, 1978). The electrogenic Na^+/K^+ pump contributes 4–8 mV (Henquin and Meissner, 1982; Ribalet and Beigelman, 1982; Meissner and Henquin, 1984), and the Na^+ (Atwater *et al.*, 1978; Meissner *et al.*, 1978) and possibly the Cl^- permeabilities may slightly influence the resting potential.

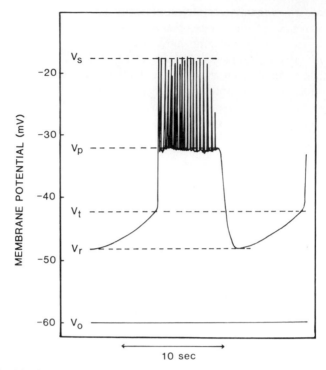

Fig. 6. Membrane potential of a β-cell in the absence of glucose and a burst of spikes at submaximal stimulation with the sugar. The approximate potentials for the resting state (V_0), maximal repolarization (V_r), the threshold (V_t), the plateau (V_p), and the reversal of the spikes (V_s) are indicated.

Figure 6 shows the membrane potential of a β-cell in the absence and presence of glucose. When the glucose concentration is raised to stimulate insulin release, there is initially a slow depolarization of about 12–15 mV (Meissner and Preissler, 1979). After reaching a threshold value (V_t), the membrane depolarizes rapidly to a plateau potential (V_p). A fast spike activity occurs at the plateau potential resulting in oscillations between V_p and an even more depolarized state, the reversal potential of the spikes (V_s). If the glucose concentration is high enough to stimulate insulin release maximally, the spike activity continues. However, at lower stimulatory concentrations of glucose, the spikes occur in bursts, which are terminated by repolarization to a new potential (V_r), a few mV more negative than V_t. From V_r the membrane depolarizes slowly to V_t, and there is a new burst of spikes. As shown in Fig. 7, the duration of the bursts is prolonged at the expense of the silent periods when the glucose concentration is increased (Meissner and Preissler, 1979). There appears to be a close correlation between the duration of the plateau phase and the amount of insulin released. At

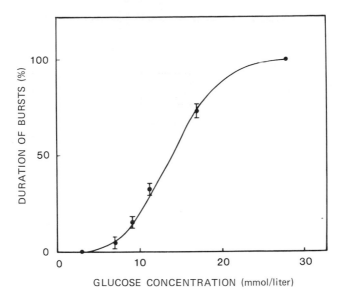

Fig. 7.   Relative duration of burst activity as a function of glucose concentration. 100% duration means continuous burst activity. Mean values ±SEM for 4–39 observations. From Meissner and Preissler (1979).

concentrations of glucose which stimulate insulin release submaximally, this correlation seems to extend to both phases of release, the initial burst of spikes being considerably longer than the following ones (Meissner and Preissler, 1979). The time course and amplitude of the spikes show no clear dependence on the glucose concentration except at very high concentrations, when the frequency of the continuous spike activity is increased (Meissner and Schmelz, 1974).

The slow depolarization from V_o to V_t initially, and from V_r to V_t between bursts, can be attributed to a decrease in the K^+ permeability (Atwater *et al.*, 1978; Meissner *et al.*, 1978). This depolarization was abolished by the K^+ ionophore valinomycin (Henquin and Meissner, 1978). Moreover, inhibitors of K^+ permeability such as tetraethylammonium, 9-aminoacridine, and quinidine cause depolarization and burst activity (Atwater *et al.*, 1979a; Henquin *et al.*, 1979; Meissner and Preissler, 1979; Ribalet and Beigelman, 1980). An increased efflux of Cl^- may possibly contribute to the depolarization (Sehlin, 1978; Meissner and Preissler, 1980). However, influxes of Ca^{2+} and Na^+ are probably insignificant, since slow depolarization occurs also in the presence of Co^{2+} or Mn^{2+} (Meissner and Preissler, 1979, 1980) or after replacement of Na^+ with choline$^+$ (Matthews and Sakamoto, 1975; Meissner and Preissler, 1980).

The concept that glucose decreases K^+ permeability is supported by measurement of the efflux of $^{42}K^+$ and its analog $^{86}Rb^+$ from prelabeled pancreatic islets (Sehlin and Täljedal, 1974a, 1975; Boschero *et al.*, 1977; Henquin, 1978a, 1979, 1980a; Malaisse *et al.*, 1978a; Carpinelli and Malaisse, 1980a,b,c). Several mechanisms have been suggested to explain the coupling between glucose stimulation and decreased K^+ permeability. The role of intracellular acidification has been emphasized (Pace and Tarvin, 1982, 1983; Pace *et al.*, 1983). However, in other studies the participation of protons has been considered unlikely (Carpinelli and Malaisse, 1980b; Henquin, 1981), and evidence for intracellular acidification are lacking (see Section VI,C). Cyclic AMP is probably not involved (Carpinelli and Malaisse, 1980c; Henquin, 1980a), but ATP was recently found to inhibit K^+ permeability by directly blocking a novel channel in the β-cell plasma membrane (Cook and Hales, 1984). Indeed, ATP is affected by glucose in a range of concentrations similar to that in which dose-related changes in ^{86}Rb fractional outflow is observed (Hellman *et al.*, 1969; Henquin, 1978a; Malaisse *et al.*, 1979a; Malaisse and Herchuelz, 1982). The coupling between metabolic and ionic events in the β-cell has also been attributed to reducing equivalents in the form of pyridine nucleotides and glutathione (Hellman *et al.*, 1974a; Ammon and Verspohl, 1979; Malaisse *et al.*, 1979a; Ammon *et al.*, 1980). Studies of ^{86}Rb efflux have indicated that the K^+ permeability depends at least in part on the redox potential (Henquin, 1980a).

The K^+ conductance of the β-cell plasma membrane has been reported to be under the control of Ca^{2+} activity in the cytoplasm (Atwater *et al.*, 1979a,b, 1980, 1983; Ribalet and Beigelman, 1979; Cook, 1984). Perhaps the primary event in the slow depolarization is a reduction of this Ca^{2+} activity. Although such an effect of glucose has been considered unlikely (Malaisse *et al.*, 1978b; Wollheim and Sharp, 1981; Prentki and Wollheim, 1984), various types of evidence including direct measurements indicate that an early result of glucose exposure is a lowering of Ca^{2+} activity in the cytoplasm (Hellman *et al.*, 1982; Gylfe *et al.*, 1983; Rorsman *et al.*, 1983, 1984; Hellman and Gylfe, 1984a,b, 1985). This reduction of Ca^{2+} activity is probably accomplished by stimulated sequestration of the cation in intracellular stores (Hellman *et al.*, 1979a,b; Kohnert *et al.*, 1979; Andersson, 1983). It is easy to envision a coupling between glucose metabolism and uptake of Ca^{2+} into mitochondria (see Section VII,C). A role of glucose-stimulated Ca^{2+} uptake into these organelles in the regulation of K^+ permeability is consistent with a relationship between changes in ^{86}Rb fluxes and production of ATP, as well as of reduced pyridine nucleotides.

Although a decreased K^+ permeability has been suggested to cause also the rapid phase of depolarization from V_t to V_p (Ribalet and Beigelman, 1979, 1980), it seems more likely that under physiological conditions it is due to an increased Ca^{2+} current after opening of voltage-dependent channels (Meissner and Preissler, 1979). This rapid depolarization does not occur after removal of extracellular Ca^{2+} (Dean and Matthews, 1970; Meissner and Schmelz, 1974; Matthews and Sakamoto, 1975; Atwater and Beigelman, 1976). Moreover, it is blocked by inhibitors of the voltage-dependent Ca^{2+} channels such as Co^{2+}, Mn^{2+} (Dean and Matthews, 1970; Meissner et al., 1979; Meissner and Preissler, 1980; Ribalet and Beigelman, 1980), and D-600 (Matthews and Sakamoto, 1975; Ribalet and Beigelman, 1980). A further argument for the importance of Ca^{2+} influx is that the magnitude of the rapid depolarization increases with the Ca^{2+} gradient across the plasma membrane (Meissner and Preissler, 1979, 1980; Meissner et al., 1979; Atwater et al., 1980; Ribalet and Beigelman, 1980).

It is likely that the rapid spike depolarization from V_p to V_s reflects the opening of the voltage-dependent Ca^{2+} channels (Atwater et al., 1980; Cook et al., 1980; Ribalet and Beigelman, 1980), since the spike peak potential follows approximately the Nernst equation as the Ca^{2+} concentration increases. The Ca^{2+} channels are not completely selective, and there may be some influx of Na^+. The repolarization of the spikes from V_s back to V_p is delayed by tetraethylammonium (Atwater et al., 1979a; Ribalet and Beigelman, 1980) and blocked by a high concentration of quinidine (Ribalet and Beigelman, 1980). It has therefore been suggested that repolarization is due either to activation of voltage-dependent K^+ channels (Atwater et al., 1979a) or to the Ca^{2+}-sensitive K^+ permeability (Ribalet and Beigelman, 1980).

The termination of bursts by repolarization from V_p to V_r is prevented by inhibitors of K^+ permeability (Atwater et al., 1979a, 1980; Meissner and Preissler, 1979; Ribalet and Beigelman, 1980) and is therefore believed to result from an increased K^+ conductance. Both voltage-dependent K^+ channels (Cook et al., 1981) and Ca^{2+}-sensitive K^+ permeability (Atwater et al., 1979a, 1980; Meissner and Preissler, 1979; Ribalet and Beigelman, 1979) have been considered. The latter alternative seems more probable, since an accumulation of Ca^{2+} and perhaps of Na^+ is to be expected during bursts. With an increased Na^+ concentration, there will also be activation of the electrogenic Na^+/K^+ pump (Henquin and Meissner, 1982; Ribalet and Beigelman, 1982; Meissner and Henquin, 1984).

The role of glucose in the regulation of the electrical activity of the β-cells may be summarized as follows. At concentrations of glucose with no effect on insulin release (3–5.5 mM), a reduction of the Ca^{2+} activity in the cytoplasm after intracellular buffering decreases the K^+ permeability, initiating a depolarization which does not reach the threshold potential (Meissner and Preissler, 1979; Henquin and Meissner, 1982; Hellman and Gylfe, 1985). When the glucose concentration is raised, the increased Ca^{2+} buffering will result in a further depolarization to V_t with the appearance of a burst pattern. The action potential may be explained in terms of an initial influx of Ca^{2+} followed by an efflux of K^+, both of which are voltage dependent. During each burst, the uptake of Ca^{2+} supersedes Ca^{2+} removal by sequestration and outward transport. Ca^{2+} in the cytoplasm consequently increases, followed by activation of Ca^{2+}-dependent K^+ permeability and termination of the burst by repolarization.

When the glucose concentration is raised further, there is a concomitant increase in the sequestration and outward transport of Ca^{2+} (see Section VI,C), resulting in prolongation of the burst. The considerably longer first burst of spikes may reflect a high initial rate of Ca^{2+} uptake into depleted stores which subsequently become saturated. At maximally stimulating concentrations of glucose, a continuous burst pattern emerges when the influx and removal of Ca^{2+} are in balance. In contrast to previous attempts to explain regulation of burst duration by Ca^{2+}-activated K^+ permeability (Atwater et al., 1979b, 1983; Ribalet and Beigelman, 1979), the time-average effect of insulin-releasing concentrations of glucose is to increase Ca^{2+} activity in the cytoplasm (see Sections VII,A and X,A). It has been possible to simulate all the basic features of the burst pattern in a model when including a Ca^{2+}-activated K^+ permeability, voltage-dependent channels for K^+ and Ca^{2+} as well as a glucose-stimulated intracellular buffering of Ca^{2+} (Chay and Keizer, 1983).

B. Calcium Influx

The influx of Ca^{2+} into the pancreatic β-cell has in most cases been estimated from measurements of the uptake of ^{45}Ca by isolated pancreatic islets. Experimental difficulties in such an approach include the presence of radioactivity in the extracellular fluid as well as a considerable binding of ^{45}Ca to the surface of cells and to the surrounding capsule (see Section IV). Extracellular space markers have been used to correct for contaminating radioactivity (Hellman et al., 1971b; Naber et al., 1977; Wollheim et al., 1977; Frankel et al., 1978b), but this technique does not cope with external binding of ^{45}Ca. Rinsing of ^{45}Ca-incubated islets with nonradioactive medium will decrease the radioactivity in the extracellular space (Wolters et al., 1982), and with extensive rinsing (Malaisse-Lagae and Malaisse, 1971) a large fraction of the externally bound ^{45}Ca will also be removed. Unfortunately, even when rinsing at reduced temperature, there will be a compromise between effective removal of superficial ^{45}Ca and loss of intracellular radioactivity. The lanthanum wash procedure represents a considerable improvement for discrimination between intracellular and extracellular ^{45}Ca (Hellman et al., 1976a). The La^{3+} ion effectively displaces extracellular Ca^{2+} and makes the β-cells almost impermeable to Ca^{2+}, particularly when washed with cold medium (Hellman, 1978).

The Ca^{2+} activity of the β-cell cytoplasm is about 10,000 times lower than that in the extracellular fluid (see Section VII,A), and with the membrane potential, there is an enormous electrochemical gradient facilitating the influx of Ca^{2+}. Such influx can be due to leakage and to gating of Ca^{2+} through specific membrane channels. As indicated from the electrophysiology (see Section VI,A), the voltage-dependent "late" Ca^{2+} channel is an important route for Ca^{2+} influx. This conclusion is supported by measurements of the effects of K^+ depolarization (Malaisse-Lagae and Malaisse, 1971; Hellman et al., 1978b;

Wollheim *et al.*, 1980; Andersson *et al.*, 1981b), and the actions of inhibitors like verapamil, D-600, and Co^{2+} on ^{45}Ca uptake (Henquin and Lambert, 1975; Malaisse *et al.*, 1976b, 1977; Donatsch *et al.*, 1977; Wollheim *et al.*, 1978b, 1980). Also the veratridine-sensitive Na^+ channel which is blocked by tetrodotoxin allows Ca^{2+} influx (Donatsch *et al.*, 1977; Pace, 1979), although this channel can only account for a minor fraction of the uptake (Wollheim and Sharp, 1981). Also, we must consider that there are receptor-operated Ca^{2+} channels specific for certain stimuli.

Under resting conditions with no net fluxes of Ca^{2+}, uptake of ^{45}Ca is probably due to leakage balanced by outward transport and to $^{40}Ca-^{45}Ca$ exchange. Since the original demonstration with double isotope (Hellman *et al.*, 1971b) and rinsing (Malaisse-Lagae and Malaisse, 1971) techniques that glucose stimulates ^{45}Ca uptake by β-cells, this has been confirmed on numerous occasions. That the ^{45}Ca taken up in response to glucose is located intracellularly was established with lanthanum washing (Hellman *et al.*, 1976a; Hellman, 1978). The time course of glucose-stimulated ^{45}Ca uptake exhibits a rapid influx followed by a slower phase approaching an apparent steady state within 60–120 min (Hellman *et al.*, 1971b, 1976a; Naber *et al.*, 1977; Wollheim *et al.*, 1977; Frankel *et al.*, 1978b). Isotopic equilibrium is not reached until after 24 hr (Ribes *et al.*, 1981).

Figure 8 shows the intracellular uptake of ^{45}Ca at different concentrations of glucose in relation to insulin release and oxidation of the sugar. Like insulin

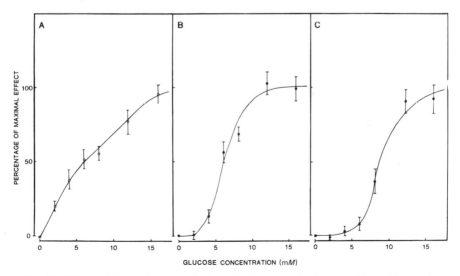

Fig. 8. Effects of increasing concentrations of glucose on *ob/ob*-mouse islets with regard to glucose oxidation, uptake of ^{45}Ca, and insulin release. The oxidation of D-(U-^{14}C)glucose was studied for 60 min (A), intracellular incorporation of ^{45}Ca for 90 min (B), and insulin release for 60 min (C). The data represent the increases above the values obtained in the absence of glucose expressed in per cent of those at 20 m*M* of the sugar. Mean values ±SEM for 6–12 experiments.

release, ^{45}Ca incorporation was related to extracellular glucose concentration sigmoidally. However, the release did not start to increase until at about 5 mmol per liter glucose, when ^{45}Ca uptake was almost half-maximal. Oxidation was even more sensitive to extracellular glucose concentration, exhibiting a hyperbolic dose–response relationship. At low glucose concentrations, there is consequently a lack of correlation between insulin release and ^{45}Ca uptake. However, as the sugar concentration is increased to depolarize the β-cells (see Section VI,A), a linear relationship becomes apparent (Malaisse-Lagae and Malaisse, 1971). There are also other situations with a considerable uptake of ^{45}Ca without a corresponding release of insulin. Fructose stimulation is one such case (Hellman *et al.*, 1978a), and other examples are exposure of pancreatic islets to glucose at reduced temperature (Hellman and Andersson, 1978) or glucose stimulation of islets from starved animals (Hellman *et al.*, 1978a). The rate of fructose oxidation is only 10% of that of glucose (Hellman *et al.*, 1978a), and reduced temperature and starvation decrease glucose metabolism (Hedeskov and Capito, 1974). A simplistic interpretation of these data is therefore that, apart from the ^{45}Ca influx through the voltage-dependent channels during depolarization, there is also an uptake which can be maintained by increased metabolism when these Ca^{2+} channels are closed. If glucose stimulates the sequestration of Ca^{2+} leaking into the β-cells, it is unnecessary to postulate the existence of voltage-independent channels for Ca^{2+} regulated by glucose (see Section VI,C).

Studies of ^{45}Ca uptake may be a very coarse and unreliable measure of the net uptake of calcium. Indeed, a major effect of glucose is to stimulate the turnover of calcium, and it has been difficult to demonstrate any change in the total amount of calcium in the β-cells after glucose exposure (see Section V). However, glucose stimulates not only the uptake of ^{45}Ca but also that of other divalent cations such as $^{28}Mg^{2+}$ (Henquin *et al.*, 1983b), $^{89}Sr^{2+}$ (Henquin, 1980b), Mn^{2+} (Rorsman *et al.*, 1982), and Cd^{2+} (Nilsson *et al.*, 1984). More important, by monitoring alterations of the Ca^{2+} concentration in a suspending medium containing micromolar concentrations of the cation, it has been possible to establish glucose-stimulated net uptake of Ca^{2+} both in β-cell-rich pancreatic islets from *ob/ob*-mice (Gylfe, 1982) and in the insulin-secreting clonal cell line RINm5F (Gylfe *et al.*, 1983). Since the RINm5F cells do not seem to be depolarized by glucose (Gylfe *et al.*, 1983), the participation of voltage-dependent Ca^{2+} channels in glucose-stimulated uptake is unlikely. This conclusion is supported by the lack of effect K^+ depolarization has on the net uptake of Ca^{2+} at micromolar concentrations (Gylfe *et al.*, 1983).

At physiological concentrations of Ca^{2+}, K^+ depolarization can mimic the stimulatory effect of glucose on ^{45}Ca uptake (Malaisse-Lagae and Malaisse, 1971; Hellman *et al.*, 1978b; Wollheim *et al.*, 1980; Andersson *et al.*, 1981b). The depolarization-independent mode of glucose-stimulated Ca^{2+} uptake may therefore be characterized by the difference between ^{45}Ca incorporation during glucose and K^+ depolarization. Such analyses have revealed a considerably

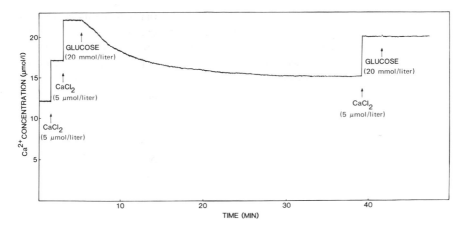

Fig. 9. Effect of glucose on the net uptake of Ca^{2+} into RINm5F cells. The cells (4.4 mg protein) were suspended in 1 ml medium (Gylfe *et al.*, 1983) containing 20 μM Arsenazo III and phenol red. The presence of these indicators enabled the simultaneous monitoring of Ca^{2+} activities and pH by recording the absorbance differences 675–685 nm and 499–525 nm, respectively. It was consequently possible to keep the pH constant at 7.4 throughout the experiment by addition of NaOH. The absolute concentration of Ca^{2+} was determined by EGTA titration (not shown).

lower mobility of the ^{45}Ca taken up in response to glucose (Andersson *et al.*, 1981b). The observations that glucose can reduce ^{45}Ca efflux (see Section VI,C), lower the cytoplasmic Ca^{2+} activity (see Section VII,A), inhibit insulin release (see Section III), decrease K^+ permeability (see Section VI,A), and enhance organelle accumulation of ^{45}Ca (see Section VII) are forceful arguments for the concept that the depolarization-independent mode of glucose-stimulated Ca^{2+} uptake represents organelle sequestration of the element. By measuring the net uptake of Ca^{2+} into RINm5F cells at micromolar extracellular Ca^{2+}, it should be possible to estimate the capacity of this glucose-stimulated buffering. Figure 9 shows that the glucose-stimulated net uptake of Ca^{2+} reaches saturation within 25–30 min, when it corresponds to 1.6 mmol/kg protein. A value of this magnitude should be expected from estimates of initial ^{45}Ca influx into normal β-cells (Wollheim and Sharp, 1981) and from glucose's lack of effect on the total content of β-cell calcium (see Section V). The saturation is not due to reduction of the extracellular Ca^{2+} concentration or to consumption of glucose, because further additions of Ca^{2+} and sugar had no effect. It is also unlikely that it represents activation of Ca^{2+} extrusion, since glucose decreases the cytoplasmic Ca^{2+} activity in the RINm5F cells (Rorsman *et al.*, 1983).

In contrast to the depolarization-independent mode of glucose-stimulated Ca^{2+} uptake, that occurring via the voltage-dependent Ca^{2+} channels will increase the Ca^{2+} activity of the cytoplasm (Rorsman *et al.*, 1984). It is not surprising that the dose–response relationship for glucose-stimulated ^{45}Ca uptake exhibits a sensitivity to glucose which is intermediate between that of insulin release and glucose oxidation (Fig. 8). Glucose-stimulated insulin release requires the opening of voltage-dependent Ca^{2+} channels, whereas the depolarization-independent uptake of Ca^{2+} is more closely associated with glucose metabolism. The time course of the depolarization-independent mode of glucose-stimulated Ca^{2+} uptake shown in Fig. 9 is consistent with the idea that the slowly rising second phase of insulin release from normal β-cells is due to a decreasing Ca^{2+} sequestration.

C. Calcium Efflux

The extrusion of Ca^{2+} from the pancreatic β-cells is mediated both by $Na^+ - Ca^{2+}$ exchange diffusion (Hellman et al., 1980c; Herchuelz et al., 1980b; Siegel et al., 1980b) and by a calmodulin-activated transport system based on a high affinity Ca^{2+},Mg^{2+}-ATPase (Pershadsingh et al., 1980). There are fundamental differences between this ATPase and that facilitating Ca^{2+} uptake into the endoplasmic reticulum (Colca et al., 1983c). In contrast to the calcium uptake by an islet endoplasmic reticulum fraction, that of plasma membrane-enriched vesicles which exhibited a higher pH optimum was not sustained by oxalate or stimulated by K^+ and had an approximately 30-fold greater affinity for ionized calcium. The apparent K_m for the ATP-stimulated uptake in the plasma membrane fraction was 50 ± 3 nM Ca^{2+}. With such a high affinity for Ca^{2+}, this calcium extrusion pump probably plays a significant role in maintaining the low cytoplasmic Ca^{2+} activity of the resting β-cell.

The exploration of Ca^{2+} extrusion from the β-cells has been considerably aided by the use of radioactive ^{45}Ca. When ^{45}Ca washout was analyzed by measuring the radioisotope remaining in islets after incubation in a nonradioac-

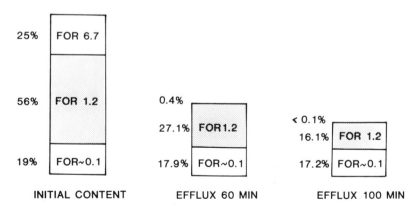

Fig. 10. Relative amounts of ^{45}Ca in different cellular pools before and after perifusion of rat islets with nonradioactive medium. Each pool is characterized by its fractional outflow rate (FOR), i.e., the ^{45}Ca mobilized per minute in percentage of the instantaneous islet content. The illustration is based on data reported by Frankel et al. (1978a).

tive medium, Naber *et al.* (1977) observed a rapid loss during the first minute followed by a slower phase with a fractional outflow rate of 3.1 ± 0.6%/min. Measurements of the effluent radioactivity in a perifusate allowed the identification of three separate cellular compartments for calcium (Frankel *et al.*, 1978a). Figure 10 shows a schematic presentation of the relative amounts of ^{45}Ca in these compartments in unstimulated rat islets after different durations of perifusion with a nonradioactive medium. It is evident that most of the ^{45}Ca appears in a slowly exchangeable pool with a fractional outflow rate of 1.2%/min and that practically all ^{45}Ca released beyond 60 min of perifusion originates from this pool.

Measurement of ^{45}Ca in perifusion media is a much more sensitive index of changes in the Ca^{2+} efflux than is the radioactivity remaining in the islets

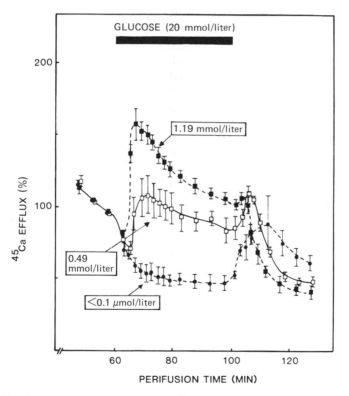

Fig. 11. Effects of glucose on the efflux of ^{45}Ca from *ob/ob*-mouse islets perifused with media containing different concentrations of Ca^{2+}. The islets were loaded with ^{45}Ca in the presence of 20 m*M* glucose. After perifusion with glucose-free medium the concentration of the sugar was increased to 20 m*M* during the period indicated by the horizontal black bar. The lowest Ca^{2+} concentration (<0.1 μ*M*) was obtained by including 0.5 m*M* EGTA in the perifusion medium. The data are given as a percentage of the average ^{45}Ca efflux during the 10 min period preceding the introduction of glucose. Mean values ±SEM for 4 experiments.

(Frankel *et al.*, 1978a; Gylfe and Hellman, 1978). Using the former approach, various researchers have found glucose to have both stimulatory and inhibitory effects on the efflux of the isotope from islets of animals (Malaisse *et al.*, 1973; Bukowiecki and Freinkel, 1976; Frankel *et al.*, 1978a; Gylfe and Hellman, 1978; Hellman *et al.*, 1980b; Abrahamsson *et al.*, 1981; Herchuelz and Malaisse, 1981) and man (Henriksson *et al.*, 1978). The effect of 20 mM glucose on ^{45}Ca efflux into a perifusion medium containing different concentrations of Ca^{2+} is shown in Fig. 11. At a physiological extracellular Ca^{2+} concentration the presence of glucose resulted in a prominent stimulation of ^{45}Ca efflux after a short initial inhibition (upper curve). The inhibitory component in the glucose action was unmasked with lowering of the medium concentration of Ca^{2+}. Only inhibition was seen when Ca^{2+} was <0.1 μM (lower curve). The inhibitory component precedes the stimulatory one, and it disappears more rapidly when glucose is omitted. The latter phenomenon explains the off-response with transiently increased ^{45}Ca efflux when glucose is omitted from medium with moderately reduced concentrations of Ca^{2+} (middle curve). Although the stimulatory and inhibitory actions of glucose differ with regard to their sensitivities to the sugar, they both depend on its metabolism (Herchuelz and Malaisse, 1980a, 1981). As

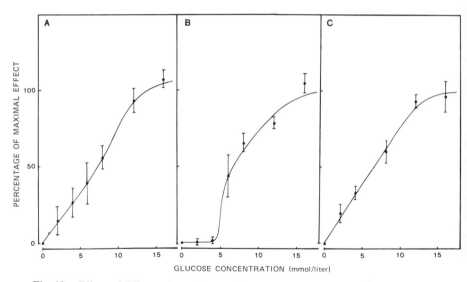

Fig. 12. Effects of different glucose concentrations on the efflux of ^{45}Ca from perifused *ob/ob*-mouse islets. Data are presented for the inhibitory (A) and stimulatory (B) effects obtained 10 min after introducing glucose into a medium deficient in or containing 1.28 mM Ca^{2+} respectively using islets loaded with ^{45}Ca in the presence of 20 mM glucose. Also, the significance of the glucose concentrations during loading for the subsequent stimulation of ^{45}Ca efflux with 20 mM glucose is illustrated (C). The data represent increases above the values obtained in the absence of glucose expressed in a percentage of those at 20 mM of the sugar. Mean values ±SEM for 5–11 experiments.

shown in Fig. 12, the dose–response relationship for the inhibitory effect (A) is similar to that for glucose oxidation (Fig. 8) in exhibiting essentially hyperbolic characteristics. However, the stimulatory action of glucose (B) mimics the intracellular uptake of ^{45}Ca and insulin release (Fig. 8) in being sigmoidal. It follows from these relationships that glucose at concentrations up to 4 mmol/liter inhibits the efflux of ^{45}Ca (Herchuelz and Malaisse, 1981; Abrahamsson et al., 1981; Hellman and Gylfe, 1984b).

The stimulatory effect of glucose on ^{45}Ca efflux resembles that on insulin release in being associated with depolarization of the β-cells and requiring the presence of extracellular Ca^{2+}. The increased ^{45}Ca efflux was at first attributed to loss of calcium with the secretory granules during exocytosis (Malaisse et al., 1973). However, later it became evident that such a mechanism could only account for a minor part of the effect, since glucose was found to stimulate ^{45}Ca efflux also under conditions of inhibited insulin release (Wollheim et al., 1977; Herchuelz and Malaisse, 1978; Gylfe and Hellman, 1978; Kikuchi et al., 1978). The magnitude of the glucose-stimulated ^{45}Ca efflux is proportional to the extracellular concentration of Ca^{2+} (Abrahamsson et al., 1981). Also, other agents which favour the entry of Ca^{2+} mobilize ^{45}Ca. In fact, stimulation of ^{45}Ca efflux provides a very sensitive index of the opening of the voltage-dependent channels in the β-cell plasma membrane (Hellman, 1981, 1982a). Even the introduction of Ca^{2+} into a perifusion medium deficient in this ion results in a marked increase in the efflux of ^{45}Ca. It is therefore likely that the stimulatory component in the glucose action is due essentially to displacement of ^{45}Ca from intracellular binding sites following increased entry of nonradioactive Ca^{2+} (Hellman et al., 1979a,b; Herchuelz et al., 1980a,c, 1981; Abrahamsson et al., 1981). The intracellular ^{40}Ca–^{45}Ca exchange is consequently sufficient to overcome the competitive inhibition of the outward transport of ^{45}Ca exerted by the entering ^{40}Ca. It has been suggested that glucose's promotion of ^{45}Ca efflux reflects also a true mobilization of Ca^{2+} from intracellular stores (Janjic et al., 1982; Prentki and Wollheim, 1984). However, this idea is difficult to reconcile with the observation of a glucose depression of the cytoplasmic Ca^{2+} activity under conditions of prevented entry of Ca^{2+} (see Section VII,A). Isolated pancreatic islets are not unique in responding with stimulation of ^{45}Ca efflux when exposed to agents which increase the entry of nonradioactive Ca^{2+}. Similar effects have been observed during perifusion of the posterior pituitary and the adrenal medulla (Gylfe and Hellman, 1982b).

The increase in ^{45}Ca efflux (and insulin release) in response to low concentrations of glucose (8.3 mM) has been found to be less sensitive to the inhibitory action of the organic calcium antagonist verapamil than to the stimulation evoked by higher concentrations or K^+ depolarization (Lebrun et al., 1982a). Since verapamil was reported to interfere also with the inhibitory component in the glucose action on ^{45}Ca efflux, it seems premature to postulate that the glucose-stimulated influx of Ca^{2+} involves channels other than those sensitive to the membrane potential.

The relative amounts of ^{45}Ca released during perifusion of the islets are markedly influenced by the presence of glucose during the loading procedure. The basal fractional outflow rate has been found to be lower when the islets are preloaded in the presence of glucose (Gylfe and Hellman, 1978; Lebrun et al., 1982b). Accordingly, ^{45}Ca taken up in response to glucose has a lower mobility than that incorporated in the absence of the sugar. Also, as shown in Fig. 12C, the glucose stimulation of ^{45}Ca efflux depends on the presence of glucose during the loading procedure. The dose–response relationship mimics that of glucose oxidation (Fig. 8A), implying that even the presence of nondepolarizing concentrations contributes significantly to the subsequent increase of ^{45}Ca efflux obtained with glucose stimulation. After a period of glucose deprivation, there is a reduction in the ability of the β-cells to metabolize the sugar (Lebrun et al., 1982b). Impairment of the metabolism may consequently be another factor when glucose fails to induce a prominent stimulation of the ^{45}Ca efflux from islets preloaded in the absence of the sugar (Lebrun et al., 1982b). The significance of glucose during the loading procedure is indicated also from the observation that both inhibitors of the metabolism (Hellman, 1979) and a rise in intracellular Na^+ (Hellman et al., 1982) result in selective mobilization of the ^{45}Ca incorporated in response to the sugar.

During perifusion with Ca^{2+}-deficient medium, glucose also inhibits the efflux of ^{45}Ca when the islets are loaded to isotopic equilibrium (Gylfe et al., 1978; Kikuchi et al., 1978). The observed inhibition can consequently be regarded as a true decrease of Ca^{2+} efflux rather than preferential mobilization of calcium stores with low specific radioactivity. Glucose interferes also with efflux of other divalent cations used as analogs of Ca^{2+} in biological systems. So far, glucose has been found to have an inhibitory action on the efflux of Sr^{2+} (Henquin, 1980b), Mn^{2+} (Rorsman et al., 1982), and Cd^{2+} (T. Nilsson, P.-O. Berggren, and B. Hellman, unpublished data). Although islets have been found to contain Ca^{2+},Mg^{2+}-ATPase activity subject to glucose inhibition (Levin et al., 1978), the significance of this observation for the outward transport of Ca^{2+} remains to be elucidated. Attention has been paid to the possibilities of a glucose interference with the Na^+,Ca^{2+} countertransport system (Siegel et al., 1980b; Herchuelz and Malaisse, 1981; Janjic and Wollheim, 1983), but there is no evidence that glucose inhibits Ca^{2+} efflux by increasing the Na^+ activity in the β-cells (Sehlin and Täljedal, 1974a,b; Kavazu et al., 1978). It has instead been postulated that the glucose inhibition of Ca^{2+} efflux reflects an increased production of H^+, competing with Ca^{2+} for exit by the Na^+,Ca^{2+} countertransport system (Malaisse et al., 1979a; Herchuelz and Malaisse, 1981; Lebrun et al., 1982c). Various observations urge for a reconsideration of this view. It has, for example, been reported that exposure to glucose makes β-cells more alkaline rather than reducing the intracellular pH (Lindström and Sehlin, 1984). Moreover, glucose is equally effective in inhibiting ^{45}Ca efflux in the absence of Na$^+$ when loss of intracellular K$^+$ is prevented by substituting Na$^+$ for K$^+$ (Hellman and Gylfe, 1984a).

The idea of a glucose interference with the active extrusion of Ca^{2+} is difficult to reconcile with the observation that the inhibitory action of 4 mM glucose remains relatively unaffected by alterations of the magnitude and even direction of the Ca^{2+} gradient across the plasma membrane (Hellman and Gylfe, 1984b). If instead the inhibitory effect reflects lowering of the cytoplasmic Ca^{2+} activity after trapping of the ion in organelles, a number of experimental pieces of evidence fall into place. After the proposal of such a mechanism (Hellman et al., 1979b), various kinds of evidence have supported its existence (see Sections VI,A and VII) culminating in the recent demonstration that glucose under certain conditions can lower cytoplasmic Ca^{2+} and inhibit insulin release (Rorsman et al., 1983, 1984; Hellman and Gylfe, 1984b, 1985; Hellman et al., 1985).

As shown in Table I, agents other than glucose also exhibit stimulatory and inhibitory actions on the efflux of ^{45}Ca during perifusion of isolated islets. In accordance with the view that the glucose effect is mediated by metabolism is the fact that other nutrients like pyruvate (Sener et al., 1978), D-glyceraldehyde (Herchuelz and Malaisse, 1978; Lebrun et al., 1982b), α-ketoisocaproic acid (Hutton et al., 1980; Lebrun et al., 1982b), and L-leucine (Malaisse et al., 1980;

TABLE I

Modifications of ^{45}Ca Efflux from Isolated Islets after Introducing Various Additives into a Perifusion Medium[a]

Additives to the medium	Inhibition of efflux	Stimulation of efflux
Pyruvate (30 mM)	+	+[b]
D-Glyceraldehyde (10 mM)	+	+
α-Ketoisocaproate (10–25 mM)	+	+
L-Leucine (20 mM)	+	+[c]
L-Alanine (20 mM)	0	+
L-Arginine (10–20 mM)	(+)	+
Sulfonylurea	0	+
Excessive K$^+$	0	+
Phosphate (10 mM)	+	0
Quinine (100 μM)	+	+
Ouabain (1 mM)	0	+
Veratridine (100 μM)	0	+

[a]The data are taken from the literature. The presence and absence of effects as well as conflicting opinions are indicated by +, 0, and (+) respectively.

[b]In the presence of D-glucose.

[c]In the presence of L-glutamine.

Charles and Henquin, 1983) have dual actions on ^{45}Ca efflux. Whereas the stimulatory effect of L-alanine can be attributed to mobilization of intracellular ^{45}Ca by Na$^+$ cotransported with the amino acid (Charles and Henquin, 1983), divergent opinions have been expressed regarding the mechanism for arginine stimulation and whether this amino acid also has an inhibitory action (Charles and Henquin, 1983; Herchuelz *et al.*, 1984). Like excessive K$^+$, hypoglycemic sulfonylureas promote ^{45}Ca efflux in the presence of physiological concentrations of Ca^{2+} but not in media deficient in the cation (Hellman, 1981, 1982a,b; Lebrun *et al.*, 1982d; Gylfe *et al.*, 1984). This observation is a strong argument for the view that the action of sulfonylureas is to increase the cytoplasmic Ca^{2+} activity by depolarizing the β-cells rather than by acting as Ca^{2+} ionophores. Because it promotes intracellular sequestration of Ca^{2+}, it is not surprising that inorganic phosphate inhibits ^{45}Ca efflux (Hellman and Andersson, 1978). The dual effects of glucose on ^{45}Ca efflux can be reproduced by quinine (Herchuelz *et al.*, 1981). However, in contradistinction to glucose the inhibition by quinine can easily be explained in terms of direct inhibition of the outward transport across the plasma membrane. Under conditions of raised intracellular Na$^+$ (addition of veratridine and ouabain or removal of K$^+$), there is a mobilization of intracellular calcium stores as indicated from stimulation of ^{45}Ca efflux during perifusion with medium deficient in Ca^{2+} (Hellman *et al.*, 1982). Evidently, glucose and Na$^+$ have opposite effects on the intracellular sequestration of calcium. A rise in the cytoplasmic Na$^+$ results in a preferential mobilization of ^{45}Ca incorporated in response to glucose, and the sugar also suppresses insulin release obtained in a Ca^{2+}-deficient medium by Na$^+$ mobilization of intracellular calcium (Hellman *et al.*, 1982). Glucose has been found to attenuate the increase of ^{45}Ca efflux associated with a rise in intracellular Na$^+$ in the presence of physiological concentrations of extracellular Ca^{2+} (Herchuelz and Malaisse, 1980b). Furthermore, there is an almost complete suppression of the alanine-induced stimulation of ^{45}Ca efflux in the presence of 7 m*M* glucose, a concentration which does not affect the cotransport of the amino acid with Na$^+$ into the β-cells (Charles and Henquin, 1983).

VII. INTRACELLULAR CALCIUM STORES

A. Cytoplasm

X-Ray microanalysis of frozen sections from unfixed rat islets indicates that the cytoplasm of the β-cells may contain substantial amounts of calcium (Howell *et al.*, 1975). However, studies of insulin release from β-cells made permeable to small molecules by electric discharges indicate that the transition from the resting to the secreting state takes place at Ca^{2+} activities as low as 100–1000 nM (Pace *et al.*, 1980; Yaseen *et al.*, 1982). The latter observation is supported by direct measurements of free ionized Ca^{2+} in clonal tumor β-cells (Rorsman *et al.*, 1983; Boyd *et al.*, 1984; Wollheim and Pozzan, 1984) and normal β-cells from *ob/ob*-mice (Rorsman *et al.*, 1984), where the resting activities were 100–175 nM. As in other cells, the cytoplasm of the β-cells thus appears to have a considerable buffering capacity for Ca^{2+}. Such buffering may be explained as binding to anions and macromolecules, some of which, like calmodulin, have specific regulatory functions. Calmodulin alone has a potential for Ca^{2+} binding corresponding to 60–200 μM (see Section VIII,C), although little of this capacity is probably used under resting conditions. The large number of binding sites available and the presence of Ca^{2+} buffering organelles will severely retard diffusion of the ion (Matthews, 1979). Considerable concentration gradients can therefore be expected within the cytoplasm after sudden changes in Ca^{2+} fluxes. The Ca^{2+} activity triggering exocytosis close to the plasma membrane may consequently be much higher than that in the rest of the cytoplasm. It has even been proposed that there is a marked reduction of the average Ca^{2+} activity in the cytoplasm during glucose-stimulated insulin release, although there would be an increase in Ca^{2+} at the plasma membrane (Atwater *et al.*, 1979b, 1983).

The Quin-2 technique has made it possible to measure changes in the cytoplasmic Ca^{2+} activity during stimulation of insulin release. Despite the fact that this method tends to underestimate the average Ca^{2+} activity when local variations occur, secretion induced by K^+ depolarization was found to be associated with a considerable rise in this activity both in clonal β-cell lines (Rorsman *et al.*, 1983; Boyd *et al.*, 1984; Wollheim and Pozzan, 1984) and normal β-cells from *ob/ob*-mice (Rorsman *et al.*, 1984). The cytoplasmic Ca^{2+} activity was also raised during glucose-stimulated insulin release (Rorsman *et al.*, 1984). After an initial delay of 1–2 min, 20 mM glucose caused a sustained 40% increase in the Ca^{2+} activity of the cytoplasm of normal β-cells (Fig. 13A). The sugar has also been reported to raise the Ca^{2+} activity by almost 30% in a hamster insulinoma cell line (Boyd *et al.*, 1984), but the extreme rapidity of the latter effect calls for confirmatory studies.

Although glucose induces a sustained increase in the average Ca^{2+} activity during insulin release, the sugar also has the ability to reduce Ca^{2+} in the cytoplasm (Rorsman *et al.*, 1983, 1984). By restricting the entry of Ca^{2+} into normal β-cells, it was possible to demonstrate a prompt 15% decrease in cytoplasmic Ca^{2+} after exposure to glucose (Fig. 13B). In the clonal rat β-cell line RINm5F, which does not release insulin in response to glucose, similar results were obtained without restriction of Ca^{2+} influx (Rorsman *et al.*, 1983). In light of these data, it is not surprising that under certain conditions glucose can inhibit insulin release (Hellman *et al.*, 1982, 1985; Hellman and Gylfe, 1984b). The timing of the two opposite effects of glucose on the Ca^{2+} activity of the cytoplasm is similar to the rapid inhibition and delayed stimulation of ^{45}Ca efflux

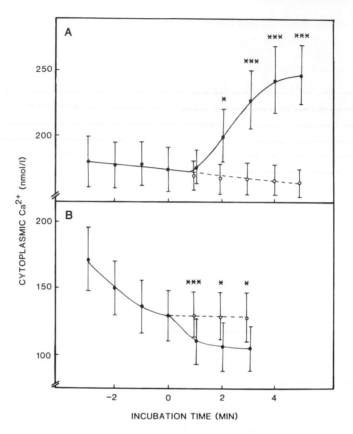

Fig. 13. Effects of glucose on the cytoplasmic-Ca^{2+} activity measured with the Quin-2 technique in suspended cells obtained from the β-cell-rich pancreatic islets of *ob/ob*-mice. At zero time, 20 mM glucose was added to a medium containing 1.20 (A) or 0.20 (B) mM Ca^{2+}. Open symbols and dotted lines indicate the Ca^{2+} activities expected without modification of the medium. Mean values ±SEM for 11 and 4 experiments, respectively. *P <0.05, and ***P <0.001. From Rorsman *et al.* (1984).

from preloaded pancreatic islets (see Section VI,C). The efflux consequently appears to reflect the Ca^{2+} activity at the sites for outward transport of the ion. If this is the case, glucose should initially lower the Ca^{2+} activity of the cytoplasm. Such an effect may be a triggering factor in membrane depolarization eventually leading to the opening of voltage-dependent channels and influx of Ca^{2+} (see Section VI,A). During stimulation of insulin release, the Ca^{2+} activity in the cytoplasm will oscillate with the membrane potential, the average activity being increased.

B. Endoplasmic Reticulum

The endoplasmic reticulum accounts for almost 20% of the ·volume of the pancreatic β-cells, providing a transport area for Ca^{2+} equivalent to about 9 times that of the plasma membrane (Dean, 1973). The presence of ATP-dependent uptake of ^{45}Ca has been demonstrated in a crude microsomal fraction obtained from β-cell-rich pancreatic islets of *ob/ob*-mice (Sehlin, 1976). In other studies, microsomal calcium metabolism was investigated by *in situ* labeling of the islets followed by isolation of the organelles under conditions minimizing the redistribution of calcium. The exposure of islets from both rats (Borowitz and Matthews, 1980) and *ob/ob*-mice (Kohnert *et al.*, 1979; Hahn *et al.*, 1980; Andersson *et al.*, 1982b) to raised concentrations of glucose has resulted in increased recovery of ^{45}Ca from the microsomal fraction. However, the effect of glucose was less pronounced than that in the mitochondria, and it was not mimicked by depolarizing the β-cells with K^+ (Andersson *et al.*, 1981b). When the glucose concentration was raised from 1 to 20 m*M* during 120 min incubation of intact *ob/ob*-mouse islets, the percentage of calcium exchanged in the microsomal fraction increased from 8.4 to 17.4 (Andersson *et al.*, 1982a).

The microsomal fraction obtained by differential centrifugation contains fragments not only of the endoplasmic reticulum but also of the plasma membrane. It is therefore important that it has been possible to obtain a relatively pure fraction of endoplasmic reticulum from batches of up to 10,000 rat islets. With access to this fraction Colca *et al.* (1982, 1983a,c) have considerably expanded our knowledge of calcium handling by the endoplasmic reticulum in the pancreatic β-cells. ATP was found to stimulate the uptake of Ca^{2+} into the membrane vesicles against a concentration gradient. The characteristics of this uptake were similar to that described for the endoplasmic reticulum in other cell types with regard to requirements for ATP, dependence on magnesium, and specific stimulation by K^+. Oxalate was required as an intravesicular trapping agent to linearize Ca^{2+} uptake over time. The uptake process differed from that in mitochondria in not being inhibited by azide or ruthenium red, and, unlike that in plasma membrane vesicles, it was not stimulated by calmodulin. Also, cyclic AMP lacked effect on Ca^{2+} uptake in the endoplasmic reticulum. Estimates of Ca^{2+} affinity of the uptake using Ca–EGTA buffers yielded an apparent K_m of 1.5 ± 0.3 μ*M*, a concentration considerably higher than that supposed to exist in the cytoplasm (see Section VII,A). Nevertheless, rat insulinoma microsomes have been reported to lower ambient Ca^{2+} to the range in the intact cells (Prentki and Wollheim, 1984). In the latter preparation inositol 1,4,5-triphosphate was found to induce a rapid and organelle-specific mobilization of Ca^{2+} (Prentki *et al.*, 1984). So far no attempts have been made to determine the steady-state ambient Ca^{2+} concentration maintained by a preparation of endoplasmic reticulum devoid of contaminating plasma membrane fragments.

Great caution is warranted when extrapolating the kinetic properties of the Ca^{2+} uptake observed in the isolated endoplasmic reticulum to that in the intact β-cells. It has for example been possible to demonstrate the existence of a cytosolic factor, which markedly increases both the rate of and capacity for Ca^{2+} uptake in the endoplasmic reticulum (Colca et al., 1983a). The observation that the islets contain endoplasmic reticulum, which in the isolated state exhibits properties for active Ca^{2+} transport similar to those in other cells, makes it likely that this membrane system is important for Ca^{2+} homeostasis also in the pancreatic β-cells.

C. Mitochondria

The pancreatic β-cell contains 400–1000 mitochondria with a total volume corresponding to 4–8% of that of the cells (Dean, 1973; Borg and Andersson, 1981). The potential significance of the mitochondria for the calcium metabolism is apparent from the area of the Ca^{2+}-transporting inner membrane, which is almost 10 times that of the plasma membrane (Borg and Andersson, 1981). Subcellular fractionation analyses of β-cell-rich pancreatic islets from ob/ob-mice under conditions preventing redistribution of calcium have indicated that the mitochondrial content of this element is 60 mmol/kg protein (Andersson et al., 1982a). Although containing only 10–14% of the total calcium in the β-cells, the concentration of mitochondrial calcium is an order of magnitude greater than that in liver and kidney (Borle, 1981). The idea that the β-cell mitochondria are rich in calcium has been supported by X-ray microanalysis of frozen sections of unfixed pancreas (Howell et al., 1975; Howell, 1977). When applying this technique for comparing the subcellular distribution of the element, it has been found that the calcium content of the mitochondria in other types of cells usually does not exceed 10 mmol/kg dry weight (Barnard, 1982).

In situ labeling of organelles in β-cell-rich pancreatic islets from ob/ob-mice indicates that the mitochondria account for a considerable part of the ^{45}Ca incorporated in response to glucose (Kohnert et al., 1979). Glucose has also been found to have a significant effect in retaining ^{45}Ca in the mitochondria when prelabeled islets are incubated in a Ca^{2+}-deficient medium (Borowitz and Matthews, 1980; Andersson, 1983). The increase in ^{45}Ca uptake after exposure to glucose was already manifested after 5 min, and reached an apparent steady state after 120 min (Andersson et al., 1982b). At the latter time, ^{45}Ca accounted for 9.5% of the mitochondrial calcium when the islets were exposed to 1 mM glucose, as compared with 27.1% when the glucose concentration was 20 mM (Andersson et al., 1982a). The glucose stimulation of the mitochondrial uptake of ^{45}Ca was dependent on an adequate supply of oxygen (Kohnert et al., 1979). The stimulatory effect of glucose became even more pronounced after prolonged starvation, which is known to abolish the secretory response to the sugar. Since starvation is associated with a lowering of cyclic AMP, it is of interest to note that conditions known to increase the concentrations of this nucleotide in the β-cell decrease the mitochondrial content of ^{45}Ca (Hahn et al., 1979, 1980; Borowitz and Matthews, 1980). Quantitative electron microscopic evaluation of the calcium distribution in rat β-cells with pyroantimonate precipitation indicated a glucose-stimulated increase of mitochondrial calcium in one experiment (Gagliardino et al., 1984) but not in another (Klöppel and Bommer, 1979). This difference may well reflect an inadequacy of the pyroantimonate approach in detecting changes of mitochondrial calcium. Even the inhibition of respiratory chain after addition of antimycin A has been found to be without effect on the number of mitochondrial pyroantimonate precipitates in the pancreatic β-cells (Lenzen and Klöppel, 1984).

The techniques for measuring the ^{45}Ca content of the mitochondria are more precise than those available for measuring their total calcium content. Accordingly, analyses with graphite furnace atomic absorption spectroscopy have so far failed to demonstrate a statistically significant net uptake of calcium in the mitochondria when the β-cells are exposed to glucose (Andersson et al., 1982a). The promotion of calcium sequestration in the mitochondria can be regarded as an effect of glucose which is independent of β-cell depolarization (Andersson, 1983). Thus, the effect of glucose in retaining ^{45}Ca in the mitochondria was not mimicked by the introduction of depolarizing concentrations of K^+ into a Ca^{2+}-deficient medium used for incubating preloaded islets. Moreover, the net uptake of ^{45}Ca into the mitochondria was found to be already activated at 4 mM glucose, a concentration which does not open the voltage-dependent Ca^{2+} channels.

The significance of glucose metabolism for mitochondrial accumulation of Ca^{2+} has also been analyzed with the use of isolated organelles. Sugden and Ashcroft (1978) studied rat islet mitochondria obtained by differential centrifugation followed by a filtration procedure permitting the passage of most of the contaminating secretory granules. The uptake of Ca^{2+} into the mitochondria was defined as the difference in the accumulation seen in the presence and absence of the proton ionophore carbonylcyanide p-trifluoromethoxyphenylhydrazone (FCCP). Even if the presence of a proton gradient also in the secretory granules may hinder the use of FCCP for discriminating between the uptakes into these organelles, the usefulness of the preparation for studying mitochondrial Ca^{2+} uptake was supported by the observation of a 95% inhibition with ruthenium red. In their tests of the effects of various glucose metabolites, Sugden and Ashcroft (1978) noted that both phosphoenolpyruvate and fructose 1,6-diphosphate decreased ^{45}Ca uptake in response to ATP. Also, methylxanthines inhibited the ATP-dependent uptake of ^{45}Ca, whereas cyclic AMP had no effect. Other studies have confirmed that phosphoenolpyruvate inhibits the uptake of ^{45}Ca in isolated rat islet mitochondria. Ewart et al. (1983) suggest that the phosphoenolpyruvate generated by glucose metabolism induces a rise of the concentration of

Ca^{2+} in the cytoplasm at the expense of that stored in the mitochondria by interacting with a mitochondrial adenine nucleotide translocase. It seems unlikely that such an action of phosphoenolpyruvate should be physiologically significant in view of the opposite effect of glucose on the mitochondria *in situ* (see above). In support of the existence of nonspecific effects of phosphoenolpyruvate, this metabolite has been reported to induce collapse of the membrane potential of liver mitochondria (Roos *et al.*, 1978) and interfere with the efflux of Ca^{2+} from isolated insulinoma mitochondria without establishing a new steady state for the element (Prentki *et al.*, 1983).

An important part in the elucidation of the role of mitochondria in the control of insulin release is the attempt to determine the set point for their buffering of cytoplasmic Ca^{2+}. So far such studies have been performed using well coupled mitochondria isolated from a rat insulinoma (Prentki *et al.*, 1983). These mitochondria maintained an extramitochondrial Ca^{2+} concentration of $0.9\ \mu M$ in the presence of $1\ mM\ Mg^{2+}$. The fact that this concentration is 5–9 times higher than the cytoplasmic Ca^{2+} activity in unstimulated β-cells (Rorsman *et al.*, 1983, 1984; Wollheim and Pozzan, 1984) may reflect limitations associated with studies of mitochondria suspended in a medium of different composition than that of the cytosol. In support of the idea that regulation of the extramitochondrial Ca^{2+} activity is mediated by independent uptake and efflux mechanisms, there was an immediate increase of the steady-state level after addition of ruthenium red. Whereas both an increase in Na^+ and a decrease in pH resulted in a rise of the extramitochondrial steady state concentration of Ca^{2+}, agents like cyclic AMP, IBMX, or NADH showed no significant effect. With the demonstration of a Na^+ effect also in the presence of ruthenium red, it seems likely that this cation stimulates the efflux rather than interferes with the uptake of Ca^{2+} into the mitochondria.

The oxidation of respiratory substrates like glucose increases the transmembrane potential in the mitochondria, favoring the uptake of Ca^{2+} into this organelle. Such an increase of the membrane potential has been postulated to activate also a component in the efflux mechanism for Ca^{2+} (Panten *et al.*, 1984), and uncertainties exist as to how glucose affects the cytoplasmic activities of Na^+ and H^+. Nevertheless, the net effect of glucose is to promote the active sequestration of calcium in the mitochondria independent of β-cell depolarization.

D. Secretory Granules

Each pancreatic β-cell contains on an average 13,000 granules, comprising 11.5% of the cell volume (Dean, 1973). The outer surface of the granule vesicle membrane carries a negative net charge, with the number of anionic sites reduced at the time of exocytosis (Howell and Tyhurst, 1977). X-Ray microanalysis of frozen sections from rat pancreas has indicated that the secretory granules are rich in calcium; the concentration is equivalent to that in mitochondria (Howell *et al.*, 1975). The pyroantimonate precipitation technique revealed deposits predominantly composed of calcium in the halos of the secretory granules. Whereas the number of these precipitates has been found to increase during glucose stimulation of insulin release (Herman *et al.*, 1973; Schäfer and Klöppel, 1974), there are divergent opinions as to whether this also applies to the initial phase of the secretory response (Klöppel and Bommer, 1979; Gagliardino *et al.*, 1984; Lenzen and Klöppel 1984). Contrary to what is seen during exposure to glucose, the stimulation of insulin release by sulfonylureas has been reported to be without effect on the number of pyroantimonate precipitates in the secretory granules (Klöppel and Schäfer, 1976; Lenzen and Klöppel, 1984). After pronounced sulfonylurea degranulation of the rat β-cells with a resulting 91% decrease in the insulin content, there was a 35% reduction in the cellular calcium content and an almost complete disappearance of the calcium histochemically detectable with the glyoxal-bis(2-hydroxyanil) technique (Wolters *et al.*, 1979, 1983). Even glucose can under certain conditions decrease the intracellular content of calcium by degranulating the β-cells (Bergsten and Hellman, 1984). The significance of calcium in the configuration of secretory granules can be demonstrated by culture of isolated islets in a medium deprived of Ca^{2+}. In this way the β-cells have been found to contain abnormally large and electron-translucent granules (Howell *et al.*, 1978).

The concept that the β-granules are organelles rich in calcium is supported by direct measurements of the element in subcellular fractions. The granule content of calcium in the β-cell-rich pancreatic islets of *ob/ob*-mice has been estimated as 50 mmol/kg dry weight by graphite furnace atomic absorption spectrophotometry (Andersson *et al.*, 1982a). In insulin-secretory granules prepared from a transplantable rat insulinoma, the same technique has indicated a calcium content of 229 mmol/kg protein; hence the secretory granules account for 26% of the total cell calcium (Hutton *et al.*, 1983). The large amount of material available from the rat insulinoma has also made it possible to predict the physicochemical form of calcium in the granules. The intragranular calcium was not sequestered as insoluble phosphates or bound in substantial amounts to insulin. On the contrary, it was rapidly solubilized by osmotic or detergent lysis in a form which dissociated readily at pH 7. After simulating the ionic composition of the granules, Hutton *et al.* (1983) concluded that up to 90% of Ca^{2+} was complexed to P_i and adenine nucleotides. The free-Ca^{2+} concentration was considered to be 10 mM, giving rise to a gradient across the granule membrane as high as 10^5. Such a gradient indicates the existence of an energy-linked Ca^{2+} transport system operating either within the granule itself or at some stage of granule morphogenesis.

The high concentration gradient for Ca^{2+} across the granule membrane may reflect the activity of a membrane-bound ATPase. When studying the subcellular distribution of Ca^{2+}-stimulated ATPase in mouse pancreatic islets, Formby *et al.* (1976) found the highest specific activity of this enzyme to be in the fraction rich in secretory granules. The data were compatible with the existence of two different forms of Ca^{2+}-ATPase with K_m for Ca^{2+} of 6.9×10^{-6} and 1.9×10^{-7} M. In support of the conclusion that the activity did not represent contamination by ATPases from other sources, it was significantly inhibited by high concentrations of Na^+ (unlike ATPase of microsomal origin) but not suppressed by caffeine (unlike mitochondrial ATPase). So far, extensively purified fractions of β-granules have not been tested for Ca^{2+}-ATPase activity. As originally suggested by Hellman *et al.* (1979b), there is also the possibility that Ca^{2+} is taken up into the β-granules in exchange for protons. An acid interior of the granule vesicle has been demonstrated both in studies employing the β-cell-rich pancreatic islets of *ob/ob*-mice (Abrahamsson and Gylfe, 1980) and a transplantable rat insulinoma (Hutton, 1982). With the access to the large amounts of insulin-secretory granules in the rat insulinoma, it has also been possible to demonstrate a proton-translocating Mg^{2+}-dependent ATPase with properties indistinguishable from those of the ATPase found in chromaffin granules of the adrenal medulla (Hutton and Peshavaria, 1982). There is no information as to how the proton translocation is related to the concentration of Mg-ATP in insulin-secretory granules. However, in chromaffin granules only micromolar concentrations of Mg-ATP are required (Baker and Knight, 1984). It is therefore difficult to envision a regulation of the granule uptake of Ca^{2+} mediated by the production of ATP. Another argument against a simple relationship between the proton gradient and the Ca^{2+} uptake into the insulin-secretory granules is that the interior has a positive potential (Hutton and Peshavaria, 1982).

The fact that the secretory granules contain large amounts of calcium does not necessarily imply that these organelles are important regulators of the Ca^{2+} activity of the aqueous cytoplasm. After exposing rat islets to ^{45}Ca and measuring the cellular distribution of label by electron microscopic autoradiography, Howell and Tyhurst (1976) concluded that the β-granules do not appear to contain a labile pool of Ca^{2+}. However, measurements of the radioactive content in secretory granules isolated from islets labeled with ^{45}Ca have provided a different picture. Whereas Borowitz and Matthews (1980) reported alterations of the granule content of ^{45}Ca in rat islets compatible with a long-term regulation of calcium, it emerged from the corresponding studies of the β-cell-rich pancreatic islets of the *ob/ob*-mice that the granules may even participate in the acute regulation of the Ca^{2+} activity in the cytoplasm (Bloom *et al.*, 1977; Hellman *et al.*, 1979a,b, 1980a,b; Kohnert *et al.*, 1979; Andersson *et al.*, 1982b). The secretory granules responded to glucose as rapidly as did the mitochondria, with a substantial increase in the incorporation of ^{45}Ca. As the granule incorporation of ^{45}Ca approached a steady state within 60 min, it seems unlikely that this process is limited to the period of granule formation (Andersson *et al.*, 1982b). Exposure of pancreatic β-cells to glucose has been reported to affect markedly the intragranular Ca^{2+} supposed to be visualized histochemically with the glyox-

al-bis(2-hydroxyanil) technique (Wolters *et al.*, 1984). The mobility of the granule calcium became evident also from chase incubation of the islets (Andersson
et al., 1982b; Andersson, 1983) and from measurement of the efflux of ^{45}Ca
from isolated granules (Abrahamsson *et al.*, 1981).

Studies of ^{45}Ca efflux from islets have indicated that the stimulatory effect of
glucose on the uptake of the radioisotope can at least to some extent be attributed
to ^{40}Ca–^{45}Ca exchange (see Section VI,C). The participation of the secretory
granules in this process is indicated both by the accelerated disappearance of
^{45}Ca from the granules of β-cells depolarized with glucose during chase incubation (Andersson *et al.*, 1982b) and by the stimulatory action of Ca^{2+} on the
efflux of ^{45}Ca from isolated granules incubated in the presence of ATP (Abrahamsson *et al.*, 1981). From the observation of stimulated ^{40}Ca–^{45}Ca exchange
it follows that glucose may increase the granule uptake of ^{45}Ca without net
accumulation of calcium. Although this possibility seems less likely, measurements with graphite furnace atomic absorption spectroscopy have so far failed to
demonstrate a statistically significant increase in the total amount of calcium in
the granules after exposure of the β-cell-rich pancreatic islets from *ob/ob*-mice to
glucose (Andersson *et al.*, 1981a, 1982a).

The secretory granules differ from other organelles in that they are directly
involved in exocytosis. It is possible that the granules have local regulatory
effects on the functionally important Ca^{2+} in the cell periphery which determine
their fusion with the plasma membrane. It is also possible that a certain concentration of Ca^{2+} in the granule sac is a prerequisite for exocytosis. The
observation that isolated granules have a very low passive permeability for Ca^{2+}
when suspended in a sucrose medium (Hutton *et al.*, 1983) does not imply that
the granules cannot function as Ca^{2+} buffers in intact β-cells. Among the
various ways in which ATP could indirectly provide energy for Ca^{2+} uptake is
the possibility of a Na^+ gradient across the β-granule membrane analogous to
that in chromaffin granules (Krieger-Brauer and Gratzl, 1981, 1982). Even if the
granules are important as local regulators of the Ca^{2+} in the vicinity of the
plasma membrane, it is unlikely that they are responsible for the glucose-induced
suppression of the Ca^{2+} activity in the aqueous cytoplasm (see Section VII,A).
The secretory granules differ from the mitochondria in not responding to glucose
with retention of ^{45}Ca during chase incubation of islets in a Ca^{2+}-deficient
medium (Andersson, 1983).

VIII. FACTORS AFFECTING THE MOVEMENTS AND ACTIONS OF CALCIUM

A. Phospholipids

In recent years increasing interest has been paid to the significance of phospholipid metabolism in the movements and actions of Ca^{2+} in the pancreatic β-

cells. Phospholipid breakdown is a prerequisite for the liberation of arachidonic acid, the parent substance for a number of important modifiers of insulin release formed metabolically via the cyclooxygenase (prostaglandins) or lipoxygenase (leukotrienes) pathways. Whereas there is evidence that the cyclooxygenase products are essentially inhibitors of insulin secretion, certain lipoxygenase products stimulate the secretory activity of the pancreatic β-cells (Yamamoto et al., 1982a,b; Metz et al., 1983; Best and Malaisse, 1983a). The arachidonic acid metabolites may not only affect the distribution of Ca^{2+} but also modify the sensitivity of the secretory apparatus to the action of the ion (Rubin, 1982). In addition to liberation of arachidonic acid, hydrolysis of the phospholipids results in production of lysophosphatides, compounds which may be directly involved in the exocytosis of the secretory granules due to their fusiogenic properties.

Most mammalian cells contain Ca^{2+}-dependent phospholipase A_2 activity with the established function of releasing arachidonic acid by breaking its C-2 ester linkage. Evidence has been presented for the existence of this type of enzyme activity also in the pancreatic islets (Laychock, 1982). In support of a role of phospholipase A_2 in insulin secretion are the findings that its activity was stimulated by glucose (Laychock, 1982) and that alleged inhibitors of phospholipase A_2 suppressed the release of insulin (Yamamoto et al., 1982a). Phospholipase A_2 activation may also be secondary to phosphatidic acid formation from phosphoinositide hydrolysis (Lapetina, 1982). However, Ca^{2+}-dependent formation of arachidonic acid does not necessarily involve activation of phospholipase A_2. Attempts to characterize enzyme activities in guinea pig islets using exogenously radiolabeled phospholipid substrates failed to demonstrate the presence of phospholipase A_2. Instead, there was clear evidence for arachidonic acid formation due to the presence of a phosphatidylinositol-specific and Ca^{2+}-regulated phospholipase C and the subsequent action of Ca^{2+}-stimulated diacylglycerol lipase (Schrey and Montague, 1983). The importance of the latter pathway was apparent from the fact that the arachidonic acid release from phosphatidylethanolamine and phosphatidylcholine was only 8 and 2.5%, respectively of that for phosphatidylinositol. In light of the observations made so far, it is reasonable to assume that an increased Ca^{2+}-activity in the pancreatic β-cells could favor the formation of arachidonic acid both by stimulating phospholipase A_2 and the phosphatidylinositol-specific phospholipase C acting in concert with diacylglycerol lipase.

Since the original report of Fex and Lernmark (1972) that glucose increases the ^{32}P-labeling of phospholipids in the ob/ob-mouse islets, evidence has been presented supporting the idea that this sugar as well as other insulin secretagogues increase the turnover of phosphatidylinositol in the pancreatic β-cells (Freinkel et al., 1975; Clements and Rhoten, 1976; Montague and Parkin, 1980; Clements et al., 1981; Axén et al., 1983; Best and Malaisse, 1983a,b). The phosphatidylinositol response can supposedly be initiated by the breakdown of the phospholipid. However, recent studies of mouse islets labeled with [³H]gly-cerol or [³H]inositol have failed to demonstrate accelerated disappearance of phosphatidylinositol-bound radioactivity in response to glucose under conditions when this process was significantly stimulated by carbamylcholine (Hallberg, 1984). Conflicting opinions have also been expressed as to whether the stimulation of the phosphoinositide turnover in the islets is a cause or effect of an increased cytoplasmic Ca^{2+}-activity. Accumulating evidence suggests that, despite Ca^{2+}'s significant role in the operation of the phosphatidylinositol cycle, the phospholipid breakdown can be initiated without promotion of the entry of Ca^{2+} into the β-cells (Best and Malaisse, 1983a).

In Fig. 14 a simplified scheme is presented indicating some alternatives for how components of the phosphoinositide cycle could affect the Ca^{2+} metabolism of the pancreatic β-cells and their secretory activity. As has been postulated for other cells (Gil et al., 1983), alterations in the phospholipid composition may influence proteins involved in Ca^{2+} transport by inducing conformational changes. Among the phosphoinositide metabolites, phosphatidic acid deserves attention both as a potential stimulator of phospholipase A_2 activity (see above) and as an agent with postulated Ca^{2+}-ionophoretic action (Dunlop et al., 1982). Diacylglycerol is important not only for the formation of arachidonic acid but also because it stimulates the Ca^{2+}-activated phospholipid-dependent protein

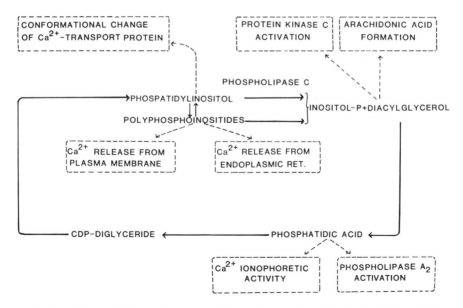

Fig. 14. Schematic diagram illustrating various mechanisms by which phospholipid metabolism may affect the secretory function of the pancreatic β-cells.

kinase referred to as C-kinase. This enzyme has been identified in the pancreatic islets (Tanigawa *et al.*, 1982; Hubinont *et al.*, 1984) and was postulated to serve as an alternative to calmodulin in the Ca^{2+}-messenger system in being responsible for the second phase of the glucose-stimulated insulin release (Zawalich *et al.*, 1983). However, it seems more likely that, under physiological conditions, activation of protein C-kinase has only a potentiatory role in glucose stimulation of insulin release (Harrison *et al.*, 1984). As in other tissues the tumor-promoting phorbol ester 12-*O*-tetradecanoylphorbol-13-acetate (TPA) can substitute for diacylglycerol as an activator of protein C-kinase (Hubinont *et al.*, 1984). Evidence has been provided that during stimulation of the β-cells, cytoplasmic protein C-kinase becomes attached to the secretory granules and phosphorylates a 29-kDa granule membrane protein (Brocklehurst and Hutton, 1983, 1984).

Studies of subcellular fractions of a rat β-cell tumor have revealed Ca^{2+}-sensitive mechanisms responsible for the metabolism of the 4-phosphate and 4,5-diphosphate derivatives of phosphatidylinositol both in the plasma membrane and in the secretory granules (Tooke *et al.*, 1984). The two types of polyphosphoinositides have been identified also in normal rat pancreatic islets and found to be temporarily diminished 30 sec after exposure to glucose (Laychock, 1983). Since the polyphosphoinositides have a high affinity for binding Ca^{2+}, breakdown of these lipids has been considered to result in a release of calcium from the plasma membrane into the cytoplasm (Fisher *et al.*, 1984). Studies of ^{45}Ca efflux indicated the presence of a distinct pool of calcium probably associated with the inner part of the plasma membrane (Gylfe and Hellman, 1982a). Moreover, electron microscopic studies with pyroantimonate provided evidence for the mobilization of calcium from the inner surface of the β-cell plasma membrane during the first 1–3 min after exposure to glucose (Lenzen and Klöppel, 1984). Also, products of the polyphosphoinositide breakdown, notably inositol 1,4,5-triphosphate, may aid in increasing the Ca^{2+} activity in the cytoplasm by mobilizing calcium stored in the endoplasmic reticulum (Prentki *et al.*, 1984; Prentki and Wollheim, 1984). However, even if the time dependency of the glucose effect is compatible with a role of the polyphosphoinositide turnover during the initial secretory peak, the physiological significance of the proposed mechanisms is obscure, considering that the overall intracellular action of glucose is to promote active sequestration of calcium (see Sections VI,C; VII,A; VII,C; X,A). The polyphosphoinositide breakdown resulting from carbamylcholine activation of muscarinic receptors was found to result in a raised cytoplasmic Ca^{2+} activity in insulin-secreting RINm5F cells suspended in a medium deprived of Ca^{2+}, but there was no increase in the presence of D-glyceraldehyde (Prentki and Wollheim, 1984).

Phospholipid methylation via the metabolic pathway described by Hirata and Axelrod (1980) is another possible mechanism for regulation of the Ca^{2+}-metabolism of the pancreatic β-cells. This process involves methylation of phosphatidylethanolamine to phosphatidylcholine with the aid of two specific methyltransferases and translocation of the methylated phospholipid from the cytoplasmic to the outer surface of the plasma membrane. Whereas the accumulation of phosphatidylcholine in the outer leaflet of the membrane bilayer may favor the entrance of Ca^{2+} by increasing membrane fluidity, methylation of the phospholipids has also been supposed to result in generation of arachidonic acid by activating phospholipase A_2. So far the significance of phospholipid methylation for the insulin secretory process remains obscure. The methyltransferase inhibitors 3-deazaadenosine and homocysteine have been reported to inhibit the glucose-stimulated insulin release without affecting the ^{45}Ca uptake into rat islets (Best and Malaisse, 1983b). Although glucose induced a transient stimulation of phospholipid methylation in rat islets, this effect was delayed as compared to the glucose-induced changes in various metabolic and ionic variables (Saceda et al., 1984).

B. Cyclic AMP

Cyclic AMP undoubtedly plays a major role as a potentiator of insulin release. The actions of the nucleotide and Ca^{2+} appear to be mutually dependent. Both the formation of cyclic AMP from ATP catalyzed by the plasma membrane-bound adenylate cyclase and the degradation by phosphodiesterase in the cytoplasm are activated by Ca^{2+}-calmodulin (see Section VIII,C). On the other hand cyclic AMP has been suggested to elevate Ca^{2+} activity in the cytoplasm by inhibiting Ca^{2+} sequestration within organelles (Howell and Montague, 1975; Howell et al., 1975; Sehlin, 1976; Hahn et al., 1979, 1980; Borowitz and Matthews, 1980; Gylfe and Hellman, 1981), by mobilizing intracellular Ca^{2+} (Brisson et al., 1972; Malaisse, 1973; Borowitz and Matthews, 1980; Siegel et al., 1980a) or by modulating the permeability of Ca^{2+} channels in the plasma membrane (Henquin and Meissner, 1983; Henquin et al., 1983a). Moreover, the insulin release mechanism has also been reported to be sensitized to Ca^{2+} by cyclic AMP (Hellman, 1976b; Hellman et al., 1979b; Siegel et al., 1979).

The catalytic subunit of adenylate cyclase at the inner surface of the plasma membrane is regulated by hormone receptors at the external surface via stimulatory or inhibitory guanine nucleotide regulatory subunits (Rodbell, 1980). There are stimulatory receptors for glucagon, vasoactive intestinal polypeptide, and β-adrenergic agents (Atkins and Matty, 1971; Rosen et al., 1971; Davis and Lazarus, 1972; Goldfine et al., 1972; Levey et al., 1972; Howell and Montague, 1973; Kuo et al., 1973) and probably also for ACTH, pancreozymin, and secretin (Kuo et al., 1973). There are also inhibitory receptors for α-adrenergic agents (Atkins and Matty, 1971; Howell and Montague, 1973; Iversen, 1973; Kuo et al., 1973) and possibly for somatostatin (Bent-Hansen et al., 1979; Leitner et al., 1980). Cholera toxin stimulates adenylate cyclase by interacting with the guanine nucleotide subunit; and forskolin, a plant diterpene, can activate the catalytic subunit directly (Seamon and Daly, 1983). A rise in the β-cell content of cyclic AMP will result not only from activation of adenylate cyclase (Hellman et al., 1974b; Niki et al., 1982) but also from inhibition of phosphodiesterase by methylxanthines (Charles et al., 1973, 1975; Hellman et al., 1974b; Grill and Cerasi, 1974). Indeed, the action of adenylate cyclase activation is greatly potentiated by phosphodiesterase inhibition (Hellman et al., 1974b).

Glucose stimulation of insulin release has repeatedly been linked to raised concentrations of cyclic AMP (Charles et al., 1973, 1975; Grill and Cerasi, 1973, 1974, 1975, 1976; Zawalich et al., 1975). Unlike the action of phosphodiesterase inhibitors or activators of adenylate cyclase, the glucose-induced increase in cyclic AMP depends on the presence of extracellular Ca^{2+} (Charles et al., 1975; Zawalich et al., 1975; Hellman, 1975a; Niki et al., 1982). Further evidence that this glucose action is mediated by an increase in the Ca^{2+} in the cytoplasm was obtained with the observation that either exposure to Ba^{2+}, which probably displaces intracellular Ca^{2+} (Berggren et al., 1983a), or exposure to a Ca^{2+} ionophore results in cyclic AMP accumulation (Hellman, 1975b; Grill, 1978). The rise in cyclic AMP concentration after stimulation with glucose alone is small in comparison to that obtained with phosphodiesterase inhibitors or with activators of adenylate cyclase. However, considerable concentrations are reached when glucose is combined with either or both of the effectors of the cyclic AMP system (Hellman et al., 1974b).

Despite the fact that intracellular cyclic AMP is a powerful potentiator of insulin release when stimulating concentrations of glucose are present, the nucleotide alone probably cannot initiate secretion (Hellman et al., 1974b). High concentrations of the phosphodiesterase inhibitor IBMX initiate insulin release (Hellman et al., 1974b; Hellman, 1975a; Rabinovitch et al., 1978; Siegel et al., 1980a), but this effect is poorly correlated to the content of cyclic AMP (Hellman et al., 1974b). Neither another phosphodiesterase inhibitor, nor activators of adenylate cyclase, nor dibutyryl cyclic AMP stimulate insulin release alone (Brisson et al., 1972; Hellman et al., 1974b; Wiedenkeller and Sharp, 1983; Ullrich and Wollheim, 1984). Indeed, IBMX may be the only one of these agents which increases the β-cell uptake of ^{45}Ca at nonstimulating concentrations of glucose (Brisson et al., 1972; Siegel et al., 1980a; Gylfe and Hellman, 1981). There are some indirect indications from studies of ^{45}Ca efflux that cyclic AMP alone may raise the cytoplasmic Ca^{2+} activity of β-cells (Brisson et al., 1972; Siegel et al., 1980a; Gylfe and Hellman, 1981). Nevertheless, the concept that cyclic AMP does not influence insulin release unless the influx of Ca^{2+} is initiated seems inconsistent with the idea that the nucleotide directly mobilizes intracellular Ca^{2+}. A more attractive alternative is that cyclic AMP raises Ca^{2+} in the cytoplasm by inhibiting the sequestration of the ion in cellular organelles. It can be predicted that such an action will potentiate but not initiate insulin secretion. In the resting β-cell the Ca^{2+} activity is regulated chiefly at the plasma membrane, but when the influx of Ca^{2+} is initiated the organelles become important sinks for Ca^{2+} (see Section X,A).

Evidence that cyclic AMP reduces the uptake of Ca^{2+} into β-cell organelles has been obtained by studying the organelles both *in situ* and after isolation. When intact *ob/ob*-mouse islets were loaded with [45]Ca, the presence of either theophylline, IBMX, or dibutyryl cyclic AMP reduced radioactivity in a subsequently isolated mitochondria fraction (Hahn *et al.*, 1979, 1980; Borowitz and Matthews, 1980). Cyclic AMP but not IBMX, theophylline, or caffeine inhibited [45]Ca uptake into a rat islet homogenate and a mitochondria-secretory granule fraction (Howell and Montague, 1975; Howell *et al.*, 1975). In a crude microsomal fraction from *ob/ob*-mouse islets both cyclic AMP and theophylline inhibited the uptake of [45]Ca (Sehlin, 1976). In contrast to these observations, it has been reported that the [45]Ca uptake into a mitochondrial fraction from rat islets is inhibited by caffeine and IBMX, but not by cyclic AMP (Sugden and Ashcroft, 1978).

The [45]Ca efflux data are consistent with the idea that cyclic AMP raises the Ca^{2+} activity in the cytoplasm in a glucose-dependent manner (Siegel *et al.*, 1980a; Gylfe and Hellman, 1981). However, the only measurements so far of this Ca^{2+} activity have failed to demonstrate any major effect from raising cyclic AMP in β-cells exposed to a maximally stimulating concentration of glucose (Abrahamsson and Rorsman, 1984). It should therefore be considered whether the cyclic AMP potentiation of [45]Ca efflux represents stimulated outward transport of Ca^{2+} or exocytosis of radioactivity with the secretory granules. The amount of [45]Ca released with the granules during glucose stimulation alone is small (Wollheim *et al.*, 1977; Gylfe and Hellman, 1978), but at higher concentrations of cyclic AMP [45]Ca exocytosis may be significant.

Even if cyclic AMP does not stimulate the β-cell uptake of Ca^{2+} either in the absence of or at maximally stimulating concentrations of glucose (Brisson *et al.*, 1972; Gylfe and Hellman, 1981), the nucleotide may still facilitate the entry of Ca^{2+} through the voltage-dependent channels at moderately stimulating concentrations of the sugar (Brisson *et al.*, 1972; Henquin and Meissner, 1983; Henquin *et al.*, 1983a). Neither can it be ruled out that cyclic AMP under such conditions may induce an increase in cytoplasmic Ca^{2+} activity by inhibiting Ca^{2+} sequestration in the organelles. Nevertheless, it may be necessary to reconsider the whole concept that the effects of cyclic AMP on insulin release are essentially mediated by Ca^{2+}. Insulin release can be stimulated by supraphysiological concentrations of extracellular Ca^{2+} alone (Hellman, 1976a,b; Devis *et al.*, 1977). Since IBMX has been found to shift the dose–response relationship for this stimulation to the left (Fig. 2), it was suggested that cyclic AMP sensitizes the β-cells to Ca^{2+} (Hellman, 1976a,b). Evidence for sensitization of the exocytotic process to Ca^{2+} has been obtained in a study of the effect of the phorbol ester TPA on catecholamine release from adrenal medullary cells made permeable to Ca^{2+} by electric discharges (Knight and Baker, 1983). Like cyclic AMP, TPA is an activator of protein kinase (see Sections VIII,A). The effect of cyclic AMP on the Ca^{2+} sensitivity of permeabilized β-cells has to our knowledge not yet been studied.

The fact that cyclic AMP promotes the formation of microtubules and microfilaments may well explain how the nucleotide can sensitize exocytosis to Ca^{2+} (see Sections IX,A and IX,B). In general the effects of cyclic AMP are mediated by activation of protein kinases. In rat islets cyclic AMP stimulates the phosphorylation of a number of proteins ranging from 15 to 138 kDa (Harrison and Ashcroft, 1982; Colca et al., 1983b, 1984; Suzuki et al., 1983; Christie and Ashcroft, 1984) and in a hamster insulinoma two cytosolic polypeptides of 16 kDa (Schubart, 1982). A 53- to 55-kDa component can be one of the subunits of tubulin (Colca et al., 1983b), and it may be worthwhile to see whether some of the other proteins are identical to actin-binding regulatory proteins (see Section IX,B). Another protein is probably the 68-kDa enzyme phosphoglucomutase, the phosphorylation of which is also stimulated by glucose (Colca et al., 1984). This observation may indicate a role for cyclic AMP in the formation of glucose 1,6-bisphosphate, a possible regulator of glycolysis in islet cells (Sener et al., 1982).

C. Calmodulin

As in other cells, certain actions of Ca^{2+} in the β-cells are apparently mediated by calmodulin, and there is indirect evidence that this protein may be involved in the initiation of insulin release (Sugden et al., 1979; Valverde et al., 1979, 1981; Gagliardino et al., 1980; Schubart et al., 1980; Krausz et al., 1980; Janjic et al., 1981). Phosphodiesterase activation assays have revealed calmodulin concentrations as high as 16–50 μM in pancreatic islets (Sugden et al., 1979; Valverde et al., 1979; Landt et al., 1982), and similar results have been obtained with radioimmunoassay (Nelson et al., 1983). With 4 Ca^{2+} sites per molecule of calmodulin the total binding capacity is probably about 1–2 orders of magnitude greater than the variations in cytoplasmic Ca^{2+} activity during insulin release (Rorsman et al., 1984). This excess may be a requisite for the high Ca^{2+} sensitivity (cf. Lin and Cheung, 1980). Although neither starvation nor exposure of pancreatic islets to glucose affects the concentrations of calmodulin measured with radioimmunoassay (Nelson et al., 1983), it is not known whether the pools of Ca^{2+}-reactive calmodulin are subject to regulation during insulin secretion. Carboxymethylation of the molecule has been suggested as a mechanism for such a regulation (Campillo and Ashcroft, 1982), and in the brain a number of proteins with the ability to inhibit calmodulin-induced activation of enzymes have been isolated (Wang et al., 1980). The elucidation of a possible modulation of calmodulin during insulin release is complicated by the lack of knowledge about its subcellular distribution in the β-cells and its Ca^{2+}-independent binding to proteins and membranes. Some effects of Ca^{2+} in insulin release may thus be mediated by calmodulin already bound at effector sites, whereas other actions would depend on a pool of free calmodulin. The specificity of this calmodulin for certain processes, such as those involved in secretion, may depend on the number of Ca^{2+} sites occupied (Klee, 1980) or on as yet unidentified regulatory molecules.

Calmodulin is known to stimulate the outward transport of Ca^{2+} from the β-cells by activating the Ca^{2+}-ATPase of the plasma membrane (Pershadsingh *et al.*, 1980; Kotagal *et al.*, 1982, 1983; Colca *et al.*, 1983b). It is unclear whether this activation has any specific function in the regulation of insulin release or whether it is merely involved in the maintenance of a low cytoplasmic Ca^{2+} activity. Although pancreatic islets have been reported to contain a Ca^{2+}-ATPase activity subject to glucose inhibition (Levin *et al.*, 1978), the significance of this observation for the outward transport of Ca^{2+} remains to be elucidated.

Other calmodulin-activated processes affect systems known or believed to be involved in the regulation of insulin release. In rat islets calmodulin activates both adenylate cyclase (Valverde *et al.*, 1979, 1981; Sharp *et al.*, 1980; Thams *et al.*, 1982) and phosphodiesterase (Sugden and Ashcroft, 1981; Sharp *et al.*, 1983). The stimulation of synthesis and degradation of cyclic AMP is not unique to β-cells. The process is thought to favor a transient increase of cyclic AMP due to activation of adenylate cyclase as Ca^{2+} passes through the plasma membrane and of phosphodiesterase when the ion diffuses through the cytoplasm (Wallace *et al.*, 1980). The initial reduction of cytoplasmic Ca^{2+} in the β-cells (see Section VII,A) would tend to amplify such an increase, and an early peak in the cyclic AMP concentrations may be a factor in the first phase of glucose-stimulated insulin release (Zawalich *et al.*, 1975). However, the reported absence of a Ca^{2+}-calmodulin-activated adenylate cyclase activity in mouse pancreatic islets (Thams *et al.*, 1982) throws doubt on whether this is a general mechanism.

Exocytosis may also be activated more directly by calmodulin. Various evidence indicates that contractile interactions between actin and myosin are involved in the translocation of the β-granule towards the plasma membrane (see Section IX,B). The presence of Ca^{2+}-calmodulin-dependent myosin light chain kinase activity in insulin-releasing cells suggests that such interactions depend on myosin light chain phosphorylation (MacDonald and Kowluru, 1982a; Penn *et al.*, 1982). The cytoskeleton is also thought to play a critical role in the release process for insulin secretory granules (see Section IX,A). It is therefore interesting that influx of Ca^{2+} into hamster insulinoma cells results in the phosphorylation of a 60-kDa protein (Schubart, 1982) identified as an islet cell intermediate filament cytokeratin (Schubart and Fields, 1984). Moreover, rat pancreatic islets contain microsomal membrane-bound Ca^{2+}-calmodulin-dependent protein kinase which has been found to catalyse the phosphorylation of a protein in the

52- to 55-kDA range (Gagliardino *et al.*, 1980; Harrison and Ashcroft, 1982; Colca *et al.*, 1983b; Nelson *et al.*, 1983) and one of 56–57 kDa (Landt *et al.*, 1982; Colca *et al.*, 1983b; Nelson *et al.*, 1983). There is evidence that these proteins are identical to the α- and β-subunits of tubulin (Boyd, 1982; Colca *et al.*, 1983b). The exact function of phosphorylation in microtubule assembly is, however, unclear. A third mechanism by which Ca^{2+}–calmodulin may affect exocytosis is the binding and/or fusion of the β-granule with the plasma membrane (Watkins and Cooperstein, 1983; Steinberg *et al.*, 1984). Model experiments with isolated insulin granules and β-cell plasma membranes have shown that Ca^{2+}–calmodulin can modulate the interaction between the granule membranes and the cytoplasmic surface of the plasma membrane (Watkins and Cooperstein, 1983). Using a model system for secretion it has even been possible to demonstrate Ca^{2+}–calmodulin-stimulated fusion of lipid vesicles with a planar bilayer (Zimmerberg *et al.*, 1980).

Ca^{2+}–calmodulin-activated processes can be manipulated pharmacologically by a number of antipsychotic drugs which inactivate calmodulin by a hydrophobic interaction (Cheung, 1982). Although such inactivation is one of the criteria for a calmodulin-dependent effect, it is important to keep in mind that these drugs also have other actions. Even the widely used trifluoperazine, which is considered to be among the more specific compounds (Weiss and Wallace, 1980), has been found to inhibit calmodulin-independent protein kinase C more than the calmodulin-dependent myosin light chain kinase (Feinstein and Hadjian, 1982). The use of "calmodulin antagonists" to confirm a role of calcium and calmodulin in the triggering of secretory exocytosis has indeed been seriously questioned (Sanchez *et al.*, 1983).

At low concentrations of "calmodulin antagonists," insulin release can be inhibited without alterations in either glucose oxidation or ^{45}Ca uptake (Valverde *et al.*, 1981). It is possible that this inhibition reflects interference with the formation of cyclic AMP and/or with the other exocytosis-related calmodulin-activated processes discussed above. However, the reduction of glucose-stimulated ^{45}Ca uptake obtained at higher concentrations of the antagonists (Janjic *et al.* 1981; Valverde *et al.*, 1981) is difficult to ascribe to inhibition of any known effect of calmodulin. Even though it is conceivable that the initial reduction of cytoplasmic Ca^{2+} supposed to start the depolarization (see Section VI,A) can be affected after inhibition of the calmodulin-dependent Ca^{2+}-ATPase of the plasma membrane, there is no evidence for such a disturbance. Trifluoperazine thus failed to interfere with the glucose reduction of ^{45}Ca efflux in Ca^{2+}-deficient media (Janjic *et al.*, 1981; Valverde *et al.*, 1981). The observation that trifluoperazine also inhibits ^{45}Ca uptake in response to a high concentration of K^+ (Janjic *et al.*, 1981) suggests that its action is related to the opening of the voltage-dependent Ca^{2+} channels rather than to depolarization per se.

Another interpretational difficulty is the lack of effect of "calmodulin antagonists'' on the potentiation of insulin release obtained after raising intracellular cyclic AMP (Schubart *et al.*, 1980; Krausz *et al.*, 1980; Valverde *et al.*, 1981; Steinberg *et al.*, 1984). Since cyclic AMP appears to act without major effects on the cytoplasmic Ca^{2+} activity (see Section VIII,B), it can be concluded that the calmodulin involved in exocytosis is either inaccessible to antagonists or that cyclic AMP can compensate for the reduction of the Ca^{2+}–calmodulin activity by sensitizing the exocytotic machinery. It is also possible that some key proteins are phosphorylated both by calmodulin and by cyclic AMP-activated protein kinases (Harrison and Ashcroft, 1982).

It is apparent from this discussion that the exact functions of calmodulin in the stimulus–secretion coupling of the β-cell are still unclear. Development of more specific antagonists would undoubtedly aid the elucidation of the mechanisms involved.

IX. PARTICIPATION OF CALCIUM IN THE PROCESS OF EXOCYTOSIS

A. Role of Microtubules

Microtubules are micrometer-long cylindrical structures with a diameter of 20–25 nm. The microtubules are formed from the subunit protein tubulin, which is a 110-kDa dimer composed of α- and β-tubulins. About 0.5% of the islet protein has been reported to be tubulin, which appears to be in dynamic equilibrium with the microtubules (Pipeleers *et al.*, 1976a). Whereas the microtubules have an established active function in the motile events of cilia, their role in secretion is more obscure. In the β-cells there is an extensive microtubular network radiating from the perinuclear region to the plasma membrane (Boyd, 1982). Although the secretory granules appear to align along these microtubules (Malaisse-Lagae *et al.*, 1971b; Somers *et al.*, 1979a), the possibility cannot be excluded that this association occurs by mere chance (Dean, 1973).

The idea that microtubules are important for insulin secretion originates from experiments with agents known to interfere with microtubule function. Insulin release was thus inhibited by colchicine (Lacy *et al.*, 1968, 1972; Malaisse *et al.*, 1971; Somers *et al.*, 1974), vincristine (Malaisse *et al.*, 1971; Malaisse-Lagae *et al.*, 1971b; Lacy *et al.*, 1972), and deuterium oxide (Malaisse *et al.*, 1971; Malaisse-Lagae *et al.*, 1971b; Lacy *et al.*, 1972). Whereas colchicine disrupts the microtubules by binding to tubulin, vincristine causes aggregation of the microtubular protein, and deuterium oxide stabilizes the microtubules. All these agents have side effects which may suppress insulin release (Boyd, 1982; Howell and Tyhurst, 1982). As a matter of fact, glucose-stimulated insulin release persisted at a reduced rate even after a complete colchicine disruption of the microtubules (Boyd, 1982). The release of insulin obtained in the presence of glucose and raised concentrations of cyclic AMP has been found to be associated with increased formation of microtubules (Montague et al., 1976; Pipeleers et al., 1976a,b). It has therefore been suggested that the microtubules are not actively involved in secretion but that they facilitate insulin discharge by directing granule movements within the β-cells (Lacy et al., 1968; Malaisse-Lagae et al., 1971b; Malaisse et al., 1975; Boyd, 1982). Such a function is supported by motion analysis of islet cell particles with time-lapse cinematography (Lacy et al., 1975; Somers et al., 1979a).

The role of calcium in microtubule function is at present unclear. Calcium could either affect the assembly and disassembly of the microtubules or regulate the interaction between the microtubules and other intracellular structures (Campbell, 1983). Extracellular Ca^{2+} has been reported both to be a prerequisite for (Montague et al., 1976), and to lack effect on (Pipeleers et al., 1976b) the microtubule formation associated with stimulation of insulin release. In vitro Ca^{2+} (Weisenberg, 1972) and Ca^{2+}–calmodulin (Marcum et al., 1978; Welsh et al., 1979; Lee and Wolff, 1982) possess well-established effects in inhibiting tubulin polymerization. However, the physiological relevance of these effects has been questioned. Even in the presence of calmodulin, concentrations of Ca^{2+} as high as 10 μM are required (Campbell, 1983). Phosphorylation processes may play an important regulatory role in microtubule assembly (Shelanski et al., 1973; Campbell, 1983). It is therefore interesting that a membrane-bound Ca^{2+}–calmodulin-activated protein kinase has been identified which appears to phosphorylate the α- and β-subunits of tubulin in pancreatic islets (Boyd, 1982; Colca et al., 1983b). Future studies may determine whether the microtubule assembly during stimulation of insulin release is a phenomenon regulated by cytoplasmic Ca^{2+}. The situation is further complicated by the fact that microtubule polymerization is increased by raised levels of cyclic AMP (Montague et al., 1976; Pipeleers et al., 1976a). A cyclic AMP-dependent protein kinase may phosphorylate tubulin subunits (Colca et al., 1983b) and play a role in microtubule assembly (Howell and Tyhurst, 1982; Colca et al., 1983b). The alternative possibility that Ca^{2+} may be important in microtubule binding to the secretory granules and other subcellular structures has so far only relatively weak support (Pipeleers et al., 1980).

B. Role of Microfilaments

The microfilaments are 4–8 nm in diameter and are distributed as a network in the β-cell cytoplasm. They are particularly numerous in the cell web, which is a layer, 30–50 nm thick, beneath the plasma membrane (Orci et al., 1972). Actin is a major component of the microfilaments, which also contain myosin that reacts with actin to form actomyosin. In rat islets the myosin ATPase activity is

70% of that in catfish islets, where the content of myosin has been estimated as 0.97% of the total protein (Ostlund *et al.*, 1978). Secretory tissue myosin is similar to smooth muscle myosin in having two heavy chains of 200 kDa and four light chains of 14 and 19 kDa. The islet content of actin has been found to be 0.7–2% of the total protein in islets from rat, mouse, and hamster (Howell and Tyhurst, 1980; Swanston-Flatt *et al.*, 1980; Boyd, 1982). Actin is a globular (G) protein of 42 kDa which in the presence of MgATP polymerizes to filamentous (F) actin.

Cytochalasins, which inhibit the formation of F-actin (Brenner and Korn, 1979), have been found to change the appearance of the cell web, to cause margination of secretory granules (Orci *et al.*, 1972), and to stimulate insulin release (Gabbay and Tze, 1972; Malaisse *et al.*, 1972a; Van Obberghen *et al.*, 1973). These observations were interpreted as suggesting that the microfilaments in the cell web determine the access of secretory granules to the cell membrane and that the cell web acts as a sphincter for exocytotic release of insulin (Orci *et al.*, 1972; Malaisse *et al.*, 1972a). In such a control mechanism for insulin release, Ca^{2+} may induce F-actin depolymerization with a resulting decrease in the viscosity of the cell web (MacDonald and Kowluru, 1982b). In a number of cells there is an increased cytoplasmic streaming due to actin depolymerization, when the cytosolic Ca^{2+} activity is raised to the micromolar range (Campbell, 1983). If the microfilaments have only a barrier function, Ca^{2+} could also facilitate interaction between the secretory granules and the plasma membrane by neutralizing the negative surface charges (Matthews, 1970; Dean, 1975).

The information available about the F/G-actin equilibrium in pancreatic islets does not support the concept of a depolymerization during stimulation of insulin release. Glucose has been reported to lack effect on actin polymerization in hamster islets (Boyd, 1982). In mouse and rat islets the fraction of F-actin even increased from 23–27% under basal conditions to 46–52% after stimulation with glucose (Howell and Tyhurst, 1980; Swanston-Flatt *et al.*, 1980). Moreover, as much as 71% was recovered as F-actin after the concentration of cyclic AMP was raised with the aid of a phosphodiesterase inhibitor (Howell and Tyhurst, 1980). Cyclic AMP has been reported to increase the phosphorylation of a number of proteins ranging from 15 to 138 kDa (see Section VIII,B). It would be interesting to compare these proteins with some actin-binding molecules believed to be involved in the regulation of microfilament formation (Korn, 1978). Since filamentous structures have been found to link the insulin secretory granule sac with the plasma membrane (Lacy *et al.*, 1968), the idea emerged that granule translocation toward the plasma membrane is due to microfilament contraction. The interactions between microfilamentous proteins and insulin secretory granules have been studied *in vitro* by observation of the sedimentation characteristics of the granules (Howell and Tyhurst, 1979, 1981, 1982). The results indicate that actomyosin is far more effective in retarding granule sedimentation than is F-actin, while G-actin and myosin have no effects. There is also ultrastructural evidence that the secretory granules are bound directly to actomyosin filaments. ATP stimulated these interactions, whereas Ca^{2+} had no effect, possibly because there was no calmodulin in the system. The role of Ca^{2+} and calmodulin in contraction-mediated granule discharge is probably to initiate the motile events by activating the myosin light chain kinase (see Section VIII,C). Myosin phosphorylation is an essential prerequisite for actomyosin ATPase activation and contraction in a number of nonmuscle cells (Adelstein and Eisenberg, 1980).

A role of microfilaments in the active granule translocation does not necessarily exclude a barrier action of the cell web. When insulin release is initiated after opening of the voltage-dependent Ca^{2+} channels, a high local Ca^{2+} activity beneath the plasma membrane may cause microfilament depolymerization within the web. However, the contractile events would be stimulated by a more modest rise of the Ca^{2+} activity in the deep cytoplasm. A dual action of Ca^{2+} on the microfilament function may account for the inhibition of insulin secretion observed at excessive concentrations of Ca^{2+} (Hellman, 1976a,b; Hellman et al., 1979b).

C. Granule Fusion

Apart from its importance for the translocation of insulin secretory granules, Ca^{2+} may also influence their interaction with the plasma membrane. The events involved include reduction of repulsion between the negatively charged membrane surfaces and the final fusion of the phospholipid structures. Calcium has thus been suggested to have the electrostatic function of diminishing the energetic barrier between the membranes (Dean, 1975), an idea supported by experiments indicating that the charge on the granule membrane is drastically reduced at the site of contact with the plasma membrane (Howell and Tyhurst, 1977). Evidence has been presented suggesting that protein kinase C may attach to insulin granules in a Ca^{2+}-dependent manner and phosphorylate a 29-kDa granule-membrane protein of possible importance for exocytosis (Brocklehurst and Hutton, 1984). Using insulin secretory granules and plasma membrane vesicles isolated from toadfish islets it has also been possible to demonstrate a calmodulin-activated interaction between the granules and the inner surface of the plasma membrane stimulated at concentrations above 100 nM Ca^{2+} (Watkins and Cooperstein, 1983). Although the latter experiments did not discriminate between binding and fusion, calmodulin in the presence of μM concentrations of Ca^{2+} has been found to stimulate fusion between lipid vesicles and a planar bilayer (Zimmerberg et al., 1980). To explain fusion of large liposomes, Papahadjopoulos (1978) suggested that a change in Ca^{2+} activity induces a transient unstable state with domains of fluid and solid lipid in the membranes. The unstable areas of the lipid bilayers would be particularly susceptible to fusion. In the living cell an alternative mechanism for fusion may involve the formation of membrane-destabilizing lysophospholipids after Ca^{2+} activation of phospholipase A_2 (see Section VIII,A).

Fusion between large liposomes (Miller *et al.*, 1976) or between liposomes and planar bilayers (Zimmerberg *et al.*, 1980) will not occur spontaneously unless an osmotic gradient is imposed across the vesicle bilayers. An increased osmotic pressure within the secretory granules has therefore been suggested to be essential also for membrane fusion during exocytosis (Pollard *et al.*, 1977; Grinstein *et al.*, 1982). The maintenance of an acidic interior in the granule (see Section VII,D) was supposed to be a basis for osmotic swelling of the granules. It was asked whether extracellular anions are driven by the electrochemical gradient into the granule when it is juxtaposed with the plasma membrane (Pollard *et al.*, 1977). Another proposal was that an electroneutral exchange of protons for cytoplasmic or extracellular monovalent cations provides the necessary increase in osmotic pressure (Grinstein *et al.*, 1982). Despite its attractiveness this chemiosmotic hypothesis for exocytosis has been the subject of severe criticism (Baker and Knight, 1984). Major arguments are that in cells made permeable to small molecules by electric discharges, exocytosis can be stimulated by Ca^{2+} even in the presence of a proton ionophore or after inhibition of the proton pump of the granule membrane.

X. MECHANISMS OF ACTION OF PRINCIPAL INITIATORS OF INSULIN RELEASE

A. Glucose

Glucose is recognized as an insulin secretagogue by virtue of its metabolism in the pancreatic β-cell (Malaisse *et al.*, 1979b; Hedeskov, 1980). It is also gener-

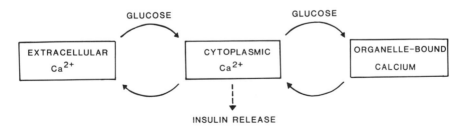

Fig. 15.   Diagram illustrating the dual action of glucose on the concentration of the cytoplasmic Ca^{2+} which regulates insulin release. The effect of glucose on this calcium pool is supposed to reflect the balance between stimulated entry of Ca^{2+} and the trapping of the ion in intracellular stores. From Hellman and Gylfe (1984b).

ally assumed that the exocytotic discharge of the insulin secretory granules is triggered by Ca^{2+}, although the exact mechanisms remain to be elucidated (see Section IX). The coupling between glucose recognition and insulin discharge is still a matter of debate. In an attempt to incorporate the different actions of glucose into a comprehensive model for stimulus–secretion coupling we have striven towards simplicity. A key feature is that glucose has dual actions on the Ca^{2+} activity in the cytoplasm (Fig. 15), a phenomenon predicted from various experimental observations (see Sections III, VI, VII), and verified by direct measurements (Rorsman *et al.*, 1983, 1984). It is implicit in the model that exposure to glucose under certain conditions results in a paradoxical inhibition of insulin release (Fig. 16).

If the initial event in stimulus–secretion coupling is a reduction of the cytoplasmic Ca^{2+} activity, it is easy to envision a strict relation between glucose metabolism and active uptake of Ca^{2+} by intracellular organelles, notably the mitochondria. The importance of the mitochondria is not only suggested from studies of the subcellular distribution of ^{45}Ca (see Section VII,C). Inhibition of ^{45}Ca efflux (Fig. 12), which probably reflects intracellular sequestration of Ca^{2+} (see Section VI,C), exhibits a hyperbolic dependence on the glucose concentra-

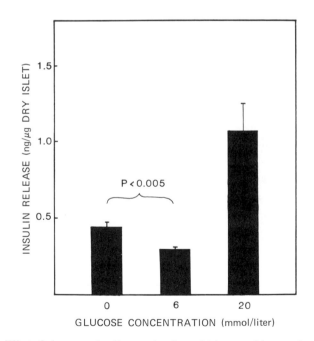

Fig. 16. Effect of glucose on insulin secretion from *ob/ob*-mouse islets previously stored for 3 days in a Ca^{2+}-deficient medium. The amounts of insulin were recorded after 60 min of incubation in RPMI 1640 medium containing 0.42 mM Ca^{2+}. Mean values ±SEM for 8 experiments.

tion, as does the oxidation of the sugar (Fig. 8). At glucose concentrations ranging from 0 to 20 mM, there appears consequently to be a gradual increase in the rate of intracellular Ca^{2+} sequestration. As discussed in Section VI,A, decreased cytoplasmic Ca^{2+} activity can be expected to result in a depolarization of the β-cells due to a reduced K^+ permeability of the plasma membrane. At glucose concentrations above 5 mM, the depolarization reaches the threshold for opening of the voltage-dependent Ca^{2+}-channels, making Ca^{2+} influx an important factor in the subsequent burst of depolarization (see Section VI,A). The influx of Ca^{2+} will eventually raise the cytoplasmic activity sufficiently for activation of K^+ permeability and termination of the burst. After repolarization with closing of the voltage-dependent channels, the raised Ca^{2+} activity of the cytoplasm will decrease as a result of outward transport and intracellular sequestration until a new cycle starts. A continuous burst pattern will emerge at high glucose concentrations when the elimination of Ca^{2+} from the cytoplasm balances the influx. It is implicit in the model that the time-average Ca^{2+} activity in the cytoplasm will increase with raised glucose concentrations. However, the highest peak activities of Ca^{2+} are reached at concentrations of glucose which stimulate secretion submaximally.

The present model is compatible with a glucose-stimulated liberation of Ca^{2+} from the inner face of the plasma membrane or from the endoplasmic reticulum (see Section VIII,A), provided that the Ca^{2+} eliminating capacity is not exceeded. Apart from the glucose-stimulated active removal of Ca^{2+}, there is an enhanced binding to proteins and other components, as well as an additional uptake into organelles when the cytoplasmic Ca^{2+} activity increases (see Section VII). Nevertheless, during prolonged exposure to glucose, it seems inevitable that these Ca^{2+} "buffers" become saturated (see Section VI,B). Therefore, it is likely that an increasing proportion of Ca^{2+} elimination is taken over by the plasma membrane. During a sustained maximal glucose stimulation, Ca^{2+} influx through the voltage-dependent Ca^{2+} channels will probably be balanced by outward transport. It is more difficult to explain a cyclic burst pattern at lower concentrations of glucose if the Ca^{2+}-pumping organelles are saturated. One possibility is that during repolarization the calmodulin-dependent Ca^{2+}-ATPase of the plasma membrane is deactivated more slowly than the cytoplasmic Ca^{2+} decreases, resulting in a reduction below the resting activity and initiation of depolarization. When glucose is omitted, release of Ca^{2+} from the sequestering organelles will prevent the Ca^{2+} activity from decreasing rapidly, and there will be time for the Ca^{2+}-ATPase to adapt to the cytoplasmic Ca^{2+}.

A number of mechanisms have been suggested to explain the biphasic nature of the insulin response to glucose stimulation (Wollheim and Sharp, 1981). Many of these do not involve any action of Ca^{2+}. However, the observation that the first phase of secretion corresponds to a prolonged burst of depolarization (see Section VI,A) is an indication that secretion may be regulated by Ca^{2+} in a biphasic manner. In the present model, a high initial rate of Ca^{2+} sequestration will prolong the initial burst (see Section VI,A), since a longer period of time is required to reach the Ca^{2+} activity at which repolarization is initiated. If there is a glucose-induced intracellular liberation of Ca^{2+}, an appropriately timed release beginning at the end of the first burst will prevent the cytoplasmic Ca^{2+} activity from reaching the nadir at which a new depolarization starts. Such a mechanism may explain the lag period between the first and second phases of secretion. It is possible to attribute the slowly rising second phase of secretion to an increase of the time-average Ca^{2+} activity in the cytoplasm as the intracellular Ca^{2+}-buffering systems become saturated (see Section VI,B). A reduced intracellular Ca^{2+} buffering is also a likely explanation for the improved secretory response after priming of the β-cells with glucose.

B. Amino Acids and Sulfonylureas

Since the original observation by Floyd et al. (1963) that L-leucine enhances the in vivo release of insulin, numerous studies have evaluated the ability of various amino acids to stimulate the secretory activity of the pancreatic β-cells. As in the case of glucose, the effects obtained with L-leucine (Milner and Hales, 1967; Malaisse-Lagae et al., 1971a) and L-arginine (Gerich et al., 1974) require the presence of extracellular Ca^{2+}. These amino acids, as well as L-alanine, have depolarizing effects on the β-cells and should therefore open the voltage-dependent Ca^{2+} channels (Henquin and Meissner, 1981; Meissner and Henquin, 1983). The idea that these amino acids favor the entry of Ca^{2+} into the β-cells is supported by measurements of the uptake of ^{45}Ca into isolated islets (Malaisse-Lagae et al., 1971a; Hellman et al., 1977; Charles and Henquin, 1983). Moreover, perifusion studies have demonstrated that the amino acids accelerate the efflux of the radioisotope, supposedly reflecting stimulated entry of nonradioactive Ca^{2+} (Table I).

It is evident from recent studies (Henquin et al., 1982; Charles and Henquin, 1983) that only those amino acids which are actively metabolized as is L-leucine stimulate secretion in a manner similar to glucose. In the case of L-arginine the stimulatory action seems to be directly related to the depolarization evoked by the entry of the positively charged molecule. Evidence has been presented indicating that alanine-induced increases in cytoplasmic Ca^{2+} activities follow the mobilization of calcium from intracellular stores by Na^+ cotransported with the amino acid into the β-cell.

The hypoglycemic sulfonylureas are examples of drugs which initiate insulin release by promoting the entry of Ca^{2+} into the pancreatic β-cells. Exposure to sulfonylureas results in depolarization and the appearance of spike activity (Dean and Matthews, 1968; Meissner et al., 1979) in combination with increased uptake of ^{45}Ca (Malaisse et al., 1972b; Hellman et al., 1977). The rapidly initiated sulfonylurea-stimulated ^{45}Ca uptake, which declines with time, fulfills the criteria for Ca^{2+} influx through voltage-dependent channels subject to progressive inactivation (Henquin, 1980c). As shown in Table I, exposure to sulfonylureas stimulates an efflux of ^{45}Ca which is secondary to increased entry of nonradioactive Ca^{2+}. In a Ca^{2+}-deficient medium sulfonylureas neither inhibit (as does glucose) nor stimulate (as do carboxylic Ca^{2+} ionophores) the efflux of ^{45}Ca (Hellman, 1982a). It is therefore likely that the sulfonylurea-induced increase in cytoplasmic Ca^{2+} which initiates insulin release reflects an enhanced membrane permeability to the ion without much involvement of intracellular stores.

In view of the depolarizing effects of sulfonylureas, it is not surprising that these compounds favor the entry of Ca^{2+} through voltage-dependent channels. However, it has also been suggested that the increased inflow of Ca^{2+} into the β-cells reflects the ability of the drugs to mediate exchange diffusion similar to that induced by carboxylic Ca^{2+} ionophores (Couturier and Malaisse, 1980a,b; Deleers et al., 1980). Major arguments against the ionophore hypothesis are that the sulfonylurea-stimulated ^{45}Ca efflux depends not only on extracellular Ca^{2+} but also on K^+ (Hellman, 1982a,b) and that even very high concentrations of sulfonylurea do not stimulate the process more than depolarizing agents do (Hellman, 1981). A more detailed account of how sulfonylureas interact with the pancreatic β-cells is given in our recent reviews on the subject (Hellman, 1982b; Gylfe et al., 1984).

XI. CALCIUM AND THE INSULIN SECRETORY DEFECT IN HUMAN DIABETES

The normal pancreatic β-cell responds rapidly and with great sensitivity to an increased glucose concentration with a biphasic release of insulin. This pattern seems to be extremely important for normal glucose homeostasis, and reduced glucose sensitivity or a delayed response will lead to deterioration of glucose tolerance or overt diabetes. It has been postulated that the early lesion of the β-cell function in noninsulin dependent diabetes is expressed as a failure of prompt stimulation of insulin release by glucose (Cerasi and Luft, 1972). In many diabetic patients there is even a temporary reduction of insulin secretion after an intravenous glucose infusion (Robertson et al., 1972; Metz et al., 1979; Hellman et al., 1985), a phenomenon which seems to have been largely overlooked. We have recently found a patient with mild diabetes, in whom the glucose depression of serum insulin was not only particularly pronounced and long-standing but also possible to induce after an oral load (Hellman et al., 1985). Figure 17 shows the results of intravenous injections of glucose and tolbutamide in this patient. The paradoxical reaction to the sugar is apparent from pronounced reductions of the

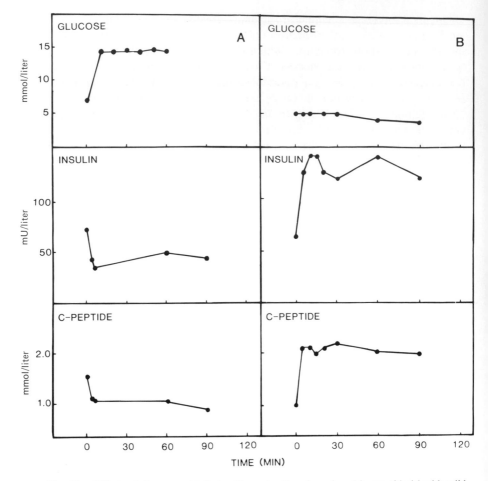

Fig. 17. Effects of glucose and tolbutamide on insulin release in a 14-year-old girl with mild diabetes. The serum concentrations of glucose, insulin, and C-peptide are shown after intravenous injection of 350 mg glucose per kilogram body weight (A) or 1 g tolbutamide (B). From Hellman *et al.* (1985).

concentrations of circulating insulin and C-peptide. Nevertheless, the response to tolbutamide was not abnormal, as indicated by the markedly raised serum concentrations of insulin and the C-peptide.

It follows from the dual actions of glucose on cytoplasmic Ca^{2+} activity that the sugar can even inhibit insulin release. Such an inhibition is seen not only in isolated pancreatic islets (Fig. 16), but also in certain cases of non-insulin-dependent diabetes. At present we cannot rule out that glucose inhibition of insulin release in man is due to the endogenous production of a nonadrenergic inhibitor of secretion such as somatostatin (Metz *et al.*, 1979; Hellman *et al.*, 1985). However, in the patient discussed above, sulfonylureas, which stimulate somatostatin release (Efendic *et al.*, 1979), enhanced insulin secretion. The limited capacity for sequestration of Ca^{2+} in the β-cells (see Section VI,B) makes it less likely that glucose-induced reduction of the cytoplasmic Ca^{2+} activity is the primary cause of the impaired insulin secretion in diabetes. The demonstration of a glucose inhibition of insulin release can instead be taken to indicate a defect in glucose-stimulated Ca^{2+} uptake into the β-cells. In patients where exposure to the depolarizing sulfonylureas results in stimulation of insulin secretion, it is likely that the impaired β-cell function is due to a failure of glucose to open the voltage-dependent Ca^{2+}-channels.

ACKNOWLEDGMENTS

We thank Ms. Kristina Linder for her excellent assistance in the preparation of the manuscript. The work from our laboratory was supported by the Swedish Medical Research Council (12x-562 and 12x-6240), the Swedish Diabetes Association, the Nordic Insulin Foundation, and the Wenner–Gren Foundation.

REFERENCES

Abrahamsson, H., and Gylfe, E. (1980). Demonstration of a proton gradient across the insulin granule membrane. *Acta Physiol. Scand.* **109**, 113–114.

Abrahamsson, H., and Rorsman, P. (1984). Cyclic AMP potentiation of insulin release from pancreatic β-cells is not mediated by an increased cytosolic Ca^{2+} activity. *Diabetologia* **27**, 250A.

Abrahamsson, H., Gylfe, E., and Hellman, B. (1981). Influence of external calcium ions on labelled calcium efflux from pancreatic β-cells and insulin granules in mice. *J. Physiol. (London)* **311**, 541–550.

Adelstein, R. S., and Eisenberg, E. (1980). Regulation and kinetics of the actin-myosin-ATP interaction. *Annu. Rev. Biochem.* **49**, 921–956.

Ammon, H. P. T., and Verspohl, E. J. (1979). Effect of methylene blue on pyridine nucleotides and insulin secretion of rat pancreatic islets. *Diabetologia* **17**, 41–44.

Ammon, H. P. T., Grimm, A., Lutz, S., Wagner-Tescher, D., Händel, M. and Hagenloh, I. (1980). Islet glutathione and insulin release. *Diabetes* **29**, 830–834.

Andersson, T. (1983). Glucose-induced retention of intracellular ^{45}Ca in pancreatic islets. *Am. J. Physiol.* **245**, C343–C347.

Andersson, T., Berggren, P.-O., and Gylfe, E. (1981a). Effects of glucose on the calcium content of intact β-cells and cellular organelles. *Upsala J. Med. Sci.* **86**, 165–170.

Andersson, T., Betsholtz, C., and Hellman, B. (1981b). Calcium and pancreatic βcell function. 12. Modification of ^{45}Ca fluxes by excess of K^+. *Acta Endocrinol. (Copenhagen)* **96**, 87–92.

Andersson, T., Berggren, P.-O., Gylfe, E., and Hellman, B. (1982a). Amounts and distribution of intracellular magnesium and calcium in pancreatic β-cells. *Acta Physiol. Scand.* **114**, 235–241.

Andersson, T., Betsholtz, C., and Hellman, B. (1982b). Granular calcium exchange in glucose-stimulated pancreatic β-cells. *Biomed. Res.* **3**, 29–36.

Atkins, T., and Matty, A. J. (1971). Adenylate cyclase and phosphodiesterase activity in the isolated islets of Langerhans of obese mice and their lean litter mates: The effect of glucose, adrenaline and drugs on adenylate cyclase activity. *J. Endocrinol.* **51**, 67–78.

Atwater, I., and Beigelman, P. M. (1976). Dynamic characteristics of electrical activity in pancreatic β-cells. I. Effects of calcium and magnesium removal. *J. Physiol. (Paris)* **72**, 769–786.

Atwater, I., Ribalet, B., and Rojas, E. (1978). Cyclic changes in potential and resistance of the β-cell membrane induced by glucose in islets of Langerhans from mouse. *J. Physiol. (London)* **278**, 117–139.

Atwater, I., Ribalet, B., and Rojas, E. (1979a). Mouse pancreatic β-cells: Tetraethylammonium blockage of the potassium permeability increase induced by depolarization. *J. Physiol. (London)* **288**, 561–574.

Atwater, I., Dawson, C. M., Ribalet, B., and Rojas, E. (1979b). Potassium permeability activated by intracellular calcium ion concentration in the pancreatic β-cell *J. Physiol. (London)* **288**, 575–588.

Atwater, I., Dawson, C. M., Scott, A., Eddlestone, G., and Rojas, E. (1980). The nature of the oscillatory behaviour in electrical activity from pancreatic β-cell. *Horm. Metab. Res., Suppl. Ser.* **10**, 100–107.

Atwater, I., Rosario, L., and Rojas, E. (1983). Properties of the Ca-activated K$^+$ channel in pancreatic β-cells. *Cell Calcium* **4**, 451–461.

Axén, K. V., Schubart, U. K., Blake, A. D., and Fleischer, N. (1983). Role of Ca^{2+} in secretagogue-stimulated breakdown of phosphatidylinositol in rat pancreatic islets. *J. Clin. Invest.* **72**, 13–21.

Baker, P. F., and Knight, D. E. (1984). Chemiosmotic hypothesis of exocytosis: A critique. *Biosci. Rep.* **4**, 285–298.

Barnard, T. (1982). Thin frozen-dried cryosections and biological X-ray microanalysis *J. Microsc. (Oxford)* **126**, 317–332.

Bent-Hansen, L., Capito, K., and Hedeskov, C. J. (1979). The effect of calcium on somatostatin inhibition of insulin release and cyclic AMP production in mouse pancreatic islets. *Biochim. Biophys. Acta* **585**, 240–249.

Berggren, P.-O., Berglund, O., and Hellman, B. (1978). Determination of calcium in microgram amounts of dried biological material by flameless atomic absorption spectrophotometry with special reference to the pancreatic islets. *Anal. Biochem.* **84**, 393–401.

Berggren, P.-O., Östenson, C.-G., Petersson, B., and Hellman, B. (1979). Evidence for divergent glucose effects on calcium metabolism in pancreatic β- and α$_2$-cells. *Endocrinology (Baltimore)* **105**, 1463–1468.

Berggren, P.-O., Andersson, T., and Hellman, B. (1983a). The interaction between barium and calcium in β-cell-rich pancreatic islets. *Biomed. Res.* **4**, 129–138.

Berggren, P.-O., Bergsten, P., Gylfe, E., Larsson, R., and Hellman, B. (1983b). Interactions between magnesium and calcium in β-cell-rich pancreatic islets. *Am. J. Physiol.* **244**, E541–E547.

Bergsten, P., and Hellman, B. (1984). Differentiation between the short and long term effects of glucose on the intracellular calcium content of the pancreatic β-cell. *Endocrinology (Baltimore)* **114**, 1854–1859.

Best, L., and Malaisse, W. J. (1983a). Phospholipids and islet function. *Diabetologia* **25**, 299–305.

Best, L., and Malaisse, W. J. (1983b). Effects of nutrient secretagogues on phospholipid metabolism in rat pancreatic islets. *Mol. Cell. Endocrinol.* **32**, 205–214.

Bloom, G. D., Hellman, B., Sehlin, J., and Täljedal, I.-B. (1977). Glucose-stimulated and La^{3+}-nondisplaceable Ca^{2+} pool in pancreatic islets. *Am. J. Physiol.* **232**, E114–E118.

Borg, L. A. H., and Andersson, A. (1981). Long-term effects of glibenclamide on the insulin production, oxidative metabolism and quantitive ultrastructure of mouse pancreatic islets maintained in tissue culture at different glucose concentrations. *Acta Diabetol. Lat.* **18,** 65–83.

Borle, A. B. (1981). Control, modulation and regulation of cell calcium. *Rev. Physiol., Biochem. Pharmacol.* **90,** 13–153.

Borowitz, J. L., and Matthews, E. K. (1980). Calcium exchange ability in subcellular fractions of pancreatic islet cells. *J. Cell Sci.* **41,** 233–243.

Boschero, A. C., Kawazu, S., Duncan, G., and Malaisse, W. J. (1977). Effect of glucose on K$^+$ handling by pancreatic islets. *FEBS Lett.* **83,** 151–154.

Boyd, A. E., III (1982). Cytoskeletal proteins and insulin secretion. *In* "Cellular Regulation of Secretion and Release" (P. M. Conn, ed.), pp. 223–263. Academic Press, New York.

Boyd, A. E., III, Zabelshansky, M., Hill, R. S., and Oberwetter, J. M. (1984). Glucose and high K$^+$ stimulate an increase in cytosolic free Ca^{2+} and insulin secretion. *Diabetes* **33,** Suppl. 1, 34A.

Brenner, S. L., and Korn, E. D. (1979). Substoichiometric concentrations of cytochalasin D inhibit actin polymerization. *J. Biol. Chem.* **254,** 9982–9985.

Brisson, G. R., Malaisse-Lagae, F., and Malaisse, W. J. (1972). The stimulus-secretion coupling of glucose-induced insulin release. VII. A proposed site of action for adenosine-3′,5′-cyclic monophosphate. *J. Clin. Invest.* **51,** 232–241.

Brocklehurst, K. W., and Hutton, J. C. (1983). Ca^{2+}-dependent binding of cytosolic components to insulin-secretory granules results in Ca^{2+}-dependent protein phosphorylation. *Biochem. J.* **210,** 533–539.

Brocklehurst, K. W., and Hutton, J. C. (1984). Involvement of protein kinase C in the phosphorylation of an insulin-granule membrane protein. *Biochem. J.* **220,** 283–290.

Bukowiecki, L., and Freinkel, N. (1976). Relationship between efflux of ionic calcium and phosphorus during excitation of pancreatic islets with glucose. *Biochim. Biophys. Acta* **436,** 190–198.

Campbell, A. K. (1983). "Intracellular Calcium. Its Universal Role as Regulator." Wiley, Chichester.

Campillo, J. E., and Ashcroft, S. J. H. (1982). Protein carboxymethylation in rat islets of Langerhans. *FEBS Lett.* **138,** 71–75.

Carpinelli, A., and Malaisse, W. J. (1980a). Regulation of ^{86}Rb$^+$ outflow from pancreatic islets. I. Reciprocal changes in the response to glucose, tetraethylammonium and quinine. *Mol. Cell. Endocrinol.* **17,** 103–110.

Carpinelli, A. R., and Malaisse, W. J. (1980b). Regulation of ^{86}Rb outflow from pancreatic islets. II. Effect of changes in extracellular and intracellular pH. *Diabete Metab.* **6,** 193–198.

Carpinelli, A. R., and Malaisse, W. J. (1980c). Regulation of ^{86}Rb outflow from pancreatic islets. IV. Effect of cyclic AMP, dibutyryl-cyclic AMP and theophylline. *Acta Diabetol. Lat.* **17,** 199–205.

Cerasi, E., and Luft, R. (1972). Pathogenesis of diabetes in man. *Isr. Med. Sci.* **8,** 207–223.

Charles, A. M., Fanska, R., Schmid, F. G., Forsham, P. H., and Grodsky, G. M. (1973). Adenosine 3′:5′-monophosphate in pancreatic islets: Glucose-induced insulin release. *Science* **179,** 569–571.

Charles, M. A., Lawecki, J., Pictet, R., and Grodsky, G. M. (1975). Insulin secretion. Interrelationships of glucose, cyclic adenosine 3′,5′-monophosphate, and calcium. *J. Biol. Chem.* **250,** 6134–6140.

Charles, S., and Henquin, J.-C. (1983). Distinct effects of various amino acids on ^{45}Ca^{2+} fluxes in rat pancreatic islets. *Biochem. J.* **214,** 899–907.

Chay, T. R., and Keizer, J. (1983). Minimal model for membrane oscillations in the pancreatic β-cells. *Biophys. J.* **42**, 181–190.

Cheung, W. Y. (1982). Calmodulin: An overview. *Fed. Proc., Fed. Am. Soc. Exp. Biol.* **41**, 2253–2257.

Christie, M. R., and Ashcroft, S. J. H. (1984). Cyclic AMP-dependent protein phosphorylation and insulin secretion in intact islets of Langerhans. *Biochem. J.* **218**, 87–99.

Clements, R. S., and Rhoten, W. B. (1976). Phosphoinositide metabolism and insulin secretion from isolated rat pancreatic islets. *J. Clin. Invest.* **57**, 684–691.

Clements, R. S., Evans, M. H., and Pace, C. S. (1981). Substrate requirements for the phosphoinositide response in rat pancreatic islets. *Biochim. Biophys. Acta* **647**, 1–9.

Colca, J. R., McDonald, J. M., Kotagal, N., Patke, C., Fink, J., Greider, M. H., Lacy, P. E., and McDaniel, M. L. (1982). Active calcium uptake by islet-cell endoplasmic reticulum. *J. Biol. Chem.* **257**, 7223–7228.

Colca, J. R., Kotagal, N., Lacy, P. E., and McDaniel, M. L. (1983a). Modulation of active Ca^{2+} uptake by the islet-cell endoplasmic reticulum. *Biochem. J.* **212**, 113–121.

Colca, J. R., Brooks, C. L., Landt, M., and McDaniel, M. L. (1983b). Correlation of Ca^{2+}- and calmodulin-dependent protein kinase activity with secretion of insulin from islets of Langerhans. *Biochem. J.* **212**, 819–827.

Colca, J. R., Kotagal, N., Lacy, P. E., and McDaniel, M. L. (1983c). Comparison of the properties of active Ca^{2+} transport by the islet-cell endoplasmic reticulum and plasma membrane. *Biochim. Biophys. Acta* **729**, 176–184.

Colca, J. R., Kotagal, N., Lacy, P. E., Brooks, C. L., Norling, L., Landt, M., and McDaniel, M. L. (1984). Glucose-stimulated protein phosphorylation in the pancreatic islet. *Biochem. J.* **220**, 529–539.

Cook, D. L. (1984). Electrical pacemaker mechanisms of pancreatic islet cells. *Fed. Proc., Fed. Am. Soc. Exp. Biol.* **43**, 2368–2372.

Cook, D. L., and Hales, C. N. (1984). Intracellular ATP directly blocks B-cell K^+ channels. *Diabetes* **33**, Suppl. 1, 35A.

Cook, D. L., Crill, W. E., and Porte, D., Jr. (1980). Plateau potential in pancreatic islet cells are voltage-dependent action potentials. *Nature (London)* **286**, 404–406.

Cook, D. L., Porte, D., Jr., and Crill, W. E. (1981). Voltage-dependence of rhythmic plateau potentials of pancreatic islet cells. *Am. J. Physiol.* **240**, E290–E296.

Couturier, E., and Malaisse, W. J. (1980a). Insulinotropic effects of hypoglycaemic and hyperglycaemic sulphonamides. The ionophoretic hypothesis. *Diabetologia* **19**, 335–340.

Couturier, E., and Malaisse, W. J. (1980b). Ionophoretic activity of hypoglycaemic sulfonylureas. *Arch. Int. Pharmacodyn. Ther.* **245**, 323–334.

Curry, D. L., Joy, R. M., Holley, D. C., and Bennett, L. L. (1977). Magnesium modulation of glucose-induced insulin secretion by the perfused rat pancreas. *Endocrinology (Baltimore)* **101**, 203–208.

Davis, B., and Lazarus, N. R. (1972). Insulin release from mouse islets. Effect of glucose and hormones on adenylate cyclase. *Biochem. J.* **129**, 373–379.

Dean, P. M. (1973). Ultrastructural morphometry of the pancreatic β-cell. *Diabetologia* **9**, 115–119.

Dean, P. M. (1975). Exocytosis modelling: An electrostatic function for calcium in stimulus-secretion coupling. *J. Theor. Biol.* **54**, 289–308.

Dean, P. M., and Matthews, E. K. (1968). Electrical activity in pancreatic islet cells. *Nature (London)* **219**, 389–390.

Dean, P. M., and Matthews, E. K. (1970). Electrical activity in pancreatic islet cells: Effect of ions. *J. Physiol. (London)* **210**, 265–275.

Deleers, M., Couturier, E., Mahy, M., and Malaisse, W. J. (1980). Calcium transport in lysosomes containing hypoglycemic and hyperglycemic sulfonamides. *Arch. Int. Pharmacodyn. Ther.* **246**, 170–172.

Devis, G., Somers, G., and Malaisse, W. J. (1975). Stimulation of insulin release by calcium. *Biochem. Biophys. Res. Commun.* **67**, 525–529.

Devis, G., Somers, G., and Malaisse, W. J. (1977). Dynamics of calcium-induced insulin release. *Diabetologia* **13**, 531–536.

Donatsch, P., Lowe, D. A., Richardson, B. P., and Taylor, P. (1977). The functional significance of sodium channels in pancreatic beta-cell membranes. *J. Physiol. (London)* **267**, 357–376.

Dunlop, M., Larkins, R. G., and Court, J. M. (1982). Endogenous ionophoretic activity in the neonatal rat pancreatic islet. *FEBS Lett.* **144**, 259–263.

Efendic, S., Enzmann, F., Nylén, A., Uvnäs-Wallensten, K., and Luft, R. (1979). Effect of glucose/sulfonylurea interaction on release of insulin, glucagon and somatostatin from isolated perfused rat pancreas. *Proc. Natl. Acad. Sci. U.S.A.* **76**, 5901–5904.

Ewart, R. B. L., Yousufzai, S. Y. K., Bradford, M. W., and Shrago, E. (1983). Rat islet mitochondrial adenine nucleotide translocase and the regulation of insulin secretion. *Diabetes* **32**, 793–797.

Feinstein, M. B., and Hadjian, R. A. (1982). Effects of the calmodulin antagonist trifluoperazine on stimulus-induced calcium mobilization, aggregation, secretion, and protein phosphorylation in platelets. *Mol. Pharmacol.* **21**, 422–431.

Fex, G., and Lernmark, Å. (1972). Effect of D-glucose on the incorporation of ^{32}P into phospholipids of mouse pancreatic islets. *FEBS Lett.* **25**, 287–291.

Fisher, S. K., Van Rooijen, L. A. A., and Agranoff, B. W. (1984). Renewed interest in the polyphosphoinositides. *Trends Biochem. Sci.* **9**, 53–56.

Flatt, P. R., and Swanston-Flatt, S. (1983). Effects of decapsulation on superficial and intracellular ^{45}Ca uptake by isolated mouse pancreatic islets. *Biomed. Res.* **4**, 557–566.

Flatt, P. R., Gylfe, E., and Hellman, B. (1979). Dissociation between ^{45}Ca uptake and insulin release: Evidence for an inhibitory pool of calcium in the β-cell plasma membrane. *Biochem. Soc. Trans.* **7**, 1097–1100.

Flatt, P. R., Berggren, P.-O., Gylfe, E., and Hellman, B. (1980a). Calcium and pancreatic β-cell function. IX. Demonstration of lanthanide-induced inhibition of insulin secretion independent of modifications in transmembrane Ca^{2+} fluxes. *Endocrinology (Baltimore)* **107**, 1007–1013.

Flatt, P. R., Boquist, L., and Hellman, B. (1980b). Calcium and pancreatic β-cell function. The mechanism of insulin secretion studied with the aid of lanthanum. *Biochem. J.* **190**, 361–372.

Flatt, P. R., Gylfe, E., and Hellman, B. (1981). Thulium binding to the pancreatic β-cell membrane. *Endocrinology (Baltimore)* **108**, 2258–2263.

Floyd, J. C., Fajans, S. S., Knopf, R. F., and Conn, J. W. (1963). Evidence that insulin release is the mechanism for experimentally induced hypoglycemia in man. *J. Clin. Invest.* **42**, 1714–1719.

Formby, B., Capito, K., Egeberg, J., and Hedeskov, C. J. (1976). Ca-activated ATPase activity in subcellular fractions of mouse pancreatic islets. *Am. J. Physiol.* **230**, 441–448.

Frankel, B. J., Imagawa, W. T., O'Connor, D. L., Lundquist, I., Kromhout, J. A., Fanska, R. E., and Grodsky, G. M. (1978a). Glucose-stimulated $^{45}calcium$ efflux from isolated rat pancreatic islets. *J. Clin. Invest.* **62**, 525–531.

Frankel, B. J., Kromhout, J. A., Imagawa, W., Landahl, H. D., and Grodsky G. M. (1978b). Glucose-stimulated ^{45}Ca uptake in isolated rat islets. *Diabetes* **27**, 365–369.

Frankel, B. J., Atwater, I., and Grodsky, G. M. (1981). Calcium affects insulin release and membrane potential in islet β-cells. *Am. J. Physiol.* **240**, C64–C72.

Freinkel, N., El Younsi, C., and Dawson, R. M. C. (1975). Interrelations between the phospholipids of rat pancreatic islets during glucose stimulation and their response to medium inositol and tetracaine. *Eur. J. Biochem.* **59**, 245–252.

Gabbay, K. H., and Tze, W. J. (1972). Cytochalasin B-sensitive emigate in the beta cell. *Diabetes* **21**, 327.

Gagliardino, J. J., Harrison, D. E., Christie, M. R., Gagliardino, E. E., and Ashcroft, S. J. H. (1980). Evidence for the participation of calmodulin in stimulus-secretion coupling in the pancreatic β-cell *Biochem. J.* **192**, 919–927.

Gagliardino, J. J., Semino, M. C., Rebolledo, O. R., Gomez Dumm, C. L., and Hernandez, R. E. (1984). Sequential determination of calcium distribution in β-cells at the various phases of glucose-induced insulin secretion. *Diabetologia* **26**, 290–296.

Gerich, J. E., Frankel, B. J., Fanska, R., West, L., Forsham, P. H., and Grodsky, G. M. (1974). Calcium dependency of glucagon secretion from the *in vitro* perfused rat pancreas. *Endocrinology (Baltimore)* **94**, 1381–1385.

Gil, D. W., Brown, S. A., Seeholzer, S. H., and Wildey, G. M. (1983). Phosphatidylinositol turnover and cellular function. *Life Sci.* **32**, 2043–2046.

Goldfine, I. D., Roth, J., and Birnbaumer, L. (1972). Glucagon receptors in β-cells. Binding of ^{125}I-glucagon and activation of adenylate cyclase. *J. Biol. Chem.* **247**, 1211–1218.

Grill, V. (1978). Ba^{++} stimulates accumulation of cyclic AMP in rat pancreatic islets. *Biochem. Biophys. Res. Commun.* **82**, 750–758.

Grill, V., and Cerasi, E. (1973). Activation by glucose of adenyl cyclase in pancreatic islets of the rat. *FEBS Lett.* **33**, 311–314.

Grill, V., and Cerasi, E. (1974). Stimulation by D-glucose of cyclic adenosine 3′:5′-monophosphate accumulation and insulin release in isolated pancreatic islets of rat. *J. Biol. Chem.* **249**, 4196–4201.

Grill, V., and Cerasi, E. (1975). Glucose-induced cyclic AMP accumulation in rat islets of Langerhans: Preferential effect of the alpha anomer. *FEBS Lett.* **54**, 80–83.

Grill, V., and Cerasi, E. (1976). Effect of hexoses and mannoheptulose on cyclic AMP accumulation and insulin secretion in rat pancreatic islets. *Biochim. Biophys. Acta* **437**, 36–50.

Grinstein, S., van der Meulen, J., and Furuya, W. (1982). Possible role of H$^+$-alkali cation countertransport in secretory granule swelling during exocytosis. *FEBS Lett.* **148**, 1–4.

Grodsky, G. M., and Bennett, L. L. (1966). Cation requirements for insulin secretion in the isolated perfused pancreas. *Diabetes* **15**, 910–912.

Gylfe, E. (1982). Glucose stimulated net uptake of Ca^{2+} in the pancreatic β-cell demonstrated with dual wavelength spectrophotometry. *Acta Physiol. Scand.* **114**, 149–151.

Gylfe, E., and Hellman, B. (1978). Calcium and pancreatic β-cell function. 2. Mobilisation of glucose-sensitive ^{45}Ca from perifused islets rich in β-cells. *Biochim. Biophys. Acta* **538**, 249–257.

Gylfe, E., and Hellman, B. (1981). Calcium and pancreatic β-cell function: Modification of ^{45}Ca fluxes by methylxanthines and dibutyryl cyclic-AMP. *Biochem. Med.* **26**, 365–376.

Gylfe, E., and Hellman, B. (1982a). Evidence for a slowly exchangeable pool of calcium in the pancreatic β-cell plasma membrane. *J. Physiol. (London)* **328**, 285–293.

Gylfe, E., and Hellman, B. (1982b). Lack of Ca^{2+} ionophoretic activity of hypoglycemic sulfonylureas in excitable cells and isolated secretory granules. *Mol. Pharmacol.* **22**, 715–720.

Gylfe, E., Buitrago, A., Berggren, P.-O., Hammarström, K., and Hellman, B. (1978). Glucose inhibition of ^{45}Ca efflux from pancreatic islets. *Am. J. Physiol.* **235**, E191–E196.

Gylfe, E., Andersson, T., Rorsman, P., Abrahamsson, H., Arkhammar, P., Hellman, P., Hellman, B., Oie, H. K., and Gazdar, A. F. (1983). Depolarization-independent net uptake of calcium into clonal insulin-releasing cells exposed to glucose. *Biosci. Rep.* **3**, 927–937.

Gylfe, E., Hellman, B., Sehlin, J., and Täljedal, I.-B. (1984). Interaction of sulfonylurea with the pancreatic β-cell. *Experientia* **40**, 1126–1134.

Hahn, H. J., Gylfe, E., and Hellman, B. (1979). Glucose-dependent effect of methylxanthines on the ^{45}Ca distribution in pancreatic β-cells. *FEBS Lett.* **103**, 348–351.

Hahn, H. J., Gylfe, E., and Hellman, B. (1980). Calcium and pancreatic β-cell function. 7. Evidence for cyclic AMP-induced translocation of intracellular calcium. *Biochim. Biophys. Acta* **630,** 425–432.

Hales, C. N., and Milner, R. D. G. (1968). Cations and the secretion of insulin from rabbit pancreas in vitro. *J. Physiol. (London)* **199,** 177–187.

Hallberg, A. (1984). Dissociation between phosphatidylinositol hydrolysis and insulin secretion of isolated mouse pancreatic islets. *Diabetologia* **27,** 284A.

Harrison, D. E., and Ashcroft, S. J. H. (1982). Effects of Ca^{2+}, calmodulin and cyclic AMP on the phosphorylation of endogenous proteins by homogenates of rat islets of Langerhans. *Biochim. Biophys. Acta* **714,** 313-319.

Harrison, D. E., Ashcroft, S. J. H., Christie, M. R., and Lord, J. M. (1984). Protein phosphorylation in the pancreatic β-cell. *Experientia* **40,** 1075–1084.

Hedeskov, C. J. (1980). Mechanisms of glucose-induced insulin secretion. *Physiol. Rev.* **60,** 442–509.

Hedeskov, C. J., and Capito, K. (1974). The effect of starvation on insulin secretion and glucose metabolism in mouse pancreatic islets. *Biochem. J.* **140,** 423–433.

Hellman, B. (1965). Studies in obese-hyperglycemic mice. *Ann. N.Y. Acad. Sci.* **131,** 541–558.

Hellman, B. (1975a). The significance of calcium for glucose stimulation of insulin release. *Endocrinology (Baltimore)* **97,** 392–398.

Hellman, B. (1975b). Modifying actions of calcium ionophores on insulin release. *Biochim. Biophys. Acta* **399,** 157–169.

Hellman, B. (1976a). Stimulation of insulin release after raising extracellular calcium. *FEBS Lett* **63,** 125–128.

Hellman, B. (1976b). Calcium and the control of insulin secretion. *In* "Diabetes Research Today. Meeting of the Minkowski Prize-Winners" (E. Lindenlaub, ed.), pp. 207–222. Schattauer Verlag, Stuttgart.

Hellman, B. (1977). Calcium and pancreatic β-cell function. 1. Stimulatory effects of pentobarbital on insulin release. *Biochim. Biophys. Acta* **497,** 766–774.

Hellman, B. (1978). Calcium and pancreatic β-cell function. 3. Validity of the La^{3+}-wash technique for discriminating between superficial and intracellular ^{45}Ca. *Biochim. Biophys. Acta* **540,** 534–542.

Hellman, B. (1979). Calcium and pancreatic β-cell function. 5. Mobilization of a glucose-stimulated pool of intracellular ^{45}Ca by metabolic inhibitors and the ionophore A-23187. *Acta Endocrinol. (Copenhagen)* **90,** 624–636.

Hellman, B. (1981). Tolbutamide stimulation of ^{45}Ca fluxes in microdissected pancreatic islets rich in β-cells. *Mol. Pharmacol.* **20,** 83–88.

Hellman, B. (1982a). Differences between the effect of tolbutamide and Ca^{2+} ionophores on Ca^{2+} efflux from pancreatic β-cells. *Pharmacol. Res. Commun.* **14,** 701–710.

Hellman, B. (1982b). The mechanism of sulfonylurea stimulation of insulin release. *Acta Biol. Med. Ger.* **41,** 1211–1219.

Hellman, B., and Andersson, T. (1978). Calcium and pancreatic β-cell function. 4. Evidence that glucose and phosphate stimulate incorporation into different intracellular pools. *Biochim. Biophys. Acta* **541,** 483–491.

Hellman, B., and Gylfe, E. (1978). Calcium and pancreatic β-cell function: Glucose stimulation of uptake of lanthanum-displaceable ^{45}Ca from low or normal calcium-containing media. *Horm. Metab. Res.* **10,** 29–31.

Hellman, B., and Gylfe, E. (1984a). Glucose inhibits ^{45}Ca efflux from pancreatic β-cells also in the absence of $Na^{+}–Ca^{2+}$ countertransport. *Biochim. Biophys. Acta* **770,** 136–141.

Hellman, B., and Gylfe, E. (1984b). Evidence for glucose stimulation of intracellular buffering of calcium in the pancreatic β-cells. *Q. J. Exp. Physiol. Cogn. Med. Sci.* **69,** 867–874.

Hellman, B., and Gylfe, E. (1985). Glucose regulation of insulin release involves intracellular sequestration of calcium. *In* "Calcium in Biological Systems" (R. P. Rubin, G. B. Weiss, J. W. Putney, Jr. eds.), pp. 93–99. Plenum, New York.

Hellman, B., Idahl, L.-Å., and Danielsson, Å. (1969). Adenosine triphosphate levels of mammalian pancreatic β cells after stimulation with glucose and hypoglycemic sulfonylureas. *Diabetes* **18**, 509–516.

Hellman, B., Sehlin, J., and Täljedal, I.-B. (1971a). Transport of α-aminoisobutyric acid in mammalian pancreatic β-cells. *Diabetologia* **7**, 256–265.

Hellman, B., Sehlin, J., and Täljedal, I.-B. (1971b). Calcium uptake by pancreatic β-cells as measured with the aid of ^{45}Ca and mannitol-^{3} H. *Am. J. Physiol.* **221**, 1795–1801.

Hellman, B., Idahl, L.-Å., Lernmark, Å., Sehlin, J., and Täljedal, I. B. (1974a). Membrane sulphydryl groups and the pancreatic beta cell recognition of insulin secretagogues. *In* "Diabetes" (W. J. Malaisse and J. Pirart, eds.), pp. 65–78. Excerpta Medica, Amsterdam.

Hellman, B., Idahl, L.-Å., Lernmark, Å., and Täljedal, I.-B. (1974b). The pancreatic β-cell recognition of insulin secretagogues: Does cyclic AMP mediate the effect of glucose? *Proc. Natl. Acad. Sci. U.S.A.* **71**, 3405–3409.

Hellman, B., Sehlin, J., and Täljedal, I. B. (1976a). Effect of glucose on $^{45}Ca^{2+}$ uptake by pancreatic islets as studied with the lanthanum method. *J. Physiol. (London)* **254**, 639–656.

Hellman, B., Sehlin, J., and Täljedal, I.-B. (1976b). Calcium and secretion: Distribution between two pools of glucose-sensitive calcium in pancreatic islets. *Science* **194**, 1421–1423.

Hellman, B., Lenzen, S., Sehlin, J., and Täljedal, I. B. (1977). Effects of various modifiers of insulin release on the lanthanum-nondisplaceable ^{45}Ca uptake by isolated pancreatic islets. *Diabetologia* **13**, 49–53.

Hellman, B., Idahl, L. Å., Lenzen, S., Sehlin, J., and Täljedal, I.-B. (1978a). Further studies on the relationship between insulin release and lanthanum-nondisplaceable $^{45}Ca^{2+}$ uptake by pancreatic islets: Effects of fructose and starvation. *Endocrinology (Baltimore)* **102**, 1856–1863.

Hellman, B., Sehlin, J., and Täljedal, I.-B. (1978b). Effects of Na^+, K^+ and Mg^{2+} on $^{45}Ca^{2+}$ uptake by pancreatic islets. *Pfluegers Arch.* **378**, 93–97.

Hellman, B., Abrahamsson, H., Andersson, T., Berggren, P.-O., Flatt, P., and Gylfe, E. (1979a). Participation of calcium pools in insulin secretion. *Int. Congr. Ser.—Excerpta Med.* **500**, 160–165.

Hellman, B., Andersson, T., Berggren, P.-O., Flatt, P., Gylfe, E., and Kohnert, K.-D. (1979b). The role of calcium in insulin secretion. *Horm. Cell Regul.* **3**, 69–96.

Hellman, B., Abrahamsson, H., Andersson, T., Berggren, P.-O., Flatt, P., Gylfe, E., and Hahn, H.-J. (1980a). Calcium movements in relation to glucose-stimulated insulin secretion. *Horm. Metab. Res., Suppl. Ser.* **10**, 122–130.

Hellman, B., Gylfe, E., Berggren, P.-O., Andersson, T., Abrahamsson, H., Rorsman, P., and Betsholtz, C. (1980b). Ca^{2+} transport in pancreatic β-cells during glucose stimulation of insulin secretion. *Upsala J. Med. Sci.* **85**, 321–329.

Hellman, B., Andersson, T., Berggren, P.-O., and Rorsman, P. (1980c). Calcium and pancreatic β-cell function. II. Modification of ^{45}Ca fluxes by Na^+ removal. *Biochem. Med.* **24**, 143–152.

Hellman, B., Honkanen, T., and Gylfe, E. (1982). Glucose inhibits insulin release induced by Na^+ mobilization of intracellular calcium. *FEBS Lett.* **148**, 289–292.

Hellman, B., Hällgren, R., Abrahamsson, H., Bergsten, P., Berne, C., Gylfe, E., Rorsman, P., and Wide, L. (1985). The dual action of glucose on the cytosolic Ca^{2+} activity in pancreatic B-cells. Demonstration of an inhibitory effect of glucose on insulin release in the mouse and man. *Biomed. Biochim. Acta* **44**, 63–70.

Henquin, J. C. (1978a). D-glucose inhibits potassium efflux from pancreatic islet cells. *Nature (London)* **271,** 271–273.

Henquin, J. C. (1978b). Relative importance of extracellular and intracellular calcium for the two phases of glucose-stimulated insulin release: Studies with theophylline. *Endocrinology (Baltimore)* **102,** 723–730.

Henquin, J. C. (1979). Opposite effects of intracellular Ca^{2+} and glucose on K^+ permeability of pancreatic islet cells. *Nature (London)* **280,** 66–68.

Henquin, J. C. (1980a). Metabolic control of the potassium permeability in pancreatic islet cells. *Biochem. J.* **186,** 541–550.

Henquin, J.-C. (1980b). Specificity of divalent cation requirement for insulin release. Effects of strontium. *Pfluegers Arch.* **383,** 123–129.

Henquin, J. C. (1980c). Tolbutamide stimulation and inhibition of insulin release: studies of the underlying ionic mechanisms in isolated rat islets. *Diabetologia* **18,** 151–160.

Henquin, J. C. (1981). The effect of pH on [86]Rubidium efflux from pancreatic islet cells. *Mol. Cell. Endocrinol.* **21,** 119–128.

Henquin, J. C., and Lambert, A. E. (1975). Cobalt inhibition of insulin secretion and calcium uptake by isolated rat islets. *Am. J. Physiol.* **228,** 1669–1677.

Henquin, J. C., and Meissner, H. P. (1978). Valinomycin inhibition of insulin release and alterations of the electrical properties of pancreatic B-cells. *Biochim. Biophys. Acta* **543,** 455–464.

Henquin, J. C., and Meissner, H. P. (1981). Effects of amino acids on membrane potential and Rb^+ fluxes in pancreatic β-cells. *Am. J. Physiol.* **240,** E245–E253.

Henquin, J. C., and Meissner, H. P. (1982). The electrogenic sodium-potassium pump of mouse pancreatic β-cells. *J. Physiol. (London)* **322,** 529–552.

Henquin, J.-C., and Meissner, H. P. (1983). Dibutyryl cyclic AMP triggers Ca^{2+} influx and Ca^{2+}-dependent electrical activity in pancreatic β-cells. *Biochem. Biophys. Res. Commun.* **112,** 614–620.

Henquin, J. C., Meissner, H. P., and Preissler, M. (1979). 9-Aminoacridine- and tetraethylammonium-induced reduction of the potassium permeability in pancreatic β-cells. *Biochim Biophys. Acta* **587,** 579–592.

Henquin, J. C., Charles, S., Nenquin, M., Mathot, F., and Tamagawa, T. (1982). Diazoxide and D-600 inhibition of insulin release: Distinct mechanisms explain the specificity for different stimuli. *Diabetes* **31,** 776–783.

Henquin, J.-C., Schmeer, W., and Meissner, H. P. (1983a). Forskolin, an activator of adenylate cyclase, increases Ca^{2+}-dependent electrical activity induced by glucose in mouse pancreatic β cells. *Endocrinology (Baltimore)* **112,** 2218–2220.

Henquin, J.-C., Tamagawa, T., Nenquin, M., and Cogneau, M. (1983b). Glucose modulates Mg^{2+} fluxes in pancreatic islet cells. *Nature (London)* **301,** 73–74.

Henriksson, C., Claes, G., Gylfe, E., Hellman, B., and Zettergren, L. (1978). Collagenase isolation and [45]Ca efflux studies of human islets of Langerhans. *Eur. Surg. Res.* **10,** 343–351.

Herchuelz, A., and Malaisse, W. J. (1978). Regulation of calcium fluxes in pancreatic islets: Dissociation between calcium and insulin release. *J. Physiol. (London)* **283,** 409–424.

Herchuelz, A., and Malaisse, W. J. (1980a). regulation of calcium fluxes in pancreatic islets: Two calcium movements dissociated response to glucose. *Am. J. Physiol.* **238,** E87–E95.

Herchuelz, A., and Malaisse, W. J. (1980b). Regulation of calcium fluxes in rat pancreatic islets: Dissimilar effects of glucose and of sodium ion accumulation. *J. Physiol. (London)* **302,** 263–280.

Herchuelz, A., and Malaisse, W. J. (1981). Calcium movements and insulin release in pancreatic islet cells. *Diabete Metab.* **7,** 283–288.

Herchuelz, A., Couturier, E., and Malaisse, W. J. (1980a). Regulation of calcium fluxes in pancreatic islets: Glucose-induced calcium-calcium exchange. *Am. J. Physiol.* **238,** E96–E103.

Herchuelz, A., Sener, A., and Malaisse, W. J. (1980b). Regulation of calcium fluxes in rat pancreatic islets: Calcium extrusion by sodium-calcium countertransport. *J. Membr. Biol.* **57,** 1–12.

Herchuelz, A., Thonnart, N., Sener, A., and Malaisse, W. J. (1980c). Regulation of calcium fluxes in pancreatic islets: The role of membrane depolarization. *Endocrinology (Baltimore)* **107,** 491–497.

Herchuelz, A., Lebrun, P., Carpinelli, A., Thonnart, N., Sener, A., and Malaisse, W. J. (1981). Quinine mimics the dual effect of glucose on calcium movements. *Biochim. Biophys. Acta* **640,** 16–30.

Herchuelz, A., Lebrun, P., Boschero, A. C., and Malaisse, W. J. (1984). Mechanism of arginine-stimulated Ca^{2+} influx into pancreatic β-cell. *Am. J. Physiol.* **246,** E38–E43.

Herman, L., Sato, T., and Hales, C. N. (1973). The electron microscopic localization of cations to pancreatic islets of Langerhans and their possible role in insulin secretion. *J. Ultrastruct. Res.* **42,** 298–311.

Hirata, F., and Axelrod, J. (1980). Phospholipid methylation and biological signal transmission. *Science* **209,** 1082–1090.

Howell, S. L. (1977). Intracellular localization of calcium in pancreatic β-cells. *Biochem. Soc. Trans.* **5,** 875–880.

Howell, S. L., and Montague, W. (1973). Adenylate cyclase activity in isolated rat islets of Langerhans. Effects of agents which alter rates of insulin secretion. *Biochim. Biophys. Acta* **320,** 44–52.

Howell, S. L., and Montague, W. (1975). Regulation by nucleotides of [45]calcium uptake in homogenates of rat islets of Langerhans. *FEBS Lett.* **52,** 48–52.

Howell, S. L., and Tyhurst, M. (1976). [45]Calcium localization in islets of Langerhans, a study by electronmicroscopic autoradiography. *J. Cell Sci.* **21,** 415–422.

Howell, S. L., and Tyhurst, M. (1977). Distribution of anionic sites on surface of β-cell granule and plasma membranes: A study using cationic ferritin. *J. Cell Sci.* **27,** 289–301.

Howell, S. L., and Tyhurst, M. (1979). Interaction between insulin-storage granules and F-actin *in vitro. Biochem. J.* **178,** 367–371.

Howell, S. L., and Tyhurst, M. (1980). Regulation of actin polymerization in rat islets of Langerhans. *Biochem. J.* **192,** 381–383.

Howell, S. L., and Tyhurst, M. (1981). Actomyosin interaction with isolated β granules. *Diabetologia* **21,** 284.

Howell, S. L., and Tyhurst, M. (1982). Microtubules, microfilaments and insulin secretion. *Diabetologia* **22,** 301–308.

Howell, S. L., Montague, W., and Tyhurst, M. (1975). Calcium distribution in islets of Langerhans: A study of calcium concentrations and of calcium accumulation in β-cell organelles. *J. Cell Sci.* **19,** 395–409.

Howell, S. L., Tyhurst, M., Duvefelt, H., Andersson, A., and Hellerström, C. (1978). Role of zinc and calcium in the formation and storage of insulin in the pancreatic β-cell. *Cell Tissue Res.* **199,** 107–118.

Hubinont, C. J., Best, L., Sener, A., and Malaisse, W. J. (1984). Activation of protein kinase C by a tumor-promoting phorbol ester in pancreatic islets. *FEBS Lett.* **170,** 247–253.

Hutton, J. C. (1982). The internal pH and membrane potential of the insulin-secretory granule. *Biochem. J.* **204,** 171–178.

Hutton, J. C., and Peshavaria, M. (1982). Proton-translocating Mg^{2+}-dependent ATPase activity in insulin-secretory granules. *Biochem. J.* **204,** 161–170.

Hutton, J. C., Sener, A., Herchuelz, A., Atwater, I., Kavazu, S., Boschero, A. C., Somers, G., Devis, G., and Malaisse, W. J. (1980). Similarities in the stimulus-secretion coupling mechanisms of glucose- and 2-keto acid-induced insulin release. *Endocrinology (Baltimore)* **106,** 203–219.

Hutton, J. C., Penn, E. J., and Peshavaria, M. (1983). Low-molecular-weight constituents of isolated insulin-secretory granules. Bivalent cations, adenine nucleotides and inorganic phosphate. *Biochem. J.* **210,** 297–305.

Iversen, J. (1973). Adrenergic receptors and the secretion of glucagon and insulin from the isolated perfused canine pancreas. *J. Clin. Invest.* **52,** 2102–2116.

Janjic, D., and Wollheim, C. B. (1983). Interactions of Ca^{2+}, Mg^{2+} and Na^+ in regulation of insulin release from rat islets. *Am. J. Physiol.* **244,** E222–E229.

Janjic, D., Wollheim, C. B., Siegel, E. G., Krausz, Y., and Sharp, G. W. G. (1981). Sites of action of trifluoperazine in the inhibition of glucose-stimulated insulin release. *Diabetes* **30,** 960–966.

Janjic, D., Wollheim, C. B., and Sharp, G. W. G. (1982). Selective inhibition of glucose-stimulated insulin release by dantrolene. *Am. J. Physiol.* **243,** E59–E67.

Kavazu, S., Boschero, A. C., Delcroix, C., and Malaisse, W. J. (1978). The stimulus-secretion coupling of glucose-induced insulin release. XXVII. Effect of glucose on Na^+ fluxes in isolated islets. *Pfluegers Arch.* **375,** 197–206.

Kikuchi, M., Wollheim, C. B., Cuendet, G. S., Renold, A. E., and Sharp, G. W. G. (1978). Studies on the dual effects of glucose on ^{45}Ca efflux from isolated islets. *Endocrinology (Baltimore)* **102,** 1339–1349.

Kikuchi, M., Wollheim, C. B., Siegel, E. G., Renold, A. E., and Sharp, G. W. G. (1979). Biphasic insulin release in rat islets of Langerhans and the role of intracellular Ca^{2+} stores. *Endocrinology (Baltimore)* **105,** 1013–1019.

Klee, C. B. (1980). Calmodulin: Structure-function relationships. *In* "Calcium and Cell Function" (W. Y. Cheung, ed.), Vol. 1, pp. 59–77. Academic Press, New York.

Klöppel, G., and Bommer, G. (1979). Ultracytochemical calcium distribution in β-cells in relation to biphasic glucose-stimulated insulin release by the perfused rat pancreas. *Diabetes* **28,** 585–592.

Klöppel, G., and Schäfer, H. J. (1976). Effects of sulfonylureas on histochemical and ultra-cytochemical calcium distribution in β-cells of mice. *Diabetologia* **12,** 227–235.

Knight, D. E., and Baker, P. F. (1983). The phorbol ester TPA increases the affinity of exocytosis for calcium in 'leaky' adrenal medullary cells. *FEBS Lett.* **160,** 98–100.

Kohnert, K.-D., Hahn, H.-J., Gylfe, E., Borg, H., and Hellman, B. (1979). Calcium and pancreatic β-cell function. 6. Glucose and intracellular ^{45}Ca distribution. *Mol. Cell. Endocrinol.* **16,** 205–220.

Korn, E. D. (1978). Biochemistry of actomyosin-dependent cell motility (A review). *Proc. Natl. Acad. Sci. U.S.A.* **75,** 588–599.

Kotagal, N., Patke, C., Landt, M., McDonald, J., Colca, J., Lacy, P., and McDaniel, M. (1982). Regulation of pancreatic islet-cell plasma membrane Ca^{2+} + Mg^{2+}-ATPase by calmodulin. *FEBS Lett.* **137,** 249–252.

Kotagal, N., Colca, J. R., and McDaniel, M. L. (1983). Activation of an islet cell plasma membrane (Ca^{2+} + Mg^{2+})-ATPase by calmodulin and CaEGTA. *J. Biol. Chem.* **258,** 4808–4813.

Krausz, Y., Wollheim, C. B., Siegel, E., and Sharp, G. W. G. (1980). Possible role of calmodulin in insulin release. *J. Clin. Invest.* **66,** 603–607.

Krieger-Brauer, H., and Gratzl, M. (1981). Influx of Ca^{2+} into isolated secretory vesicles from adrenal medulla. Influence of external K^+ and Na^+. *FEBS Lett.* **133,** 244–246.

Krieger-Brauer, H., and Gratzl, M. (1982). Uptake of Ca^{2+} by isolated secretory vesicles from adrenal medulla. *Biochim. Biophys. Acta* **691,** 61–70.

Kuo, W.-N., Hodgins, D. S., and Kuo, J. F. (1973). Adenylate cyclase in islets of Langerhans. Isolation of islets and regulation of adenylate cyclase activity by various hormones and agents. *J. Biol. Chem.* **248,** 2705–2711.

Lacy, P. E., Howell, S. L., Young, D. A., and Fink, C. J. (1968). New hypothesis of insulin secretion. *Nature (London)* **219**, 1177–1179.

Lacy, P. E., Walker, M. M., and Fink, C. J. (1972). Perifusion of isolated rat islets in vitro. Participation of the microtubular system in the biphasic release of insulin. *Diabetes* **21**, 987–998.

Lacy, P. E., Finke, E. H., and Codilla, R. C. (1975). Cinemicrographic studies on β granule movement in monolayer culture of islet cells. *Lab. Invest.* **33**, 570–576.

Landt, M., McDaniel, M. L., Bry, C. G., Kotagal, N., Colca, J. R., Lacy, P. E., and McDonald, J. M. (1982). Calmodulin-activated protein kinase activity in rat pancreatic islet cell membranes. *Arch. Biochem. Biophys.* **213**, 148–154.

Lapetina, E. G. (1982). Regulation of arachidonic acid production: Role of phospholipases C and A_2. *Trends Pharmacol. Sci.* **3**, 115–118.

Laychock, S. G. (1982). Phospholipase A_2 activity in pancreatic islets is calcium-dependent and stimulated by glucose. *Cell Calcium* **3**, 43–54.

Laychock, S. G. (1983). Identification and metabolism of polyphosphoinositides in isolated islets of Langerhans. *Biochem. J.* **216**, 101–106.

Lebrun, P., Malaisse, W. J., and Herchuelz, A. (1982a). Evidence for two distinct modalities of Ca^{2+} influx into pancreatic β-cell. *Am. J. Physiol.* **242**, E59–E66.

Lebrun, P., Malaisse, W. J., and Herchuelz, A. (1982b). Nutrient-induced intracellular calcium movement in rat pancreatic β-cell. *Am. J. Physiol.* **243**, E196–E205.

Lebrun, P., Malaisse, W. J., and Herchuelz, A. (1982c). Effect of the absence of bicarbonate upon intracellular pH and calcium fluxes in pancreatic islet cells. *Biochim. Biophys. Acta* **721**, 357–363.

Lebrun, P., Malaisse, W. J., and Herchuelz, A. (1982d). Modalities of gliclazide-induced Ca^{2+} influx into the pancreatic β-cell. *Diabetes* **31**, 1010–1015.

Lee, Y. C., and Wolff, J. (1982). Two opposing effects of calmodulin on microtubule assembly depend on the presence of microtubule-associated proteins. *J. Biol. Chem.* **257**, 6306–6310.

Leitner, J. W., Rifkin, R. M., Maman, A., and Sussman, K. E. (1980). The relationship between somatostatin binding and cyclic AMP-stimulated protein kinase inhibition. *Metab., Clin. Exp.* **29**, 1065–1074.

Lenzen, S., and Klöppel, G. (1984). Intracellular localization of calcium in pancreatic β-cells in relation to insulin secretion by the perfused ob/ob mouse pancreas. *Endocrinology (Baltimore)* **114**, 1012–1020.

Levey, G. S., Schmidt, W. M. I., and Mintz, D. H. (1972). Activation of adenylate cyclase in a pancreatic islet cell adenoma by glucagon and tolbutamide. *Metab., Clin Exp.* **21**, 93–98.

Levin, S. R., Kasson, B. G., and Driessen, J. F. (1978). Adenosine triphosphatases of rat pancreatic islets. Comparison with those of rat kidney. *J. Clin. Invest.* **62**, 692–701.

Lin, Y. M., and Cheung, W. Y. (1980). Ca^{2+}-dependent cyclic nucleotide phosphodiesterase. *In* "Calcium and Cell Funtion" (W. Y. Cheung, ed.), Vol. 1, pp. 79–107. Academic Press, New York.

Lindh, U., Juntti-Berggren, L., Berggren, P. -O., and Hellman, B. (1985). Proton microprobe analysis of pancreatic β-cells. *Biomed. Biophys. Acta* **44**, 55–62.

Lindström, P., and Sehlin, J. (1984). Effect of glucose on the intracellular pH of pancreatic islet cells. *Biochem. J.* **218**, 887–892.

MacDonald, M. J., and Kowluru, A. (1982a). Calcium-calmodulin-dependent myosin phosphorylation by pancreatic islets. *Diabetes* **31**, 566–570.

MacDonald, M. J., and Kowluru, A. (1982b). Calcium-activated factors in pancreatic islets that inhibit actin polymerization. *Arch. Biochem. Biophys.* **219**, 459–462.

Malaisse, W. J. (1973). Theophylline-induced translocation of calcium in the pancreatic β-cell: Inhibition by deuterium oxide. *Nature (London), New Biol.* **242,** 189–190.

Malaisse, W. J., and Herchuelz, A. (1982). Nutritional regulation of K⁺ conductance: An unsettled aspect of pancreatic B cell physiology. *In* "Biochemical Actions of Hormones" (G. Litwack, ed.), Vol. 9, pp. 69–92. Academic Press, New York.

Malaisse, W. J., Malaisse-Lagae, F., Walker, M. O., and Lacy, P. E. (1971). The stimulus-secretion coupling a glucose-induced insulin release. V. The participation of a microtubular-microfilamentous system. *Diabetes* **20,** 257–265.

Malaisse, W. J., Hager, D. L., and Orci, L. (1972a). The stimulus-secretion coupling of glucose-induced insulin release. IX. The participation of the beta cell web. *Diabetes* **21,** 594–604.

Malaisse, W. J., Mahy, M., Brisson, G. R., and Malaisse-Lagae, F. (1972b). The stimulus-secretion coupling of glucose-induced insulin release. VIII. Combined effects of glucose and sulfonylureas. *Eur. J. Clin. Invest.* **2,** 85–90.

Malaisse, W. J., Brisson, G. R., and Baird, L. E. (1973). Stimulus-secretion coupling of glucose-induced insulin release. X. Effect of glucose on ⁴⁵Ca efflux from perifused islets. *Am. J. Physiol.* **224,** 389–394.

Malaisse, W. J., Herchuelz, A., Levy, J., Somers, G., and Devis, G. (1975). Insulin release and the movements of calcium in pancreatic islets. *In* "Calcium Transport in Contraction and Secretion" (E. Carafoli, F. Clementi, W. Drabikowski, and A. Margreth, eds.), pp. 211–226. North-Holland Publ., Amsterdam.

Malaisse, W. J., Devis, G., Herchuelz, A., Sener, A., and Somers, G. (1976a). Calcium antagonists and islet function. VIII. The effect of magnesium. *Diabete Metab.* **2,** 1–4.

Malaisse, W. J., Devis, G., Pipeleers, D. G., and Somers, G. (1976b). Calcium-antagonists and islet function. IV. Effect of D600. *Diabetologia* **12,** 77–81.

Malaisse, W. J., Herchuelz, A., Levy, J., and Sener, A. (1977). Calcium antagonists and islet function. III. The possible site of action of verapamil. *Biochem. Pharmacol.* **26,** 735–740.

Malaisse, W. J., Boschero, A. C., Kawazu, S., and Hutton, J. C. (1978a). The stimulus secretion coupling of glucose-induced insulin release. XXVII. Effect of glucose on K⁺ fluxes in isolated islets. *Pfluegers Arch.* **373,** 237–242.

Malaisse, W. J., Herchuelz, A., Devis, G., Somers, G., Boschero, A. C., Hutton, J. C., Kawazu, S., Sener, A., Atwater, I. J., Duncan G., Ribalet, B., and Rojas, E. (1978b). Regulation of calcium fluxes and their regulatory roles in pancreatic islets. *Ann. N.Y. Acad. Sci.* **307,** 562–582.

Malaisse, W. J., Hutton, J. C., Sener, A., Levy, J., Herchuelz, A., Devis, G., and Somers, G. (1978c). Calcium antagonists and islet function. VII. Effect of calcium deprivation. *J. Membr. Biol.* **38,** 193–198.

Malaisse, W. J., Hutton, J. C., Kawazu, S., Herchuelz, A., Valverde, I., and Sener, A. (1979a). The stimulus-secretion coupling of glucose-induced insulin release. XXXV. Links between metabolic and cationic events. *Diabetologia* **16,** 331–334.

Malaisse, W. J., Sener, A., Herchuelz, A., and Hutton, J. C. (1979b). Insulin release: The fuel hypothesis. *Metab., Clin. Exp.* **28,** 373–386.

Malaisse, W. J., Hutton, J. C., Carpinelli, A. R., Herchuelz, A., and Sener, A. (1980). The stimulus-secretion coupling of amino acid-induced insulin release. 1. Metabolism and cationic effects of L-leucine. *Diabetes* **29,** 431–437.

Malaisse-Lagae, F., and Malaisse, W. J. (1971). Stimulus-secretion coupling of glucose-induced insulin release. III. Uptake of ⁴⁵calcium by isolated islets of Langerhans. *Endocrinology (Baltimore)* **88,** 72–80.

Malaisse-Lagae, F., Brisson, G. R., and Malaisse, W. J. (1971a). The stimulus-secretion coupling of glucose-induced insulin release. VI. Analogy between the insulinotropic mechanisms of sugars and amino acids. *Horm. Metab. Res.* **3**, 374–378.

Malaisse-Lagae, F., Greider, M. H., Malaisse, W. J., and Lacy, P. E. (1971b). The stimulus-secretion coupling of glucose-induced insulin release. IV. The effect of vincristine and deuterium oxide on the microtubular system of the pancreatic beta cell. *J. Cell Biol.* **49**, 530–535.

Marcum, J. M., Dedman, J. R., Brinkley, B. R., and Means, A. R. (1978). Control of microtubule assembly-disassembly by calcium-dependent regulator protein. *Proc. Natl. Acad. Sci. U.S.A.* **75**, 3771–3775.

Matthews, E. K. (1970). Calcium and hormone release. *In* "Calcium and Cellular Function" (A. W. Cuthbert, ed.), pp. 163–182. Macmillan, London.

Matthews, E. K. (1979). Calcium translocation and control mechanisms for endocrine secretion. *Symp. Soc. Exp. Biol.* **33**, 225–249.

Matthews, E. K., and Sakamoto, Y. (1975). Electrical characteristics of pancreatic islet cells. *J. Physiol. (London)* **246**, 421–437.

Meissner, H. P. (1976). Electrical characteristics of the beta-cell in pancreatic islets. *J. Physiol. (Paris)* **72**, 757–767.

Meissner, H. P., and Henquin, J. C. (1983). Comparison of the effects of glucose, amino acids and sulphonamides on the membrane potential of pancreatic β-cells. *Int. Congr. Ser.—Excerpta Med.* **500**, 166–171.

Meissner, H. P., and Henquin, J. C. (1984). The sodium pump of mouse pancreatic β-cells: Electrogenic properties and activation by intracellular sodium. *In* "Electrogenic Transport: Fundamental Principles and Physiological Implications" (M. P. Blaustein and M. Liberman, eds.), pp. 295–306. Raven Press, New York.

Meissner, H. P., and Preissler, M. (1979). Glucose-induced changes of the membrane potential of pancreatic β-cells: Their significance for the regulation of insulin release. *In* "Treatment of Early Diabetes" (R. A. Camerini-Davalos and B. Hanover, eds.), pp. 97–107. Plenum, New York.

Meissner, H. P., and Preissler, M. (1980). Ionic mechanisms of the glucose-induced membrane potential changes in B-cells. *Horm. Metab. Res., Suppl. Ser.* **10**, 91–99.

Meissner, H. P., and Schmelz, H. (1974). Membrane potential of beta-cells in pancreatic islets. *Pfluegers Arch.* **351**, 195–206.

Meissner, H. P., Henquin, J. C., and Preissler, M. (1978). Potassium dependence of the membrane potential of pancreatic B-cells. *FEBS Lett.* **94**, 87–89.

Meissner, H. P., Preissler, M., and Henquin, J. C. (1979). Possible ionic mechanisms of the electrical activity induced by glucose and tolbutamide in pancreatic B cells. *In* "Diabetes 1979" (W. K. Waldhäusl, ed.), pp. 166–171. Excerpta Medica, Amsterdam.

Metz, S. A., Halter, J. B., and Robertson, R. P. (1979). Paradoxical inhibition of insulin secretion by glucose in human diabetes mellitus. *J. Clin. Endocrinol. Metab.* **48**, 827–835.

Metz, S. A., Fujimoto, W. Y., and Robertson, P. (1983). A role for the lipoxygenase pathway of arachidonic acid metabolism in glucose and glucagon-induced insulin secretion. *Life Sci.* **32**, 903–910.

Miller, C., Avran, P., Telford, J. M., and Racker, E. (1976). Ca^{++}-induced fusion of proteoliposomes: Dependence on transmembrane osmotic gradient. *J. Membr. Biol.* **30**, 271–282.

Milner, R. D. G., and Hales, C. N. (1967). The role of calcium and magnesium in insulin secretion from rabbit pancreas studied in vitro. *Diabetologia* **3**, 47–49.

Montague, W., and Parkin, E. N. (1980). Changes in membrane lipids of the β-cell during insulin secretion. *Horm. Metab. Res., Suppl. Ser.* **10**, 153–157.

Montague, W., Howell, S. L., and Green, I. C. (1976). Insulin release and the microtubular system of the islets of Langerhans: Effects of insulin secretagogues on microtubule subunit pool size. *Horm. Metab. Res.* **8**, 166–169.

Naber, S. P., McDaniel, M. L., and Lacy, P. E. (1977). The effect of glucose on the acute uptake and efflux of calcium-45 in isolated rat islets. *Endocrinology (Baltimore)* **101**, 686–693.

Naber, S. P., McDonald, J. M., Jarett, L., McDaniel, M. L., Ludvigsen, C. W., and Lacy, P. E. (1980). Preliminary characterization of calcium binding in islet-cell plasma membranes. *Diabetologia* **19**, 439–444.

Nelson, T. Y., Oberwetter, J. M., Chafouleas, J. G., and Boyd, A. E., III (1983). Calmodulin-binding proteins in a cloned rat insulinoma cell line. *Diabetes* **32**, 1126–1133.

Niki, A., Niki, H., and Koide, T. (1982). Effects of forskolin on insulin release and cAMP levels in rat pancreatic islets. *Int. Congr. Ser.—Excerpta Med.* **577**, 171.

Nilsson, T., Rorsman, F., Berggren, P.-O., and Hellman, B. (1984). Glucose stimulates the accumulation of cadmium in pancreatic B-cells. *Diabetologia* **27**, 315A.

Orci, L., Gabbay, K. H., and Malaisse, W. J. (1972). Pancreatic beta-cell web: Its possible role in insulin secretion. *Science* **175**, 1128–1130.

Ostlund, R. E., Jr., Leung, J. T., and Kipnis, D. M. (1978). Myosins of secretory tissues. *J. Cell Biol.* **77**, 827–836.

Pace, C. S., (1979). Activation of Na channels in islet cells: metabolic and secretory effects. *Am. J. Physiol.* **237**, E130–E135.

Pace, C. S., and Tarvin, J. T. (1982). Influence on anion transport on glucose-induced electrical activity in the β-cell. *Diabetes* **31**, 653–655.

Pace, C. S., and Tarvin, J. T. (1983). pH modulation of glucose-induced electrical activity in B-cells: Involvement of Na/H and HCO_3/Cl antiporters. *J. Membr. Biol.* **73**, 39–49.

Pace, C. S., Tarvin, J. T., Neighbors, A. S., Pirkle, J. A., and Greider, M. H. (1980). Use of a high voltage technique to determine the molecular requirements for exocytosis in islet cells. *Diabetes* **29**, 911–918.

Pace, C. S., Tarvin, J. T., and Smith, J. S. (1983). Stimulus-secretion coupling in β-cells: modulation by pH. *Am. J. Physiol.* **244**, E3–E18.

Panten, U., Zielmann, S., Langer, J., Zünkler, B.-J., and Lenzen, S. (1984). Regulation of insulin secretion by energy metabolism in pancreatic β-cell mitochondria. Studies with a non-metabolizable leucine analogue. *Biochem. J.* **219**, 189–196.

Papahadjopoulos, D. (1978). Calcium-induced phase changes and fusion in natural and model membranes. *In* "Membrane Fusion" (G. Poste and G. L. Nicolson, eds.), pp. 765–790. Elsevier/North-Holland Biomedical Press, Amsterdam.

Penn, E. J., Brocklehurst, K. W., Sopwith, A. M., Hales, C. N., and Hutton, J. C. (1982). Ca^{2+}-calmodulin dependent myosin light-chain phosphorylating activity in insulin-secreting tissues. *FEBS Lett.* **139**, 4–8.

Pershadsingh, H. A., McDaniel, M. L., Landt, M., Bry, C. G., Lacy, P. E., and McDonald, J. M. (1980). Ca^{2+}-activated ATPase and ATP-dependent calmodulin-stimulated Ca^{2+} transport in islet cell plasma membrane. *Nature (London)* **288**, 492–495.

Pipeleers, D. G., Pipeleers-Marichal, M. A., and Kipnis, D. M. (1976a). Regulation of tubulin synthesis in islets of Langerhans. *Proc. Natl. Acad. Sci. U.S.A.* **73**, 3188–3191.

Pipeleers, D. G., Pipeleers-Marichal, M. A., and Kipnis, D. M. (1976b). Microtubule assembly and the intracellular transport of secretory granules in pancreatic islets. *Science* **191**, 88–90.

Pipeleers, D. G., Harnie, N., Heylen, L., and Wauters, G. (1980). Microtubule interactions in islets of Langerhans. *Horm. Metab. Res., Suppl. Ser.* **10**, 163–167.

Pollard, H. B., Pazoles, C. J., Creutz, C. E., Ramu, A., Strott, C. A., Ray, P., Brown, E. M., Aurbach, G. D., Tack-Goldman, K. M., and Shulman, R. (1977). A role of anion transport in the regulation of release from chromaffin granules and exocytosis from cells. *J. Supramol. Struct.* **7**, 277–285.

Prentki, M., and Wollheim, C. B. (1984). Cytosolic free Ca^{2+} in insulin secretory cells and its regulation by isolated organelles. *Experientia* **40**, 1052–1060.

Prentki, M., Janjic, D., and Wollheim, C. B. (1983). The regulation of extramitochondrial steady state free Ca^{2+} concentration by rat insulinoma mitochondria. *J. Biol. Chem.* **258**, 7597–7602.

Prentki, M., Biden, T. J., Janjic, D., Irvine, R. F., Berridge, M. J., and Wollheim, C. B. (1984). Rapid mobilisation of Ca^{2+} from rat insulinoma microsomes by inositol-1,4,5-trisphosphate. *Nature (London)* **309**, 562–564.

Rabinovitch, A., Cuendet, G. S., Sharp, G. W. G., Renold, A. E., and Mintz, D. H. (1978). Relation of insulin release to cyclic AMP content in rat pancreatic islets maintained in tissue culture. *Diabetes* **27**, 766–773.

Ribalet, B., and Beigelman, P. M. (1979). Cyclic variation of K^+ conductance in pancreatic β-cells: Ca^{2+} and voltage dependence. *Am. J. Physiol.* **237**, C137–C146.

Ribalet, B., and Beigelman, P. M. (1980). Calcium action potentials and potassium permeability activation in pancreatic β-cells. *Am. J. Physiol.* **239**, C124–C133.

Ribalet, B., and Beigelman, P. M. (1982). Effects of sodium on β-cell electrical activity. *Am. J. Physiol.* **242**, C296–C303.

Ribes, G., Siegel, E., Wollheim, C. B., Renold, A. E., and Sharp, G. W. G. (1981). Rapid changes in calcium content of rat pancreatic islets in response to glucose. *Diabetes* **30**, 52–55.

Robertson, R. P., Brunzell, J. D., Hazzard, W. R., Lerner, R. L., and Porte, D., Jr. (1972). Paradoxical hypoinsulinaemia: An alpha-adrenergic-mediated response to glucose. *Lancet* **2**, 787–789.

Rodbell, M. (1980). The role of hormone receptors and GTP-regulatory proteins in membrane transduction. *Nature (London)* **284**, 17–22.

Roos, I., Crompton, M., and Carafoli, E. (1978). The effect of phosphoenol-pyruvate on the retention of calcium by liver mitochondria. *J. Biol. Chem.* **258**, 7597–7602.

Rorsman, P., Berggren, P. O., and Hellman, B. (1982). Manganese accumulation in pancreatic β-cells and its stimulation by glucose. *Biochem. J.* **202**, 435–444.

Rorsman, P., Berggren, P.-O., Gylfe, E., and Hellman, B. (1983). Reduction of the cytosolic calcium activity in clonal insulin-releasing cells exposed to glucose. *Biosci. Rep.* **3**, 939–946.

Rorsman, P., Abrahamsson, H., Gylfe, E., and Hellman, B. (1984). Dual effects of glucose on the cytosolic Ca^{2+} activity of mouse pancreatic β-cells. *FEBS Lett.* **170**, 196–200.

Rosen, O. M., Hirsch, A. H., and Goren, E. N. (1971). Factors which influence cyclic AMP formation and degradation in an islet cell tumor of the Syrian hamster. *Arch. Biochem. Biophys.* **146**, 660–663.

Rubin, R. P. (1982). "Calcium and Cellular Secretion." Plenum, New York.

Saceda, M., Garcia-Morales, P., Mato, J. M., Malaisse, W. J., and Valverde, I. (1984). Phospholipid methylation in pancreatic islets. *Biochem. Int.* **8**, 445–452.

Sanchez, A., Hadlam, T. J., and Rink, T. J. (1983). Trifluoperazine and chlorpromazine block secretion from human platelets evoked at basal cytoplasmic free calcium by activators of C-kinase. *FEBS Lett.* **164**, 43–46.

Schäfer, H. J., and Klöppel, G. (1974). The significance of calcium in insulin secretion. Ultrastructural studies on identification and localization of calcium in activated and inactivated β-cells of mice. *Virchows Arch. A: Pathol. Anat. Histol.* **362**, 231–245.

Schrey, M. P., and Montague, W. (1983). Phosphatidylinositol hydrolysis in isolated guinea-pig islets of Langerhans. *Biochem. J.* **216**, 433–441.

Schubart, U. K. (1982). Regulation of protein phosphorylation in hamster insulinoma cells. *J. Biol. Chem.* **257**, 12231–12238.

Schubart, U. K., and Fields, K. L. (1984). Identification of a calcium-regulated insulinoma cell phosphoprotein as an islet cell keratin. *J. Cell Biol.* **98**, 1001–1009.

Schubart, U. K., Erlichman, J., and Fleischer, N. (1980). The role of calmodulin in the regulation of protein phosphorylation and insulin release in hamster insulinoma cells. *J. Biol. Chem.* **255**, 4120–4124.

Seamon, K. B., and Daly, J. W. (1983). Forskolin, cyclic AMP and cellular physiology. *Trends Pharmacol. Sci.* **4**, 120–123.

Sehlin, J. (1976). Calcium uptake by subcellular fractions of pancreatic islets. Effects of nucleotides and theophylline. *Biochem. J.* **156**, 63–69.

Sehlin, J. (1978). Interrelationship between chloride fluxes in pancreatic islets and insulin release. *Am. J. Physiol.* **235**, E501–E508.

Sehlin, J., and Täljedal, I.-B. (1974a). Transport of rubidium and sodium in pancreatic islets. *J. Physiol. (London)* **242**, 505–515.

Sehlin, J., and Täljedal, I. B. (1974b). Sodium uptake by microdissected pancreatic islets: Effects of ouabain and chloromercuribenzene-p-sulphonic acid. *FEBS Lett.* **39**, 209–213.

Sehlin, J., and Täljedal, I.-B., (1975). Glucose-induced decrease in Rb+ permeability in pancreatic β cells. *Nature (London)* **253**, 635–636.

Sener, A., Kawazu, S., Hutton, J. C., Boschero, A. C., Devis, G., Somers, G., Herchuelz, A., and Malaisse, W. J. (1978). Stimulus-secretion coupling of glucose-induced insulin release. Effect of exogenous pyruvate on islet function. *Biochem. J.* **176**, 217–232.

Sener, A., Malaisse-Lagae, F., and Malaisse, W. J. (1982). Glucose-induced accumulation of glucose 1,6-biphosphate in pancreatic islets: Its possible role in the regulation of glycolysis. *Biochem. Biophys. Res. Commun.* **104**, 1033–1040.

Sharp, G. W. G., Widenkeller, D. E., Kaelin, D., Siegel, E. G., and Wollheim, C. B. (1980). Stimulation of adenylate cyclase by Ca^{2+} and calmodulin in rat islets of Langerhans. Explanation for the glucose-induced increase in cyclic AMP levels. *Diabetes* **29**, 74–77.

Sharp, G. W. G., Wiedenkeller, D. E., Lipson, L. G., Oldham, S. B., Krausz, Y., Janjic, D., Pian-Smith, M. C. M., and Wollheim, C. B. (1983). The multiple roles of calmodulin in the endocrine pancreas and the control of insulin release. *In* "Diabetes 1982" (E. N. Mngola, ed.), pp. 329–336. Excerpta Medica, Amsterdam.

Shelanski, M. L., Gaskin, F., and Cantor, C. R. (1973). Microtubule assembly in the absence of added nucleotides. *Proc. Natl. Acad. Sci. U.S.A.* **70**, 765–768.

Siegel, E. G., Wollheim, C. B., Sharp, G. W. G., Herberg, L., and Renold, A. E. (1979). Defective calcium handling in insulin release in islets from diabetic chinese hamsters. *Biochem. J.* **180**, 233–236.

Siegel, E. G., Wollheim, C. B, Kikuchi, M., Renold, A. E., and Sharp, G. W. G. (1980a). Dependency of cyclic AMP-induced insulin release on intra- and extracellular calcium in rat islets of Langerhans. *J. Clin. Invest.* **65**, 233–241.

Siegel, E. G., Wollheim, C. B., Renold, A. E., and Sharp, G. W. G. (1980b). Evidence for involvement of Na/Ca exchange in glucose-induced insulin release from rat pancreatic islets. *J. Clin. Invest.* **66**, 996–1003.

Siegel, E. G., Wollheim, C. B., and Sharp, G. W. G. (1980c). Glucose-induced first phase insulin release in the absence of extracellular Ca^{2+} in rat islets. *FEBS Lett.* **109**, 213–215.

Somers, G., Van Obberghen, E., Devis, G., Ravazzola, M., Malaisse-Lagae, F., and Malaisse, W. J. (1974). Dynamics of insulin release and microtubular-microfilamentous system. III. Effect of colchicine upon glucose-induced insulin secretion. *Eur. J. Clin. Invest.* **4**, 299–305.

Somers, G., Blondel, B., Orci, L., and Malaisse, W. J. (1979a). Motile events in pancreatic endocrine cells. *Endocrinology (Baltimore)* **104**, 255–263.

Somers, G., Devis, G., and Malaisse, W. J. (1979b). Calcium antagonists and islet function. IX. Is extracellular calcium required for insulin release. *Acta Diabetol. Lat.* **16**, 9–18.

Steinberg, J. P., Leitner, J. W., Draznin, B., and Sussman, K. E. (1984). Calmodulin and cyclic AMP. Possible different sites of action of these two regulatory agents in exocytotic hormone release. *Diabetes* **33**, 339–345.

Sugden, M. C., and Ashcroft, S. J. H. (1978). Effects of phosphoenolpyruvate, other glycolytic intermediates and methylxanthines on calcium uptake by a mitochondrial fraction from rat pancreatic islets. *Diabetologia* **15**, 173–180.

Sugden, M. C., and Ashcroft, S. J. H. (1981). Cyclic nucleotide phosphodiesterase of rat pancreatic islets. *Biochem. J.* **197**, 459–464.

Sugden, M. C., Christie, M. R., and Ashcroft, S. J. H. (1979). Presence and possible role of calcium-dependent regulator (calmodulin) in rat islets of Langerhans. *FEBS Lett.* **105**, 95–100.

Suzuki, S., Oka, H., Yasuda, H., Ikeda, M., Cheng, P. Y., and Oda, T. (1983). Effect of glucagon and cyclic adenosine 3′,5′-monophosphate on protein phosphorylation in rat pancreatic islets. *Endocrinology (Baltimore)* **112**, 348–352.

Swanston-Flatt, S. K., Carlsson, L., and Gylfe, E. (1980). Actin filament formation in pancreatic β-cells during glucose stimulation of insulin secretion. *FEBS Lett.* **117**, 299–302.

Täljedal, I.-B. (1978). Chlorotetracycline as a fluorescent Ca^{2+} probe in pancreatic islet cells. Methodological aspects and effects of alloxan, sugars, methylxanthines and Mg^{2+}. *J. Cell Biol.* **76**, 652–674.

Täljedal, I.-B. (1979). Polarization of chlorotetracycline fluorescence in pancreatic islet cells and its response to calcium ions and D-glucose. *Biochem. J.* **178**, 187–193.

Tanigawa, K., Kuzuya, H., Imura, H., Taniguchi, H., Baba, S., Takai, Y., and Nishizuka, Y. (1982). Calcium-activated, phospholipid-dependent protein kinase in rat pancreatic islets of Langerhans. *FEBS Lett.* **138**, 183–186.

Thams, P., Capito, K., and Hedeskov, C. J. (1982). Differential effects of Ca^{2+}-calmodulin on adenylate cyclase activity in mouse and rat pancreatic islets. *Biochem. J.* **206**, 97–102.

Tooke, N. E., Hales, C. N., and Hutton, J. C. (1984). Ca^{2+}-sensitive phosphatidylinositol 4-phosphate metabolism in a rat β-cell tumour. *Biochem. J.* **219**, 471–480.

Ullrich, S., and Wollheim, C. B. (1984). Islet cyclic AMP levels are not lowered during α_2-adrenergic inhibition of insulin release. *J. Biol. Chem.* **259**, 4111–4115.

Valverde, I., Vandermeers, A., Anjaneyulu, R., and Malaisse, W. J. (1979). Calmodulin activation of adenylate cyclase in pancreatic islets. *Science* **206**, 225–227.

Valverde, I., Sener, A., Lebrun, P., Herchuelz, A., and Malaisse, W. J. (1981). The stimulus-secretion coupling of glucose-induced insulin release. XLVII. The possible role of calmodulin. *Endocrinology (Baltimore)* **108**, 1305–1312.

Van Obberghen, E., Somers, G., Devis, G., Vaughan, G. D., Malaisse-Lagae, F., Orci, L., and Malaisse, W. J. (1973). Dynamics of insulin release and microtubular-microfilamentous system. I. Effect of cytochalasin B. *J. Clin. Invest.* **52**, 1041–1051.

Wallace, R. W., Tallant, E. A., and Cheung, W. Y. (1980). Assay, preparation and properties of calmodulin. *In* "Calcium and Cell Function" (W. Y. Cheung, ed.), Vol. 1, pp. 13–40. Academic Press, New York.

Wang, J. H., Sharma, R. K., and Tam, S. W. (1980). Calmodulin-binding proteins. *In* "Calcium and Cell Function" (W. Y. Cheung, ed.), Vol. 1, pp. 305–328. Academic Press, New York.

Watkins, D. T., and Cooperstein, S. J. (1983). Role of calcium and calmodulin in the interaction between islet cell secretion granules and plasma membranes. *Endocrinology (Baltimore)* **112**, 766–768.

Weisenberg, R. C. (1972). Microtubule formation in vitro in solution containing low calcium concentrations. *Science* **177**, 1104–1105.

Weiss, B., and Wallace, T. L. (1980). Mechanisms and pharmacological implications of altering calmodulin activity. *In* "Calcium and Cell Function" (W. Y. Cheung, ed.), Vol. 1, pp. 329–379. Academic Press, New York.

Welsh, M. J., Dedman, J. R., Brinkley, B. R., and Means, A. R. (1979). Tubulin and calmodulin. Effects of microtubule and microfilament inhibitors on localization in the mitotic apparatus. *J. Cell Biol.* **81**, 624–634.

Wiedenkeller, D. E., and Sharp, G. W. G. (1983). Effects of forskolin on insulin release and cyclic AMP content in rat pancreatic islets. *Endocrinology (Baltimore)* **113**, 2311–2313.

Wollheim, C. B., and Pozzan, T. (1984). Correlation betwen cytosolic free Ca^{2+} and insulin release in an insulin-secreting cell line. *J. Biol. Chem.* **259**, 2262–2267.

Wollheim, C. B., and Sharp, G. W. G. (1981). Regulation of insulin release by calcium. *Physiol. Rev.* **61**, 914–973.

Wollheim, C. B., Kikuchi, M., Renold, A. E., and Sharp, G. W. G. (1977). Somatostatin- and epinephrine-induced modifications of $^{45}Ca^{++}$ fluxes and insulin release in rat pancreatic islets maintained in tissue culture. *J. Clin. Invest.* **60**, 1165–1173.

Wollheim, C. B., Blondel, B., Kikuchi, M., and Sharp, G. W. G. (1978a). Inhibition of insulin release by somatostatin: No evidence for interaction with calcium. *Metab., Clin. Exp.* **27**, 1303–1307.

Wollheim, C. B., Kikuchi, M., Renold, A. E., and Sharp, G. W. G. (1978b). The roles of intracellular and extracellular Ca^{2+} in glucose-stimulated biphasic insulin release by rat islets. *J. Clin. Invest.* **62**, 451–458.

Wollheim, C. B., Siegel, E. G., Kikuchi, M., Renold, A. E., and Sharp, G. W. G. (1980). The role of extracellular Ca^{2+} and islet cell calcium stores in the regulation of biphasic insulin release. *Horm. Metab. Res., Suppl. Ser.* **10**, 108–115.

Wollheim, C. B., Janjic, D., Siegel, E. G., Kikuchi, M., and Sharp, G. W. G. (1981). Importance of cellular calcium stores in glucose-stimulated insulin release. *Upsala J. Med. Sci.* **86**, 149–164.

Wolters, G. H. J., Pasma, A., Konijnendijk, W., and Boom, G. (1979). Calcium, zinc and other elements in islet and exocrine tissue of the rat pancreas as measured by histochemical methods and electron-probe microanalysis. Effects of fasting and tolbutamide. *Histochemistry* **62**, 1–17.

Wolters, G. H. J., Wiegman, J. B., and Konijnendijk, W. (1982). The effect of glucose stimulation on ^{45}calcium uptake of rat pancreatic islets and their total calcium content as measured by a fluorometric micromethod. *Diabetologia* **22**, 122–127.

Wolters, G. H. J., Pasma, A., Wiegman, J. B., and Konijnendijk, W. (1983). Changes in histochemically detectable calcium and zinc during tolbutamide-induced degranulation and subsequent regranulation of rat pancreatic islets. *Histochemistry* **78**, 325–338.

Wolters, G. H. J., Pasma, A., Wiegman, J. B., and Konijnendijk, W. (1984). Glucose-induced changes in histochemically determined Ca^{2+} in β-cell granules, ^{45}Ca uptake and total Ca^{2+} of rat pancreatic islets. *Diabetes* **33**, 409–414.

Yamamoto, S., Nakadate, T., Nakaki, T., Ishii, K., and Kato, R. (1982a). Tumor promoter 12-0-tetradecanoylphorbol-13-acetate-induced insulin secretion: Inhibition by phospholipase A_2- and lipoxygenase inhibitors. *Biochem. Biophys. Res. Commun.* **105**, 759–765.

Yamamoto, S., Nakadate, T., Nakaki, T., Ishii, K., and Kato, R. (1982b). Prevention of glucose-induced insulin secretion by lipoxygenase inhibitor. *Eur. J. Pharmacol.* **78**, 225–227.

Yaseen, M. A., Pedley, K. C., and Howell, S. L. (1982). Regulation of insulin secretion from islets of Langerhans rendered permeable by electric discharge. *Biochem. J.* **206**, 81–87.

Zawalich, W. S., Karl, R. C., Ferrendelli, J. A., and Matschinsky, F. M. (1975). Factors governing glucose induced elevation of cyclic 3'5'AMP levels in pancreatic islets. *Diabetologia* **11**, 231–235.

Zawalich, W. S., Brown, C., and Rasmussen, H. (1983). Insulin secretion: combined effects of phorbol ester and A23187. *Biochem. Biophys. Res. Commun.* **117,** 448–455.
Zimmerberg, J., Cohen, F. S., and Finkelstein, A. (1980). Micromolar Ca^{2+} stimulates fusion of lipid vesicles with planar bilayers containing calcium-binding protein. *Science* **210,** 906–908.

Chapter 9

Roles of Calcium in Photosynthesis

CLANTON C. BLACK, JR.

Biochemistry Department
University of Georgia
Athens, Georgia

JERRY J. BRAND

Department of Botany
University of Texas
Austin, Texas

CALCIUM AND CELL FUNCTION, VOL. VI
Copyright © 1986 by Academic Press, Inc.

I. INTRODUCTION

The ubiquitous presence of the divalent cation Ca^{2+} in photosynthetic tissues has been recognized for several decades. However, until recently the roles or functions of Ca^{2+} in photosynthesis were unknown; general bridging roles such as cross-linking and membrane binding were postulated. A resurgence of interest in Ca^{2+} arose from the discovery of Ca^{2+}-binding proteins such as calmodulin in plants (Anderson and Cormier, 1978) and regulation of the cloroplast NAD kinase by a calmodulin-like protein (Jarrett *et al.,* 1982), plus the potential functioning of Ca^{2+} in the chloroplast O_2 evolution process and in the operation of the reaction center of Photosystem II (Brand and Becker, 1984). The importance of Ca^{2+} in intracellular regulation of cellular processes is widely acknowledged. But for Ca^{2+} to be an effective intracellular regulator its localization within an organism is critical, as is the flux of Ca^{2+} between cellular compartments and between the organism and its environment. Within each responding cellular compartment there also must be a Ca^{2+} receptor or some change such as voltage change which regulates the cellular process. Here we will present our current understanding of these subjects and other roles for Ca^{2+} in photosynthesis while simultaneously recognizing the primitive nature of our efforts in unraveling the precise roles of Ca^{2+} in photosynthesis.

To assist general readers in understanding this chapter, we will give a brief description of the subcellular compartments in photosynthesis where Ca^{2+} may function. The higher plant chloroplast consists of a double membrane, the envelope that surrounds a soluble matrix, and the stroma that surrounds the thylakoid membranes containing the pigments of photosynthesis. An intrathylakoid compartment, the lumen, is surrounded by the thylakoid. In higher plants the soluble chloroplast compartments that may contain free Ca^{2+} are the stroma and the lumen; they occupy about 87 and 13%, respectively, of the chloroplast's soluble volume. The outer membrane of the envelope is freely permeable to ions, but the inner membrane is not; thus, we will not consider the compartment between the double envelope membranes. The stroma contains soluble enzymes for processes such as CO_2 fixation and reduction, while the lumen is involved as a proton transfer site in support of photosynthetic phosphorylation and as the site of O_2 evolution. The photochemical portion of photosynthesis occurs on or in the thylakoids. Higher plant photosynthesis occurs via two photosystems within the thylakoid, and bacterial photosynthesis occurs via one photosystem which is also in a pigment-containing membrane. We will examine the compartmentation and roles of Ca^{2+} in the stroma, on or within the pigmented thylakoid membrane, in the photosystems, and in the lumen.

II. Ca^{2+} CONTENT AND COMPARTMENTATION IN PHOTOSYNTHETIC TISSUES

At the outset one needs to establish how much Ca^{2+} is present in photosynthetic tissues and in what form. Classical quantitative analysis of the roles of individual components or processes in photosynthesis, revolves around the photosynthetic unit. The photosynthetic unit was originally defined as the number of chlorophyll molecules required to evolve a molecule of O_2. Thus, it was realized that 500–600 chlorophyll molecules comprise a complete O_2-evolving electron transport unit in plant chloroplasts to support CO_2 fixation. For example, with chloroplasts, 2 photosystems are present per 500–600 chlorophylls or 4 Mn are present per 200 chlorophylls in a Photosystem II subchloroplast particle that can evolve O_2 through the photolysis of water.

By such reasoning one can determine how two photosystems function in plant photosynthesis, how 4 Mn molecules function in the evolution of O_2, or how Ca^{2+} is stoichiometric with known components of photosynthesis. Knowing stoichiometries allows one to infer functions. Table I shows known Ca^{2+} contents in photosynthetic tissues, chloroplasts, and subchloroplast particles. The amount of Ca^{2+} found in leaf chloroplasts can be a major percentage of the total leaf Ca^{2+}; indeed perhaps 60% of the leaf Ca^{2+} is in the chloroplast. The amount of Ca^{2+} in photosynthetic tissues is high but lies within a fairly narrow range relative to the pigments of photosynthesis. The range of 0.2 to 0.7 Ca^{2+} per chlorophyll molecule in bacteria is remarkably similar to the range of 0.26 to 0.83 Ca^{2+} per chlorophyll molecule reported in leaves and algae. Hence Ca^{2+} is present in a roughly half-stoichiometric ratio to photosynthetic pigments in a variety of organisms. One can conclude from these stoichiometric relationships that Ca^{2+} is in great excess over catalytic components of photosynthesis. For example we find 4 Mn per O_2 evolving unit of 200 chlorophylls, and 2 ferredoxin molecules per 500–600 chlorophyll molecules.

The fact that Ca^{2+} is nearly half-stoichiometric with chlorophyll possibly has hindered research into understanding the exact functions of Ca^{2+} in photosynthesis. Perhaps an instructive analogy can be made with the multiple functions of chlorophyll in photosynthesis, where the abundance of light-harvesting chlorophyll for many years obscured the detection of the reaction center chlorophylls, P_{700} and P_{680}. Indeed, multiple functions for Ca^{2+} can be inferred from the data in Table I. Combining the evidence from the data in Table I yields the conclusion that most of the chloroplast Ca^{2+} is bound to the thylakoids. Therefore, the old idea that Ca^{2+} functions in membrane organization as it does

TABLE I

Ca²⁺ Content of Photosynthetic Cells, Isolated Whole Chloroplasts, and Subchloroplast Particles

Organism or photosynthetic structure	Ca²⁺/chlorophyll (Chl) or tissue Ca²⁺ level	References
Cyanobacteria		
Anacystis cells	0.46 Ca²⁺/Chl	Becker and Brand (1984)
Phormidium cells[a]	0.2–0.4 Ca²⁺/Chl; 20 mM Ca²⁺ total in thylakoids	Piccioni and Mauzerall (1978)
PS bacteria		
Chromatium vinosum cells	0.7 Ca²⁺/bacteria Chl in light; 1.5 Ca²⁺/bacteria Chl in dark	Davidson and Knaff (1981)
Leaves		
Tobacco chloroplasts isolated nonaqueously	63% of the total leaf Ca²⁺	Stocking and Ongun (1962)
Bean chloroplasts isolated nonaqueously	58% of the total leaf Ca²⁺	
Isolated whole chloroplasts		
Pea Leaf	0.76 Ca²⁺/Chl; 15 mM Ca²⁺	Nakatani *et al.* (1979); Nobel (1969)
Bryopsis, a marine alga[b]	0.26 Ca²⁺/Chl; 13 ± 2 mM Ca²⁺	Yamagishi *et al.* (1981)
Spinach leaf	0.77 Ca²⁺/Chl	Miginiac-Maslow and Hoarew (1977)
Isolated broken chloroplasts		
Spinach fragments	0.70 Ca²⁺/Chl; 20 to 30 mM Ca²⁺	O'Keefe and Dilley (1977)
Unwashed fragments	0.47 Ca²⁺/Chl	Yamashita and Tomita (1975)
0.8 M KCl-washed fragments	0.06 Ca²⁺/Chl	Yamashita and Tomita (1975)
Isolated Photosystem II preparation:		
Spinach PSII, up to 2 M NaCl in washing	0.52–0.84 Ca²⁺/Chl	Ghanotakis *et al.* (1984a,b)
Spinach PSII, chemically stripped with 1 mM EDTA + 20 μM A23187	0.05 Ca²⁺/Chl	F. Callahan and G. M. Cheniae (1984)[c]
Light-harvesting chlorophyll–protein Complex prepared with sodium dodecyl sulfate	9–13 Ca²⁺/Chl	Davis and Gross (1975)

[a]2.2 × 10⁻⁵ M total Ca²⁺ in whole cells.
[b]93–95% of the total chloroplast Ca²⁺ is bound.
[c]Personal communication; Callahan (1984).

in the chloroplast certainly retains credence, though this topic will not be pursued here.

Two studies described in Table I show that much of the chloroplast Ca^{2+} can be chemically washed or stripped from thylakoids while O_2 evolution or electron transport remains intact (Yamashita and Tomita, 1975; F. Callahan and G. M. Cheniae, personal communication, 1984). Ca^{2+} can be depleted from chloroplast to near 1 Ca^{2+} per 20 chlorophylls without apparent deleterious effects on O_2 evolution or on the associated electron transport. Further Ca^{2+} depletion results in the loss of both activities; restoration of activity upon the addition of Ca^{2+} will be treated under the topic of Photosystem II of chloroplasts (see Section VIII). Thus, nearly 90% of the thylakoid Ca^{2+} can be removed without appreciable damage to electron transport or O_2 evolution. The function of this excess Ca^{2+} is unknown. How much free or soluble Ca^{2+} is in the intact chloroplast is also unknown, as is the function of free Ca^{2+}.

Based on the total amount of Ca^{2+} in a typical chloroplast volume, Ca^{2+} concentrations could range from 15 to 30 mM (Table I). However, since most of the chloroplast Ca^{2+} is complexed with the thylakoids, the free Ca^{2+} in soluble compartments such as the stroma and lumen must be low. Unfortunately few measurements have been made of the concentration of free Ca^{2+} in chloroplasts, and none have been made in green leaf cells. In other eukaryotic cells the cytoplasmic soluble Ca^{2+} concentration is low, 10^{-8} to $10^{-7} M$; hence we assume that leaf cell cytoplasmic Ca^{2+} is similar. With aequorin injections into resting algae cells, measurement of the free cytoplasmic Ca^{2+} levels gave values of 10^{-7} to $5.6 \times 10^{-7} M$ for *Chara* and 7.9×10^{-7} to $1.6 \times 10^{-6} M$ for *Nitella* (Williamson and Ashley, 1982). By comparison, though no measurements are available, we will assume that the soluble Ca^{2+} levels in the chloroplast stroma also are near 10^{-7}.

The soluble Ca^{2+} levels in the chloroplast lumen also have not been measured, but a variety of experiments show that isolated thylakoids take up Ca^{2+} via a reversible light-driven process (Nobel and Packer, 1964; Dilley and Vernon, 1965; Nobel, 1967), which leads us to believe that the lumen has higher levels of soluble Ca^{2+} than the stroma; perhaps the range of Ca^{2+} in the lumen is near $10^{-4} M$.

III. Ca^{2+} FLUX BY ISOLATED WHOLE CHLOROPLASTS AND BY NAKED THYLAKOIDS

As stated above, chloroplasts *in vivo* have two soluble compartments, the stroma and the thylakoid lumen, in which soluble Ca^{2+} may function. Experiments have been conducted on the light-dependent and dark flux of Ca^{2+} with (a) isolated whole chloroplasts with both soluble compartments intact, and (b) isolated naked thylakoids with only the lumen compartment intact. Using such

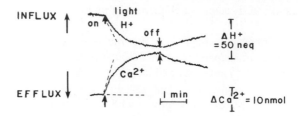

Fig. 1. Light-induced Ca²⁺ and H⁺ fluxes with isolated whole chloroplasts from wheat leaves. The ion changes were recorded in the light and dark with ion-selective electrodes in a chloroplast suspension medium containing 0.33 M sorbitol, 0.2 mM CaCl₂, 1 mM HEPES/bis-Tris-propane (pH 7.6), plus chloroplasts at the level of 40 μg chlorophyll per milliliter. Adapted from Muto *et al.* (1982).

chloroplast preparations isolated from wheat leaf protoplasts and from spinach leaves, Muto *et al.* (1982) performed a set of experiments on chloroplast Ca²⁺ fluxes that can be integrated with other research (Table I) to yield an understanding of how soluble Ca²⁺ levels can change in the chloroplast stroma and lumen under conditions of light versus dark.

Using ⁴⁵Ca²⁺ and an ion-specific electrode, Muto *et al.* (1982) demonstrated a light-dependent uptake of Ca²⁺ by whole chloroplasts (from both wheat and spinach) which was complete in less than 5 min. As illustrated in Fig. 1 for wheat, the Ca²⁺ uptake from the surrounding medium was accompanied by a stoichiometric release of protons from whole chloroplasts. From the initial light on-slope, the rate of Ca²⁺ uptake and H⁺ release was 21 μmol and 38 neq per milligram chlorophyll per hour, respectively, while the maximum extent was 205 nmol per milligram chlorophyll and 390 neq per milligram chlorophyll, respectively. In the dark both Ca²⁺ and H⁺ decayed toward their initial levels, and this decay was accelerated by adding K⁺ to the external medium. One can conclude that with their envelopes intact, illuminated whole chloroplasts take up Ca²⁺ and release H⁺ in a 2:1 stoichiometric fashion, which is reversible in the dark.

If the envelopes of the same whole chloroplasts are ruptured osmotically in a hypotonic medium, the direction of both Ca²⁺ and H⁺ movements is reversed. In the light these naked thylakoids release Ca²⁺ and take up H⁺ (Fig. 2). Therefore, opposite ion movements occur with naked thylakoids as with whole chloroplasts in the presence of the same external or media Ca²⁺ level (0.2 mM) (compare Figs. 1 and 2). In the dark the decay kinetics for both ions are rapid and fully reversible, occurring with equivalent stoichoimetries in the light and dark (Fig. 2). Thus, thylakoids without their normal surrounding envelopes release Ca²⁺ from their lumen into what would be the stroma of a whole chloroplast *in vivo*.

Collectively, the results illustrated in Figs. 1 and 2 show that the stroma may have Ca²⁺ influx in the light both from the cytoplasm and from the lumen. The Ca²⁺ from the stroma is removed in the dark. These Ca²⁺-ion movements occur

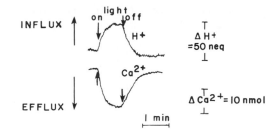

Fig. 2. Light-induced Ca^{2+} and H^+ fluxes with isolated naked thylakoids from wheat leaves. The measurements were made as in Fig. 1 in the same medium but omitting the sorbitol; this medium ruptured the whole chloroplast envelopes to yield naked thylakoids with their lumen intact. Adapted from Muto *et al.* (1982).

with opposite and equivalent changes in H^+. If we combine these results with the data on soluble Ca^{2+} of chloroplasts and green cells (Table I), we can present a model for changes in the levels of soluble Ca^{2+} in the chloroplast stroma and lumen. We propose that these light and dark changes result in reversibly providing Ca^{2+} in the stroma for Ca^{2+}-binding proteins such as calmodulin, thereby modulating photosynthetic activities.

Figure 3 is a model for the light regulation of soluble Ca^{2+} levels in the

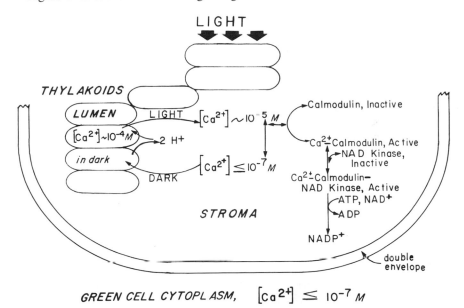

Fig. 3. Unified model for the light and dark regulation of Ca^{2+} levels in the stroma and lumen of leaf chloroplasts plus the subsequent activation/inactivation of NAD kinase by a soluble Ca^{2+}-calmodulin-like protein in the stroma in the light and dark.

stroma and lumen of intact chloroplasts. Most data indicate that the plant cell cytoplasm has low levels ($\leq 10^{-7} M$) of free Ca^{2+}. In this model the chloroplast uptake of Ca^{2+} from the cytoplasm is not considered a requirement for the normal operation of photosynthesis. The chloroplast is capable of taking up Ca^{2+} from the cytoplasm (Fig. 1), though a $K_{m,app}$ of 0.18 mM (Muto et al., 1982) is much higher than the level of Ca^{2+} in the cytoplasm (Fig. 3). Therefore, rather than utilizing Ca^{2+} from its surrounding cytoplasm, the chloroplast internally modulates its free Ca^{2+} pools between the lumen and the stroma.

Knowing the approximate volumes of stroma and lumen plus their resting Ca^{2+} levels, we can determine the light–dark reversible changes in both chloroplast compartments. The light-dependent release of Ca^{2+} from the lumen (Fig. 2) leads to an increase in the Ca^{2+} level in the stroma of an illuminated chloroplast (Fig. 3). As a consequence, the Ca^{2+} level in the stroma changes from $\leq 10^{-7} M$ in the dark to $\sim 10^{-5} M$ in the light. Ca^{2+} is then available for binding to Ca^{2+}-sensitive proteins such as calmodulin in the stroma, which activates the NAD kinase to produce $NADP^{+}$ and thereby regulates photosynthesis (Fig. 3). The Ca^{2+} is reversibly released and moved to the lumen in the dark.

IV. Ca^{2+}-BINDING AND Ca^{2+}-SENSITIVE PROTEINS IN CHLOROPLASTS

In recent years Ca^{2+} has been implicated as a regulatory molecule or messenger for the transduction of stimuli in plants via Ca^{2+}-binding proteins. Within the chloroplast, Ca^{2+} moves in response to illumination and darkness (Figs. 1 and 2), but for Ca^{2+} to induce a change in a cellular reaction within the chloroplast, Ca^{2+}-binding or Ca^{2+}-sensitive proteins must be present.

The discovery of such proteins in plants has already been described in this series (Cormier et al., 1980; Roux and Slocum, 1983; Marmé and Dieter, 1983). Since the pioneering work of Anderson and Cormier in 1978, the detection of Ca^{2+}-binding proteins such as calmodulin in plants with a binding affinity for soluble $Ca^{2+} \geq 1$ μM has become a common occurrence. As has been abundantly illustrated in this series, calmodulin binds up to 4 Ca^{2+} molecules and then binds to a target enzyme, thereby altering its activity. In plants, three target enzymes are known: NAD kinase, Ca^{2+}-transport ATPase, and perhaps a protein kinase which activates quinate : NAD^{+}-oxidoreductase. We will consider only the NAD kinase and its regulation in chloroplasts, since it is likely that in plant cells the Ca^{2+}-transport ATPase is near the plasmalemma, and the oxidoreductase is not in the chloroplast.

Plant calmodulins have been intensively studied since 1978, including a partial sequencing of the proteins from spinach, barley (Schleicher et al., 1983; Lukas *et al.,* 1984), and mung bean (M. J. Cormier and M. Charbanneau, personal communication, 1984). The amino acid composition of calmodulin from numerous organisms had earlier shown much similarity in the protein from animals, plants, and fungi (Cormier *et al.,* 1980; Marmé, 1982). Current work continues to confirm this similarity through demonstrating similar peptide maps, electrophoretic mobilities, activator activities, and immunoreactivities (Schleicher *et al.,* 1983; Lukas *et al.,* 1984; Roberts *et al.,* 1984).

A detailed analytical comparison with calmodulin isolated from six diverse organisms in highly purified forms quantified these similarities and differences based on the calmodulin-stimulated activities of pea seedling NAD kinase, bovine brain cyclic nucleotide phosphodiesterase, and human erythrocyte Ca^{2+}-ATPase (Harmon *et al.,* 1984). A comparison of pea NAD kinase activation in the presence of varying concentrations of calmodulin (Fig. 4) shows similar stimulation as well as a high sensitivity of the kinase to low levels of calmodulin. The amount of calmodulin required for 50% activation of the pea seedling NAD kinase ($K_{0.5}$) ranged from a low of 0.52 ± 0.07 ng/ml with the *Tetrahymena* calmodulin to a high of 2.2 ± 0.82 ng/ml with the bovine brain calmodulin (Harmon *et al.,* 1984). Clearly these diverse calmodulins are similar. The calmodulin concentration range is narrow and the pea NAD kinase was the most sensitive calmodulin-stimulated enzyme assayed. These results are relevant to

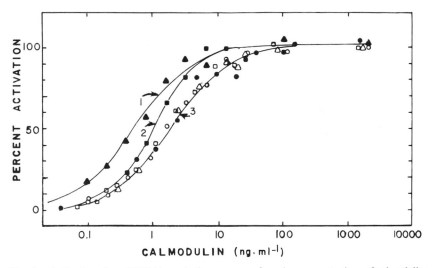

Fig. 4. Activation of pea NAD kinase in the presence of varying concentrations of calmodulins isolated from the following organisms: human erythrocyte (□); bovine brain (△); sea pansy (○); mung beans (■); *Tetrahymena* (▲); and mushroom (●). The dose–response curves were fitted as follows: curve 1, *Tetrahymena* calmodulin; curve 2, mung bean calmodulin; and curve 3, brain, erythrocyte, sea pansy, and mushroom calmodulin. Assays were performed in the presence of 1 m*M* $CaCl_2$ (Harmon *et al.,* 1984).

the Ca^{2+}–calmodulin activation of the NAD kinase (Jarrett et al., 1982) found in the chloroplast which we will soon consider.

For Ca^{2+} to be a chloroplast messenger a receptor protein must be present; thus the question, where is calmodulin localized in the plant cell? So far the answer to this question is controversial, but we find general agreement that the protein is ubiquitous in plants and that it has both an inter- and intracellular localization in plant tissues. Certainly a portion of the controversy about calmodulin in plants is the presence of other Ca^{2+}-binding proteins which are as yet very poorly defined. Calmodulin has been reported to be associated with the cell wall, plasmalemma, microsomal fractions, mitochondria, chloroplasts, etioplasts, and nuclei (Cormier et al., 1980; Roux and Slocum, 1983; Marmé, 1982; Jarrett et al., 1982; Muto et al., 1982; Biro et al., 1984; Stosic et al., 1983). The evidence for these associations has accumulated primarily from either the detection of calmodulin activity in the isolated fraction or organelle, or by radioimmunoassays with its antibody. We shall concentrate on the evidence for a subcellular localization of calmodulin in the chloroplast.

In 1982, a detailed study was presented on the localization of a calmodulin-like protein in the stroma of pea chloroplasts (Jarrett et al., 1982) which was immediately substantiated by Muto (1982) with isolated wheat leaf chloroplasts. These studies were aimed at understanding the roles of Ca^{2+} and calmodulin in activating NAD kinase (Anderson et al., 1980), which will be considered next. Both studies found a low amount ($<1\%$) of the total Ca^{2+}-sensitive calmodulin activity in this organelle, but concluded there is a Ca^{2+}-sensitive protein in the chloroplast stroma which is like calmodulin and activates the chloroplast NAD kinase. These conclusions were challenged, however, by other work (Simon et al., 1982, 1984; Roberts et al., 1983). Simon et al. (1982) reported that spinach chloroplast NAD kinase was not affected by Ca^{2+} or calmodulin; and recently Simon et al. (1984) concluded that there is a Ca^{2+}–calmodulin sensitive NAD kinase on the chloroplast envelope, but not in the stroma. Roberts et al. (1983) had reported earlier that there was a calmodulin-binding protein in the pea chloroplast envelope, but that no detectable calmodulin-binding protein was present in the stroma. Collectively, all of this work indicates that a Ca^{2+}-sensitive, calmodulin-like protein exists in the chloroplast; the work disagrees as to whether it is in the stroma or in the envelope, and whether or not NAD kinase in the chloroplast is activated by Ca^{2+}.

The resolution of these disagreements probably resides in the high affinity of NAD kinase for Ca^{2+}–calmodulin (Fig. 4, Harmon et al., 1984). Plant extracts form a Ca^{2+}–calmodulin/NAD kinase complex which does not bind to a calmodulin-Sepharose affinity column (Harmon et al., 1984); indeed 98% of the chloroplast NAD kinase passed through this column in the experiments by Simon et al. (1982). Harmon et al. (1984) show that this complex must be separated as in an anion exchange step (e.g., with DEAE-cellulose chromatography) in order to detect the Ca^{2+}–calmodulin dependency of NAD kinase. Similar explanations probably apply to the Roberts *et al.* (1983) study on stromal proteins. We can reasonably conclude that there is a calmodulin-like protein in the chloroplast which activates NAD kinase, but the degree of association with the envelope is unknown.

Historically the interest in NAD kinase in the chloroplast arose from the observation that light increases the ratio of $NADP^+$ to NAD^+ in photosynthetic tissues (Oh-Hama and Miyachi, 1959). Since $NADP^+$ is the preferred nucleotide for photosynthesis, what was demonstrated may be a photosynthetic regulatory mechanism. Indeed an NAD kinase was found in photosynthetic organisms which required calmodulin (reviewed in Cormier *et al.*, 1980; Anderson *et al.*, 1980; Roux and Slocum, 1983).

Muto *et al.* (1981) and Muto and Miyachi (1981) found the NAD kinase almost exclusively in the chloroplast. However, there is an NAD kinase in the cytoplasm and in other organelles, particularly in nonphotosynthetic plant tissues (Cormier *et al.*, 1980; Roux and Slocum, 1983; Simon *et al.*, 1982, 1984; Dieter and Marmé, 1984). A large body of data is now available showing NAD kinase activity in the leaf chloroplast (Jarrett *et al.*, 1982; Muto, 1982, 1983; Marmé, 1982). As just discussed, the Ca^{2+} or calmodulin stimulation of the chloroplast NAD kinase *in vitro* requires first breaking the complex between these proteins, thus rendering the chloroplast NAD kinase sensitive to exogenous calmodulin (Harmon *et al.*, 1984).

Based on the above data, Fig. 3 presents a model for the light–dark regulation of chloroplast Ca^{2+} levels in the stroma and lumen. Upon illumination Ca^{2+}, which can bind to calmodulin, is pumped from the lumen to the stroma. Subsequently, NAD kinase is complexed with calmodulin, resulting in an increase in the chloroplast $NADP^+ : NAD^+$ ratio. Thus, $NADP^+$ is made available for the continuous operation of photosynthesis. In other words, the light-induced regulation of Ca^{2+} levels in the stroma regulates photosynthesis through the synthesis of $NADP^+$.

V. OTHER Ca^{2+}-SENSITIVE PROTEINS AND PROCESSES IN THE CHLOROPLAST

A. Fructose-1,6-biphosphatase (FBPase) and Seduheptulose-1,7-bisphosphatase (SBPase)

There are reports that Ca^{2+} functions in the activation/inactivation of FBPase and SBPase in the chloroplast stroma (Hertig and Wolosiuk, 1980; Rosa, 1981; Charles and Halliwell, 1980) with $K_{Act_{0.5}} \approx 55$ μM and $KI_{act_{0.5}}$, 7–40 μM. However, Mg^{2+} or Mn^{2+} also serve the same function. The activation/inactivation of these proteins is dependent upon pH and Mg^{2+}, both of which change in the stroma in light and dark, as well as upon reduced thioredoxin. In view of the high stroma level of Mg^{2+}, e.g., ~23 mM in the light and ~16 mM in the dark (Miginiac-Maslow and Hoavau, 1977), there seems to be little reason to speculate on participation from Ca^{2+} with these stromal enzymes. Rather the evidence is quite convincing that these activations/inactivations occur through thioredoxin, with Mg^{2+} serving as the activating metal.

B. Ca^{2+}-ATPase of Chloroplasts

About two decades ago, a Ca^{2+}-dependent ATPase activity was uncovered by a partial digestion of chloroplast thylakoids with trypsin (Selman and Selman–Reimer, 1981). As a result, a massive research effort was launched to understand the ATPase and its role in ATP synthesis by chloroplasts. Does Ca^{2+} play a role *in vivo* in photophosphorylation? The answer seems to be no. The Ca^{2+}-ATPase probably is a laboratory expression of a change *in vitro* in the topographical arrangement of the chloroplast ATP synthesis proteins. No role for Ca^{2+} in the synthesis or hydrolysis of ATP by intact chloroplasts can be reasonably proposed from the experiments performed *in vitro* on the Ca^{2+}-ATPase (Selman and Selman-Reimer, 1981).

C. Ca^{2+} and Photosynthesis Ontogeny

Plant research workers have known for a half century that Ca^{2+} is required for cell division and growth in plants (Burstrom, 1968). However, a specific role for Ca^{2+} during chloroplast development has not been defined, though there is much interest currently in the relation of Ca^{2+} to plant growth (Roux and Slocum, 1983). Some evidence is accumulating on the role of Ca^{2+} in the synthesis and degradation of chlorophyll (Tanaka and Tsuji, 1980, 1981; Ferguson *et al.*, 1983) and in senescence (Leshem *et al.*, 1984), but these subjects have not been studied sufficiently on a biochemical level to warrant a critical review.

VI. CALCIUM AND PROKARYOTIC PHOTOSYNTHESIS

A. Photosynthetic Bacteria

Two distinct groups of photosynthetic prokaryotes have been characterized extensively: photosynthetic bacteria and cyanobacteria (blue–green algae). Neither group contains distinct chloroplasts, but both possess internal photosynthetic membranes which separate the cytosol from an intravesicular space. The membrane-associated light reactions of cyanobacterial photosynthesis differ greatly from those of photosynthetic bacteria, but are very similar to those of higher plants. Photosynthetic bacteria possess only a single photoreaction, which drives the (sometimes indirect) reduction of NAD^+ and synthesis of ATP. The reductant is generally a reduced sulfur or organic compound, but never water, and photosynthesis is obligately anaerobic (Clayton, 1980).

Photosynthetic vesicles are easily isolated from broken cells of photosynthetic bacteria such as *Rhodopseudomonas*. These chromatophores have an inside-out orientation with respect to the plasma membrane of intact cells (Michels and Konings, 1978). The chromatophores are capable of Ca^{2+} uptake but "rightside-out" vesicles are not. Energy-dependent Ca^{2+} uptake does not occur in intact cells (Jasper and Silver, 1978).

Davidson and Knaff (1981) demonstrated a light-dependent Ca^{2+}/H^+ antiport in *Chromatium*. This antiport system apparently pumps Ca^{2+} into the intravesicular spaces and out of the cell during photosynthesis, thus diminishing the cytosol concentration and the total cell content of Ca^{2+}. One might predict that Ca^{2+} content of the periplasmic space would increase during illumination as Ca^{2+} traverses the plasma membrane. These authors proposed that the Ca^{2+}/H^+ antiport system in chromatophores serves to dissipate much of the ΔpH generated during illumination without losing the membrane potential.

Recently a cytochemical staining method for localizing intracellular Ca^{2+} has been employed with *Chromatium* (Petruyaka and Vesil'ev, 1984). The pyroantimonate-osmium stain accumulated only within the intravesicular membranes during illumination. Surprisingly, no periplasmic Ca^{2+} was seen. Vesicular Ca^{2+} accumulation occurred in the dark when O_2 was present, suggesting that energy-dependent uptake of Ca^{2+} did not require light. The function of Ca^{2+} which accumulates within chromatophores and intracellular photosynthetic vesicles is by no means clear. It may serve as an antiport system to adjust pH values as has been proposed for other bacterial systems (Jasper and Silver, 1978), but other roles cannot be ruled out.

B. Cyanobacteria—Whole Cells

Except in endospore formation (Rosen, 1982) direct roles for Ca^{2+} in photosynthetic and nonphotosynthetic bacteria have not been clearly shown. In contrast, there is increasing evidence that Ca^{2+} affects directly several activities of cyanobacteria (Becker and Brand, 1984). Phosphate accumulation is enhanced by Ca^{2+} (Faulkner et al., 1980; Rigby et al., 1980), and preliminary evidence indicates that a calmodulin-like molecule may facilitate phosphate uptake in *Oscillatoria* (Kerson et al., 1984).

Many filamentous cyanobacteria exhibit some type of motility, such as oscillation or gliding, when in contact with a solid substrate. Ca^{2+} appears to be required for this motility (Castenholz, 1973). The rapid change in gliding direction which follows decreased light intensity in *Phormidium* probably involves a gated influx of Ca^{2+} (Hader, 1982). This effect involves photosynthesis only in that the cytoplasmic pH rises during photosynthetic proton pumping into intrathylakoid spaces, which also affects the cytoplasmic membrane potential. A sudden decrease in light intensity diminishes the membrane potential, triggering gated Ca^{2+} channels, which in turn permit Ca^{2+} to flood in and bind to membrane sites which cause reversed direction of motility. The Ca^{2+} is then pumped back out until another trigger again opens the channels. This proposed mechanism now has considerable support in *Phormidium* (Murvanidze and Glagolev, 1981; Hader and Poff, 1982). Recently, Ca^{2+}-dependent depolarization waves have been shown to move from cell to cell along the trichome after a light–dark stimulus in *Phormidium*. An externally applied voltage could be substituted for the change in light intensity (Murvanidze and Glagolev, 1983). However, gliding in *Spirulina* also requires Ca^{2+}, but in this case a pH-dependent Ca^{2+}-mediated trigger may utilize only external Ca^{2+} (Abeliovich and Gan, 1982).

When *Anacystis nidulans* is grown on medium devoid of Ca^{2+}, then photosynthetic O_2 evolution declines, and within a few hours ceases altogether (Becker and Brand, 1982) (Fig. 5). This loss in activity is light-dependent and follows a loss in approximately two-thirds of the intracellular Ca^{2+}. Besides light, the

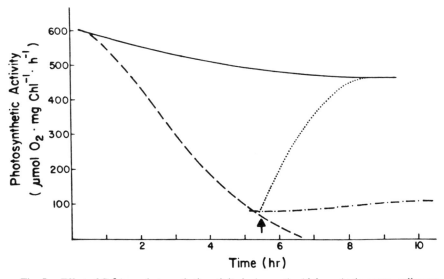

Fig. 5.   Effect of Ca^{2+} on photosynthetic activity in *Anacystis nidulans*. At time zero, cells were transferred from a complete medium to a medium devoid of Ca^{2+}. Incubation was resumed in light (- - - -) or darkness (——). Samples were collected at frequent intervals for measurements of photosynthetic activity in saturating white light. After 5.5 hr, 0.35 mM $Ca(NO_3)_2$ was added to a portion of the Ca^{2+}-deficient culture. This culture was then divided into two aliquots. One aliquot was incubated in light (····) and the other in darkness (·–·–). Samples were collected at frequent intervals for photosythentic measurements as before. From Becker and Brand (1982).

loss in activity requires a divalent cation chelator such as glycylglycine, and the medium must not contain Na^+. Cell growth ceases as soon as photosynthesis is impaired. Lipophilic electron donors and acceptors penetrate the cells, demonstrating that Ca^{2+} depletion inhibits Photosystem II but has no effect on Photosystem I activities (Becker and Brand, 1982). When Ca^{2+} is added to medium of cells depleted of Ca^{2+}, then O_2 activity recovers, and normal growth is restored within a few hours (Becker and Brand, 1984) (Fig. 5). Full recovery requires that at least some residual Photosystem II activity ($>5\%$) be retained.

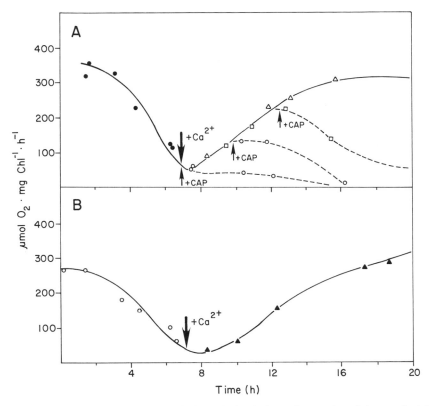

Fig. 6. Effect of chloramphenicol (CAP) on the incubation and restoration of photosynthesis by Ca^{2+} in *Anacystis nidulans*. Cells were placed in Ca^{2+}-deficient medium at time zero and then incubated in light. Samples were collected at times indicated to measure photosynthetic activity under saturating white light. (A) 0.35 mM $Ca(NO_3)_2$ was added as indicated; then the culture was divided into four aliquots. Incubation in light was resumed, and 6.0 μM CAP added to individual aliquots at times indicated by upward pointing arrows. Photosynthetic activity was measured in samples taken from each aliquot before (——) and after (- - - -) addition of CAP. (B) Immediately after cells were placed in the Ca^{2+}-deficient medium, 0.6 μM CAP was added to the culture. At 7 hr cells were washed twice by centrifugation in growth medium to remove residual CAP. Incubation in light was then resumed in medium containing 0.35 mM $Ca(NO_3)_2$ and samples taken for photosynthetic measurements at times indicated (▲). From S.-H. Kim and J. Brand (unpublished).

Recovery of activity also requires light and is impaired when the depletion is performed with Ca^{2+} chelators such as EGTA or EDTA.

Anacystis cells can be disrupted under conditions which retain their photochemical activities (Yu and Brand, 1980). When intact cells are depleted of Ca^{2+} and then broken to measure photochemical activities of the photosynthetic membranes, no Photosystem II activity is detected, even when Ca^{2+} or other cations are present. Photochemical electron transfer from diphenylcarbazide to silicomolybdate includes only reactions very near the reaction center of Photosystem II (Izawa, 1980). These electron transfer reactions do not occur in cells depleted of Ca^{2+} prior to disruption (Becker, 1983).

Chloramphenicol also prevents the restoration of Photosystem II activity in these Ca^{2+}-depleted cells (Fig. 6). Thus, we conclude that new protein synthesis is required for restoration of activity lost upon Ca^{2+} depletion, and that the light-dependent loss of activity may involve the destruction of one or more polypeptides.

A large number of cations, including Li^+, Na^+, K^+, NH^{4+}, Mg^{2+}, Sr^{2+}, Ba^{2+}, Ca^{2+}, Mn^{2+}, Zn^{2+}, La^{3+}, Fe^{3+}, and Al^{3+} were tested to determine if they would restore activity in substitution for Ca^{2+}. The Cl^-, SO_4^{2-}, acetate and NO_3^- salts of several of these ions, including Ca^{2+}, and Sr^{2+} partially, restored activity while the accompanying anion made no difference. Of the monovalent cations tested, only Na^+ restored activity as did Ca^{2+} (Becker, 1983). Thus, in whole *Anacystis* cells the requirement is apparently met by Ca^{2+} or Na^+, and each restores activity fully at about 0.1 mM cation in the external medium. However, isolated photosynthetic membrane preparations from cells not depleted of Ca^{2+} still require Ca^{2+} in the assay medium for high Photosystem II activity, and Na^+ does not replace Ca^{2+} in the assay. Furthermore, other strains of cyanobacteria also display the Ca^{2+} requirement and, at least in *Agmenellum* (PR-6), *Anabaena flos-aquae,* and *Fremyella diplosiphon,* Na^+ did not appear to substitute for Ca^{2+}. The reason for Na^+ replacement of Ca^{2+} in *Anacystis* cells is unknown, but it is interesting that these two ions have virtually identical ionic radii.

In an effort to determine the site or mechanism of Ca^{2+} action, the maximum turnover rates of photosynthetic units were measured in whole cells using bright flashes of short duration (1 μsec) light. Turnover rates did not diminish as Ca^{2+} depletion became manifest. This indicates that Ca^{2+} depletion switches off photosynthetic units one at a time rather than gradually slowing the rate at which each functions (Becker and Brand, 1984).

Variable fluorescence also is lost in parallel with the loss in Photosystem II activity during Ca^{2+} depletion; and in cells devoid of activity the fluorescence remains near F_0 (Brand *et al.,* 1983a). This suggests that the site of inhibition is on or very near the oxidizing side of Photosystem II. Millisecond-delayed fluorescence also is eliminated in Ca^{2+}-depleted cells (Brand *et al.,* 1983b). This signal is attributed to recombination of primary charge separation in Photosystem II, strengthening the hypothesis that the effect is very near the reaction center. Recently ESR Signal II was examined in control cells and in cells at various stages of Ca^{2+} depletion (M. Boska, J. Brand, and K. Sauer, unpublished data). The Signal II spectrum (measured in darkness immediately after illuminations to eliminate Signal I) was easily seen in control cells but not detectable in Ca^{2+}-depleted cells. Signal II_f is manifest in cells heated to 55°C for 10 min. The heating inactivates the O_2-evolving mechanism, but not other portions of the photosynthetic light reactions. This signal, when resolved kinetically, reflects the oxidation state of Z, the primary electron donor to Photosystem II. Ca^{2+} depletion diminishes the amplitude of this signal without changing its kinetics. Thus, it appears that Z^+ formation is precluded in Ca^{2+}-depleted cells, supporting the proposal that the site of Ca^{2+} function is at or immediately adjacent to the reaction center of Photosystem II. Signal II_f and all measured fluorescence parameters return to normal in cells with restored Photosystem II activity by the addition of Ca^{2+} to the external medium.

Cells which have lost all Photosystem II activity by incubation without Ca^{2+} for many hours are not restored by addition of Ca^{2+} to the external medium. During this extended incubation a new fluorescing component appears (Mohanty et al., 1983). The peak (at 683 nm when measured at 77 K) appears upon phycocyanin excitation, but not when only chlorophyll is excited (Mohanty et al., 1985). This secondary effect of Ca^{2+} depletion apparently results from phycobilin emission, probably from an allophycocyanin species. Thus, Ca^{2+} depletion in intact cells results in an initial loss in Photosystem II activity at a site near the reaction center. This loss is fully reversible. A secondary, later effect in inhibited cells is an irreversible disconnection of phycobilisomes from the membranes.

C. Cyanobacteria—Subcellular Preparations

Cells of *Anacystis* can be made permeable to polar electron donors or acceptors without disrupting the cell structure by treatment with lysosyme and EDTA (Ward and Myers, 1972). These permeaplasts showed a significant stimulation of ferricyanide-dependent Hill activity by Ca^{2+}, but very little Ca^{2+} stimulation when the lipophilic electron acceptor phenylenediamine was used (Sotiropoulou et al., 1984). The effect was specific for Ca^{2+}, but a similar set of experiments with *Synechococcus* did not show specificity for Ca^{2+} (Wavare and Mohanty, 1983).

Isolated membrane preparations derived from cyanobacteria perform photosynthetic reactions much like chloroplasts isolated from plants, but special precautions are required. In particular, divalent cations are often reported as a component of Photosystem II assays in cyanobacteria (Black et al., 1963; Fredricks and Jagendorf, 1964; Susor and Krogmann, 1964; Binder et al., 1976;

McSwain *et al.*, 1976). In these early studies, when various ions were compared, Ca^{2+} usually produced the highest levels of activity. However, a specific Ca^{2+} effect could not be claimed, since other ions also stimulated activity. When negatively charged ferricyanide was used as the electron acceptor in sonicated preparations, cation effectiveness in Photosystem II increased with the charge, suggesting a general charge screening effect was involved (DeRoo and Yocum, 1981).

Recently, studies of photosynthetic membrane preparations and Photosystem II particles derived from cyanobacteria suggest that Ca^{2+} plays a more specific role in Photosystem II. Piccioni and Mauzerall (1976) demonstrated that membrane preparations from *Phormidium* have a Ca^{2+} requirement for ferricyanide-mediated O_2 evolution which is not replaced by other cations. Piccioni and Mauzerall (1978) showed that the Ca^{2+} binding was freely reversible, with a first-order dissociation constant of 3 mM. The major effect of added Ca^{2+} was to increase the number of active photosynthetic units. Several other cyanobacteria also showed the Ca^{2+} requirement. *Anacystis* membranes prepared from French pressure cell-disrupted cells retained high Photosystem II activity only when Ca^{2+} was present during cell disruption (Brand, 1979; England and Evans, 1981). Other cations substituted only poorly for Ca^{2+}. Photosystem II activity required divalent cations during assays also, but other cations would substitute partially for Ca^{2+} (Yu and Brand, 1980; Pistorius and Voss, 1982). Photochemical activity also required osmotic support during membrane preparation. This led to the proposal that Ca^{2+} binds a protein to the inner face of the thylakoid membranes. When cells are broken without Ca^{2+} in the medium, membrane Ca^{2+} is released, which also releases the protein into the medium. If Ca^{2+} is present during cell disruption, the required protein remains firmly bound. When vesicles again seal in the presence of osmotic support, then Ca^{2+} may be removed, and the released protein is retained within the vesicles. Evidence for a required protein factor had also been reported earlier (Fredricks and Jagendorf, 1964).

The isolated membranes from cyanobacteria can be further fractionated by exposure to a nonionic detergent followed by density gradient centrifugation. One component fraction is highly enriched in Photosystem II activity, including O_2 evolution, when proper precautions are taken. Such Photosystem II preparations from *Anacystis* show a pronounced requirement for Ca^{2+} (Pistorius, 1983; England and Evans, 1983). Both Ca^{2+} and Mn^{2+} were required for optimum activity, which was inhibited by chlorpromazine.

The site of Ca^{2+} action was not at the O_2-evolving site, since the Ca^{2+} stimulation of Photosystem II activity was still pronounced when O_2 evolution was chemically blocked. The absence of Ca^{2+} during preparation of the active fraction resulted in a loss of three polypeptides of approximate MW 30,000, 33,000, and 39,000, respectively (England and Evans, 1983). It is reasonable to expect that one or more of these peptides may facilitate Photosystem II near the reaction center, while Ca^{2+} serves a stabilizing or regulatory function by holding the peptides(s) in place.

Pistorius and co-workers have examined properties of Photosystem II in several kinds of preparations from *Anacystis*. A bound flavoprotein, a dimer of MW 98,000, acts as an amino acid oxidase (Pistorius and Voss, 1980). Its physiological function is by no means clear, but Ca^{2+} appears to repress its activity approximately to the same extent that it stimulates Photosystem II. These effects of Ca^{2+} were especially manifest in Photosystem II "particles" washed with EDTA (Pistorius, 1983). It was proposed that the effect of chlorpromazine on Photosystem II is at the amino acid oxidase, thus implicating this protein directly in the central reactions of photosynthesis. Pistorius and Schmid (1984) also noted that *Anacystis* membrane preparations treated with EDTA demonstrate only F_0-level fluorescence. A variable component could be restored only in the presence of Mn^{2+} and Ca^{2+}. Of various other cations tested, only $Sr2^+$ substituted partially for Ca^{2+}. These workers also examined the flash yield of O_2 evolution in EDTA-washed particles before and after the addition of Ca^{2+} and/or Mn^{2+}. The flash yield in isolated chloroplasts and in isolated cyanobacteria is well known to oscillate with a frequency of 4 (Kok and Joliot, 1975). The absence of Ca^{2+} resulted in much lower yield on each flash, but did not alter the flash yield pattern in any obvious way. One might conclude from these experiments that Ca^{2+} is acting as a switching mechanism, facilitating Photosystem II on an all-or-none basis.

Not all subcellular membrane preparations from cyanobacteria require Ca^{2+} for Photosystem II activity. Brief sonication of *Anacystis* cells subsequent to lysosyme digestion produces a lamellar fraction with high activities (Ono and Murata, 1978). Although phycobilins are largely removed, electron micrographs indicate that the membranes remain largely intact after this treatment. Membranes prepared by osmotic shock of lysozyme-treated cells without any mechanical disruption retain photochemical O_2 evolution, as do Photosystem II "particles" derived from these membranes (Stewart and Bendall, 1979; Smutzer and Wang, 1984). Also, Photosystem II preparations in which phycobilisomes remain attached do not require added Ca^{2+} for activity, even when French pressure cell disruption is used (Pakrasi and Sherman, 1984).

Polypeptide profiles following SDS–polyacrylamide gel electrophoresis were examined for several of the Photosystem II preparations described above and in other preparations (e.g., Koenig and Vernon, 1981). Quite different patterns are seen among different procedures although polypeptides in the 47–50 kDa, 30–36 kDa, and 18–27 kDa ranges are a common feature. It appears to be easier to selectively remove peptides from membranes of cyanobacteria without massive membrane disruption than from chloroplast thylakoids. Thus, it may soon be possible to relate the Ca^{2+} effect to specific components in cyanobacteria.

VII. MODEL TO EXPLAIN THE ROLE OF Ca^{2+} IN PHOTOSYSTEM II OF CYANOBACTERIA

A polypeptide bound near the reaction center of Photosystem II is required for primary charge separation or for the secondary $Z \rightarrow P_{680+}$ reaction. This polypeptide is bound in place by Ca^{2+}, which requires a critical-Ca^{2+} concentration within the intralamellar space. In isolated lamellar fractions the Ca^{2+} is usually lost, resulting in release of the polypeptide. In whole cells energy-dependent Ca^{2+} pumps are directed away from the cytoplasm, toward the lamellar lumen and the cell exterior, while gated channels maintain low, constant cytoplasmic Ca^{2+} concentrations.

Under illumination in medium devoid of Ca^{2+}, a mass-action effect causes accumulation of Ca^{2+} outside the cells, eventually depleting the lamellar lumen. This results in irreversible loss from the membranes of the polypeptide which facilitates Photosystem II. Addition of Ca^{2+} to the external medium results in its redistribution within the cells, but restoration of Photosystem II activity requires new protein synthesis as well. In cyanobacteria most energy for metabolic processes comes from photosynthetic ATP synthesis. This explains why light is required for restoration of activity and why a small residual photosynthetic activity must be retained in Ca^{2+}-depleted cells to permit restoration.

Under nonstress conditions the cells may maintain a Ca^{2+} level within the lamellar lumen to stabilize the essential polypeptide in place. Alternatively, cells may vary their intralamellar Ca^{2+} concentration to modulate the rate of Photosystem II activity. The latter alternative is attractive since it would provide a way to regulate a balance between cyclic and noncyclic electron flow without generating a potentially damaging strong oxidant when cyclic flow predominates.

Two observations remain unexplained by this hypothesis. First, Na^+ appears to substitute for Ca^{2+} in some (but not all) cyanobacteria in facilitating Photosystem II in whole cells. This might be explained by a Na^+ displacement of Ca^{2+} at some other site, making the freed Ca^{2+} available for the Photosystem II site. Alternatively, Na^+ might substitute directly for Ca^{2+} at the active site, but this would be surprising since no other ion (except Sr^{2+}, slightly) will substitute for Ca^{2+} in restoring activity to Ca^{2+}-depleted cells.

Second, some submembrane preparations from cyanobacteria retain Photosystem II activity without a Ca^{2+} requirement. Apparently, these preparations stabilize the Ca^{2+}–polypeptide-membrane complex in place, as in photosynthetic lamellar preparations from green plant chloroplasts. Future comparative studies of these different photosynthetic preparations should help to identify the key polypeptide(s).

VIII. CALCIUM AND PHOTOSYSTEM II OF CHLOROPLASTS

Chloroplasts can be prepared from a number of plant sources under conditions in which their envelope remains intact for several hours. Without special precautions, however, isolated chloroplasts lose at least a portion of their surrounding membrane, resulting in equilibration of stroma components with the suspending medium. The vast majority of studies of the light reactions of photosynthesis have employed these "broken chloroplasts" or "thylakoid preparations." In order to better characterize specific components, photosynthetic membranes have been fractionated into simpler systems by a variety of physical and chemical treatments which retain photochemical or biochemical activities. With broken chloroplasts or partial membrane fractions, it appears that practically any function is affected by almost any ion tested, if the right concentration of ion is used. Thus, a bewildering array of effects of cations are known for light-driven processes, many of which are seen with Ca^{2+}. We will make no attempt to interpret or even catalog these diverse effects, but will confine this discussion to effects where a physiological role for Ca^{2+} appears likely.

The O_2-evolving component of the photosynthetic electron transport chain requires bound maganese, probably four ions per reaction center or per 200 chlorophylls (Radmer and Cheniae, 1977). Treatment with NH_2OH (Cheniae and Martin, 1970) or alkaline Tris buffer at high concentration (Yamashita and Butler, 1968) inactivates photosynthetic O_2 evolution and partially removes the bound manganese. Activity is restored in inactivated broken chloroplasts by illumination in the presence of exogenous Mn^{2+}, Ca^{2+} and reductant (Yamashita and Tomita, 1975). Calcium is absolutely required for $^{54}Mn^{2+}$ incorporation and simultaneous development of O_2 evolution (Yamashita and Tomita, 1976).

Photosystem II is inactive in angiosperm leaves developed in flashing light, but then occurs normally after continuous light is provided (Remy, 1973). Intact chloroplasts could be isolated from leaves of wheat seedlings grown in flashing light which were inactive in photochemical O_2 evolution but became active upon illumination, with no need for external ion addition (Ono and Inoue, 1982). The divalent cation ionophore A23187 prevented photoactivation, but did not inhibit chloroplasts already photoactivated. EDTA had also previously been shown to preclude photoactivation in broken chloroplasts, but it did not inhibit those already activated (Yamashita and Tomita, 1974). The A23187 inhibition of photoactivation could be prevented only by the addition of both Mn^{2+} and Ca^{2+} to the intact wheat chloroplast preparation (Ono and Inoue, 1983). The antagonistic effect of excessive Mn^{2+} or Ca^{2+} and the competitive inhibition by other ions strongly suggest that both Ca^{2+} and Mn^{2+} must bind to facilitate photoactivation, and that the sites of binding are different for these two ions.

Barr et al. (1980) demonstrated that exposure of broken chloroplasts to a high concentration of EGTA caused inhibition of Photosystem II activity at a site close to the reaction center. Calcium addition to these inhibited chloroplasts restored activity better than did any of a number of other cations tested. These authors (Barr et al., 1982) proposed that a calcium-binding molecule facilitates Photosystem II, since phenothiazine drugs blocked Photosystem II at a site near the reaction center, and Ca^{2+} afforded protection from the inhibition (also see Nakatani, 1984b). Acid treatment of broken chloroplasts caused up to 75% inhibition of Photosystem II activity; Ca^{2+} (or Sr^{2+}) restored activity much better than did other cations (Barr et al., 1983). The acid treatment stripped several polypeptides from the membranes. Best restoration was achieved when these polypeptides along with Ca^{2+} were added back to the inactivated thylakoid membranes (Barr and Crane, 1984). The authors proposed that a 16-kDa polypeptide removed by acid treatment was a calmodulin-like molecule.

"Inside-out" thylakoid membranes (Andersson and Akerlund, 1978) and Photosystem II "particles" prepared by nonionic detergent treatment of chloroplasts (Dunahay et al., 1984) have become important research tools for examining structural requirements for functional Photosystem II. These membrane preparations can be depleted of polypeptides of approximately 34, 23, and 17 kDa, each of which appears to function at the oxidizing side of Photosystem II (see Barber, 1984, for a brief review). Removal of the 23- and 17-kDa peptides do not fully inhibit Photosystem II activity; typically 25–40% of photochemical O_2 evolution remains (Kuwabara and Murata, 1983). The extracted peptides restore nearly fully activity (Wensink et al., 1984) but, curiously, so does Ca^{2+} (Miyao and Murata, 1984; Ghanotakis et al., 1984a; Nakatani, 1984a; Ono and Inoue, 1984). Other cations do not substitute well for Ca^{2+}. Although there is no timidity in providing models (see, for example, Barber, 1984, and articles in Sybesma, 1984), the physiological function of these peptides is not established. A currently popular view is that the extrinsic 23-kDa polypeptide serves a regulatory or structural role in facilitating Photosystem II. It may do so by facilitating the tight binding of Ca^{2+} to the photosynthetic membranes (Ghanotakis et al., 1984b).

Very recently a Ca^{2+}-binding protein was extracted from Photosystem II particle preparations from lettuce chloroplasts (Sparrow and England, 1984). The 14-Kda polypeptide had some calmodulin-like characteristics, and clearly was not identical to the 16-kDa protein extracted by salt treatment in other studies.

A specific calcium effect was seen on the EPR signal II_f of Tris-washed chloroplasts (Yerkes and Babcock, 1981). Calcium appeared to facilitate efficient electron transfer from an anionic donor to Z^+, indicating that Z may be more exposed to this (and other) hydrophobic components when Ca^{2+} is bound.

Broken chloroplasts from Avicennia (mangrove) require Ca^{2+} for high Photosystem II activity; other cations did not appreciably substitute (Critchley *et al.*, 1982). It may be that these chloroplasts had lost the 23-kDa and 17-kDa polypeptides without the special treatments which are required for expression of the calcium effect in nonhalophilic plants. The effect of chlorpromazine (Burris and Black, 1983) on photosynthesis in a coral ecosystem also suggested a calcium requirement in these halophytes.

It is tempting to extend the proposed model for Ca^{2+} function in cyanobacteria to Photosystem II in green plants. However, the diverse and incomplete observations reported thus far are difficult to accommodate within the model. The whole chloroplast experiments of Ono and Inoue (1983, 1984) and the photoactivation experiments of Yamashita and Tomita (1976) perhaps provide the most convincing support for this hypothesis.

IX. CONCLUSIONS

Substantial amounts of calcium are present in all photosynthetic tissues and cells; indeed over 50% of leaf calcium is localized in the chloroplast. One function of Ca^{2+} in the chloroplast is a light-dependent transfer of Ca^{2+} from its thylakoid lumen to its stroma, stoichiometric with an opposite movement of H^+ (Figs. 1 and 2). In the dark, equal amounts of both ions move in reverse directions. Evidence also is accumulating for the existence in chloroplasts of a calmodulin-like Ca^{2+}-binding protein that binds to NAD kinase to result in the synthesis of $NADP^+$ in the light. We propose that a light-dependent Ca^{2+} flux from the chloroplast lumen to the chloroplast stroma regulates the continuous operation of photosynthesis by providing $NADP^+$ (Fig. 3).

In the O_2-evolving Photosystem II of cyanobacteria, Ca^{2+} binds a polypeptide within the intralamellar space. This polypeptide is required for the primary charge separation which permits the reaction center, P_{680}, to function. Upon Ca^{2+} depletion, O_2 evolution activity is lost and the polypeptide is released. With Ca^{2+} restoration the polypeptide is synthesized and bound by Ca^{2+}, and O_2-evolution is restored (Figs. 5 and 6). There may be a similar function for Ca^{2+} in Photosystem II of green-plant chloroplasts, although convincing data for this role have not been presented. Nevertheless, these roles for Ca^{2+} can account for only a small fraction of the total Ca^{2+} in photosynthetic tissues; the roles for the remaining Ca^{2+} are unknown.

ACKNOWLEDGEMENTS

This work was partially supported by National Science Foundation grant No. PCM 8023949 to C. C. Black. Work performed in the laboratory of J. J. Brand described here was supported by a grant from the Robert A. Welch Foundation.

REFERENCES

Abeliovich, A., and Gan, J. (1982). Site of Ca^{2+} action in triggering motility in the cyanobacterium *Spirulina subsalsa*. *Cell Motil.* **2**, 393.

Anderson, J. M., and Cormier, M. J. (1978). Calcium-dependent regulator of NAD kinase in higher plants. *Biochem. Biophys. Res. Commun.* **84**, 595–602.

Anderson, J. M., Charbonneau, H., Jones, H. P., McCann, R. O., and Cormier, M. J. (1980). Characterization of plant nicotinamide dinucleotide kinase activator protein and its identification as calmodulin. *Biochemistry* **19**, 3113–3120.

Andersson, B., and Akerlund, H.-E. (1978). Inside-out membrane vesicles isolated from spinach thylakoids. *Biochim. Biophys. Acta* **503**, 462–472.

Barber, J. (1984). Has the mangano-protein of the water splitting reaction of photosynthesis been isolated? *Trends Biochem. Sci.* **9**, 79–80.

Barr, R., and Crane, F. L. (1984). A Photosystem II-specific calcium site in spinach chloroplasts. *In* "Advances in Photosynthesis Research" (C. Sybesma, ed.), Vol. 1, pp. 441–444. Nijhoff, The Hague.

Barr, R., Troxel, K. S., and Crane, F. L. (1980). EGTA, a calcium chelator, inhibits electron transport in photosystem II of spinach chloroplasts at two different sites. *Biochem. Biophys. Res. Commun.* **92**, 206–212.

Barr, R., Troxel, K. S., and Crane, F. L. (1982). Calmodulin antagonists inhibit electron transport in photosystem II of spinach chloroplasts. *Biochem. Biophys. Res. Commun.* **104**, 1182–1188.

Barr, R., Troxel, K. S., and Crane, F. L. (1983). A calcium-selective site in photosystem II of spinach chloroplasts. *Plant Physiol.* **73**, 309–315.

Becker, D. (1983). An *in vivo* calcium/sodium requirement for Photosystem II activity in *Anacystis nidulans*. Ph.D. Dissertation, University of Texas at Austin.

Becker, D. W., and Brand, J. J. (1982). An *in vivo* requirement for calcium in Photosystem II of *Anacystis nidulans*. *Biochem. Biophys. Res. Commun.* **109**, 1134–1139.

Becker, D. W., and Brand, J. J. (1984). A calcium/sodium requirement near reaction center II in *Anacystis nidulans*. *In* "Advances in Photosynthetic Research" (C. Sybesma, ed.), Vol. 2, pp. 659–762. Nijhoff, The Hague.

Binder, A., Tel-Or, E., and Avron, M. (1976). Photosynthetic activities of membrane preparations of the blue-green alga *Phormidium luridum*. *Eur. J. Biochem.* **67**, 187–196.

Biro, R. L., Daye, S., Serlin, B. S., Terry, M. E., Datta, N., Sopory, S. K., and Roux, S. J. (1984). Characterization of oat calmodulin and radioimmunoassay of its subcellular distribution. *Plant Physiol.* **75**, 382–386.

Black, C. C., Fewson, C. A., and Gibbs, M. (1963). Photochemical reduction of triphosphopyridine nucleotide by cell-free extracts of blue-green algae. *Nature (London)* **198**, 88.

Brand, J. J. (1979). The effect of Ca^{2+} on oxygen evolution in membrane preparations from *Anacystis nidulans*. *FEBS Lett.* **103**, 114–117.

Brand, J. J., and Becker, D. W. (1984). Evidence for direct roles of calcium in photosynthesis. *J. Bioenerg. Biomembr.* **16**, 239–249.

Brand, J. J., Mohanty, P., and Fork, D. C. (1983a). Reversible inhibition of Photosystem II photochemistry in *Anacystis nidulans* by removal of Ca^{2+}. In *Year Book—Carnegie Inst. Washington, Dep. Plant Biol.*, *1982–1983*, pp. 72–75.

Brand, J. J., Mohanty, P., and Fork, D. C. (1983b). Reversible inhibition of photochemistry of Photosystem II by Ca^{2+} removal from intact cells of *Anacystis nidulans*. *FEBS Lett.* **155**, 120–124.

Burris, J. E., and Black, C. C., Jr. (1983). Inhibition of coral and algal photosynthesis by Ca^{2+}-antagonist phenothiazine drugs. *Plant Physiol.* **71**, 712–715.

Burstrom, H. G. (1968). Calcium and plant growth. *Biol. Rev. Cambridge Philos. Soc.* **43**, 287–316.

Callahan, F. (1984). Ph.D. Thesis, University of Kentucky, Lexington.

Castenholz, R. W. (1973). Movements. *In* "The Biology of Blue-Green Algae" (N. G. Carr and B. A. Whitten, eds.), pp. 320–339. Blackwell, Oxford.

Charles, S. A., and Halliwell, B. (1980). Action of calcium ions on spinach (*Spinacia oleracea*) chloroplast fructose bisphosphatase and other enzymes of the Calvin cycle. *Biochem. J.* **188**, 775–779.

Cheniae, G. M., and Martin, I. F. (1970). Sites of function of manganese within photosystem II. Roles in O_2 evolution and system II. *Biochim. Biophys. Acta* **197**, 219–239.

Clayton, R. K. (1980). "Photosynthesis: Physical Mechanisms and Chemical Patterns," IUPAB Ser. Cambridge Univ. Press, London and New York.

Cormier, M. J., Anderson, J. M., Charbonneau, M., Jones, H. P., and McCann, R. (1980). Plant and fungal calmodulin and the regulation of plant NAD kinase. *In* "Calcium and Cell Function" (W. Y. Cheung, ed.), Vol. 1, pp. 201–218. Academic Press, New York.

Critchley, D., Baianu, I. C., Govindjee, and Gutowsky, H. S. (1982). The role of chloride in O_2 evolution by thylakoids from salt-tolerant higher plants. *Biochim. Biophys. Acta* **682**, 436–445.

Davidson, V. L., and Knaff, D. B. (1981). Calcium-proton antiports in photosynthetic purple bacteria. *Biochim. Biophys. Acta* **637**, 53–60.

Davis, D. J., and Gross, E. L. (1975). Protein-protein interactions of light-harvesting pigment protein from spinach chloroplasts. I. Ca^{2+} binding and its relation to protein association. *Biochim. Biophys. Acta* **387**, 557–567.

DeRoo, C. L. S., and Yocum, C. F. (1981). Cation-induced, inhibitor-resistant Photosystem II reactions in cyanobacterial membranes. *Biochem. Biophys. Res. Commun.* **100**, 1025–1031.

Dieter, P., and Marmé, D. (1984). A Ca^{2+}, calmodulin-dependent NAD kinase from dark-grown corn is located in the outer mitochondrial membrane. *J. Biol. Chem.* **259**, 184–189.

Dilley, R. A., and Vernon, L. P. (1965). Ion and H_2O transport processes related to the light-dependent shrinkage of spinach chloroplasts. *Arch. Biochem. Biophys.* **111**, 365–375.

Dunahay, T. G., Staehelin, L. A., Seibert, M., Ogilvie, P. D., and Berg, S. P. (1984). Structural, biochemical and biophysical characterization of four oxygen-evolving photosystem II preparations from spinach. *Biochim. Biophys. Acta* **764**, 179–193.

England, R. R., and Evans, E. H. (1981). A rapid method for extraction of oxygen-evolving Photosystem II preparations from cyanobacteria. *FEBS Lett.* **134**, 175–177.

England, R. R., and Evans, E. H. (1983). A requirement for Ca^{2+} in the extraction of O_2-evolving Photosystem II preparations from the cyanobacterium *Anacystis nidulans*. *Biochem. J.* **210**, 473–476.

Faulkner, G., Horner, F., and Simonis, W. (1980). The regulation energy-dependent phosphate uptake by the blue-green alga, *Anacystis nidulans*. *Planta* **149**, 138–143.

Ferguson, I. B., Watkins, C. B., and Harman, J. E. (1983). Inhibition by calcium of senescence of detached cucumber cotyledons. *Plant Physiol.* **71**, 182–186.

Fredericks, W. W., and Jagendorf, A. T. (1964). A soluble component of the Hill reaction in *Anacystis nidulans*. *Arch. Biochem. Biophys.* **104**, 39–49.

Ghanotakis, D. F., Babcock, G. T., and Yocum, C. F. (1984a). Calcium reconstitutes high rates of oxygen evolution in polypeptide depleted Photosystem II preparations. *FEBS Lett.* **167**, 127–130.

Ghanotakis, D. F., Topper, J. N., Babcock, G. T., and Yocum, C. F. (1984b). Water soluble 17 and 23 kDa polypeptides restore oxygen evolution activity by creating a high-affinity binding site for Ca^{2+} on the oxidizing side of Photosystem II. *FEBS Lett.* **170**, 169–173.

Hader, D.-P. (1982). Gated ion fluxes involved in photophobic responses in the blue-green alga, *Phormidium uncinatum. Arch. Microbiol.* **131**, 77–80.

Hader, D.-P., and Poff, K. L. (1982). Dependence of the photophobic response of the blue-green algae, *Phormidium uncinatum,* on cations. *Arch. Microbiol.* **132**, 345–348.

Harmon, A. C., Jarrett, H. W., and Cormier, M. J. (1984). An enzymatic assay for calmodulins based on plant NAD kinase activity. *Anal. Biochem.* **141**, 168–178.

Hertig, C., and Wolosiuk, R. A. (1980). A dual effect of Ca^{2+} on chloroplast fructose-1,6-bisphosphatase. *Biochem. Biophys. Res. Commun.* **97**, 325–333.

Izawa, S. (1980). Aceptors and donors for chloroplast electron transport. *In* "Methods in Enzymology" (A. San Pietro, ed.), Vol. 69, pp. 413–434. Academic Press, New York.

Jarrett, H. W., Brown, C. J., Black, C. C., and Cormier, M. J. (1982). Evidence that calmodulin is in the chloroplast of peas and serves a regulatory role in photosynthesis. *J. Biol. Chem.* **257**, 13795–13804.

Jasper, P., and Silver, S. (1978). Divalent cation transport systems of *Rhodopseudomonas capsulata. J. Bacteriol.* **133**, 1323–1328.

Kerson, W. K., Miernyk, J. A., and Budd, K. (1984). Evidence for the occurrence of, and possible physiological role for, cyanobacterial calmodulin. *Plant Physiol.* **75**, 222–224.

Koenig, F., and Vernon, L. P. (1981). Which polypeptides are characteristic for photosystem II? Analysis of active photosystem II particles from the blue-green alga *Anacystis nidulans. Z. Naturforsch., C. Biosci.* **36C**, 295–304.

Kok, B., and Joliot, P. (1975). Oxygen evolution in photosynthesis. *In* "Bioenergetics of Photosynthesis" (Govindjee, ed.), pp. 388–410. Academic Press, New York.

Kuwabara, T., and Murata, N. (1983). Quantitative analysis of the inactivation of photosynthetic oxygen evolution and the release of polypeptides and mangenese in the Photosystem II particles of spinach chloroplast. *Plant Cell Physiol.* **24**, 741–747.

Leshem, Y. Y., Sridhara, S., and Thompson, J. E. (1984). Involvement of calcium and calmodulin in membrane deterioration during senescence of pea foliage. *Plant Physiol.* **75**, 329–335.

Lukas, T. J., Iverson, D. B., Schleicher, M., and Watterson, D. M. (1984). Structural characterization of a higher plant calmodulin. *Plant Physiol.* **75**, 788–795.

McSwain, B. D., Tsujimoto, H. Y., and Arnon, D. I. (1976). Effects of magnesium and chloride ions on light-induced electron transport in membrane fragments from a blue-green alga. *Biochim. Biophys. Acta* **423**, 313–318.

Marmé, D. (1982). The role of Ca^{2+} and calmodulin in plants. *What's New Plant Physiol.* **13**, 37–40.

Marmé, D., and Dieter, P. (1983). Role of Ca^{2+} and calmodulin in plants. *In* "Calcium and Cell Function (W. Cheung, ed.), Vol. 4, pp. 263–311. Academic Press, New York.

Matsumura-Kadota, H., Muto, S., and Miyachi, S. (1982). Light-induced conversion of NAD^+ to $NADP^+$ in *Chlorella* cells. *Biochim. Biophys. Acta* **679**, 300–307.

Michels, P. A. M., and Konings, W. N. (1978). Structural and functional properties of chromatophores and membrane vesicles from *Rhodopseudomonas sphaeroides. Biochim. Biophys. Acta* **507**, 353–368.

Miginiac-Maslow, M., and Hoarau, A. (1977). The effect of the ionophore A23187 on Mg^{2+} and Ca^{2+} movements, and internal pH of isolated intact chloroplasts. *Plant Sci. Lett.* **9**, 7–15.

Miyao, M., and Murata, N. (1984). Calcium ions can be substituted for the 24-kD polypeptide in photosynthetic oxygen evolution. *FEBS Lett.* **168**, 118–120.

Mohanty, P., Brand, J., and Fork, D. C. (1983). Calcium depletion affects energy transfer and alters the fluorescence yield and emission of Photosystem II in *Anacystis nidulans. Year Book Carnegie Inst. Washington, Dep. Plant Biol. 1982–1983,* pp. 75–80.

Mohanty, P., Brand, J., and Fork, D. C. (1985). Calcium depletion alters energy transfer and prevents state changes in intact *Anacystis. Photosynth. Res.* **6**, 349–361.

Murvanidze, G. V., and Glagolev, A. N. (1981). Calcium ions regulate reverse motion in photoactically active *Phormidum uncinatum* and *Halobacterium halobium*. *FEMS Microbiol.* **12**, 3–6.

Murvanidze, G. V., and Glagolev, A. N. (1983). Role of calcium concentration graident and proton electrochemical potential in cyanobacterium phot taxis. *Biophysics* **28**, 837–891.

Muto, S. (1982). Distribution of calmodulin within wheat leaf cells. *FEBS Lett.* **147**, 161–164.

Muto, S. (1983). Kinetic nature of calmodulin-dependent NAD kinase from pea seedlings. *Z. Pflanzenphysiol.* **109**, 385–393.

Muto, S., and Miyachi, S. (1981). Light-induced conversion of nicotinamide adenine dinucleotide to nicotinamide adenine dinucleotide phosphate in higher plant leaves. *Plant Physiol.* **68**, 324–328.

Muto, S., Miyachi, S., Usuda, H., Edwards, G. E., and Bassham, J. A. (1981). Light-induced conversion of nicotinamide adenine dinucleotide to nicotinamide adenine dinucleotide phasphate in higher plant leaves. *Plant Physiol.* **68**, 324–328.

Muto, S., Izawa, S., and Miyachi, S. (1982). Light-induced Ca^{2+} uptake by intact chloroplasts. *FEBS Lett.* **139**, 250–254.

Nakatani, H. Y. (1984a). Photosynthetic oxygen evolution does not require the participation of polypeptides of 16 and 24 kilodaltons. *Biochem. Biophys. Res. Commun.* **120**, 299–304.

Nakatani, H. Y. (1984b). Inhibition of photosynthetic oxygen evolution by calmodulin-type inhibitors and other calcium-antagonists. *Biochem. Biophys. Res. Commun.* **121**, 626–633.

Nakatani, H. Y., Barber, J., and Minski, M. J. (1979). The influence of the thylakoid membrane surface properties on the distribution of ions in chloroplasts. *Biochim. Biophys. Acta* **545**, 24–35.

Nobel, P. S. (1967). Calcium uptake, ATPase and photophosphorylation by chloroplasts *in vitro*. *Nature (London)* **214**, 875–877.

Nobel, P. S. (1969). Light-induced changes in the ionic content of chloroplasts in *Pisum sativum*. *Biochim. Biophys. Acta* **172**, 134–143.

Nobel, P. S., and Packer, L. (1964). Energy-dependent ion uptake in spinach chloroplasts. *Biochim. Biophys. Acta* **88**, 453–455.

Oh-Hama, T., and Miyachi, S. (1959). Effects of illumination and oxygen supply upon the levels of pyridine nucleotides in *Chlorella* cells. *Biochim. Biophys. Acta* **34**, 202–210.

O'Keefe, D. P., and Dilley, R. A. (1977). The effect of chloroplast coupling factor removal on thylakoid membrane ion permeability. *Biochim. Biophys. Acta* **461**, 48–60.

Ono, T.-A., and Inoue, Y. (1982). Photoactivation of the water-oxidation system in isolated intact chloroplasts prepared from wheat leaves grown under intermittent flash illumination. *Plant Physiol.* **69**, 1418–1422.

Ono, T.-A., and Inoue, Y. (1983). Requirement of divalen cations for photoactivation of the latent water-oxidation system in intact chloroplast from flashed leaves. *Biochim. Biophys. Acta* **723**, 191–201.

Ono, T.-A., and Inoue, Y. (1984). Ca^{2+}-dependent restoration of O_2-evolving activity in $CaCl_2$-washed PS II particles depleted of 33, 24 and 16 kDa. *FEBS Lett.* **168**, 281–286.

Ono, T.-A., and Murata, N. (1978). Photosynthetic electron transport and phosphorylation reactions in thylakoid membranes from the blue-green alga *Anacystis nidulans*. *Biochim. Biophys. Acta* **502**, 477–485.

Pakrasi, H. B., and Sherman, L. A. (1984). A highly active oxygen-evolving photosystem II preparation from the cyanobacterium *Anacystis nidulans*. *Plant Physiol.* **74**, 742–745.

Petrunyaka, V. V., and Vasil'ev, B. G. (1984). Ultrastructural localization of calcium in purple sulfur bacteria. *Microbiology (Engl. Transl.)* **52**, 655–658.

Piccioni, R. G., and Mauzerall, D. C. (1976). Increase effected by calcium ion in the rate of oxygen evolution from preparations of *Phormidium luridum*. *Biochim. Biophys. Acta* **423**, 605–609.

Piccioni, R. G., and Mauzerall, D. C. (1978). Calcium and photosynthetic oxygen evolution in cyanobacteria. *Biochim. Biophys. Acta* **504**, 384–397.

Pistorius, E. K. (1983). Effects of Mn^{2+}, Ca^{2+} and chlorpromazine on photosystem II of *Anacystis nidulans*. *Eur. J. Biochem.* **135**, 217–222.

Pistorius, E. K., and Schmid, G. H. (1984). Effect of Mn^{2+} and Ca^{2+} on O_2 evolution and on the variable fluorescence yield associated with Photosystem 2 in preparations of *Anacystis nidulans*. *FEBS Lett.* **171**, 173–178.

Pistorius, E. K., and Voss, H. (1980). Some properties of a basic L-amino-acid oxidase from *Anacystis nidulans*. *Biochim. Biophys. Acta* **611**, 227–240.

Pistorius, E. K., and Voss, H. (1982). Presence of an amino acid oxidase in photosystem II of *Anacystis nidulans*. *Eur. J. Biochem.* **126**, 203–209.

Radmer, R., and Cheniae, G. (1977). Mechanisms of oxygen evolution. *In* "Primary Processes of Photosynthesis" (J. Barber, ed.), pp. 303–350. Elsevier, Amsterdam.

Remy, R. (1973). Appearance and development of photosynthetic activities in wheat etioplasts greened under continuous or intermittent light-evidence for water-side photosystem II deficiency after greening under intermittent light. *Photochem. Photobiol.* **18**, 409–416.

Rigby, C. H., Craig, S. R., and Budd, K. (1980). Phosphate uptake by *Synechococcus leopoliensis* (Cyanophyceae): Enhancement by calcium ion. *J. Physcol.* **16**, 389–393.

Roberts, D. M., Zielinski, R. E., Schleicher, M., and Watterson, D. M. (1983). Analysis of suborganellar fractions from spinach and pea chloroplasts for calmodulin-binding proteins. *J. Cell Biol.* **97**, 1644–1647.

Roberts, D. M., Burgess, W. H., and Watterson, D. M. (1984). Comparison of the NAD kinase and myosin light chain kinase activator properties of vertebrate, higher plant, and algal calmodulins. *Plant Physiol.* **75**, 796–798.

Rosa, L. (1981). A rapid activation *in vitro* of the chloroplast fructose-1,6-bisphosphatase following a new assay procedure. *FEBS Lett.* **134**, 151–154.

Rosen, B. P. (1982). Calcium transport in microorganisms. *In* "Membrane Transport of Calcium" (E. Carafoli, ed.), pp. 187–216. Academic Press, New York.

Roux, S. J., and Slocum, R. D. (1983). Role of calcium in mediating cellular functions important for growth and development in higher plants. *In* "Calcium and Cell Function" (W. Y. Cheung, ed.), Vol. 3, Chapter 13, pp. 409–453. Academic Press, New York.

Schleicher, M., Lukas, T. J., and Watterson, D. M. (1983). Further characterization of calmodulin from the monocotyledon barley (*Hordeum vulgare*). *Plant Physiol.* **73**, 666–670.

Selman, B. R., and Selman-Reimer, S., eds. (1981). "Energy Coupling in Photosynthesis." Am. Elsevier, New York.

Simon, P., Dieter, P., Bonzan, M., Greppin, H., and Marmé, D. (1982). Calmodulin-dependent and independent NAD kinase activities from cytoplasmic and chloroplastic fractions of spinach (*Spinacia oleracea* L.). *Plant Cell Rep.* **1**, 119–122.

Simon, P., Bonzon, M., Greppin, H., and Marmé, D. (1984). Subchloroplastic localization of NAD kinase activity: Evidence for a Ca^{2+}, calmodulin-dependent activity at the envelope and for a Ca^{2+}, calmodulin-dependent activity in the stroma of a pea chloroplasts. *FEBS Lett.* **167**, 332–337.

Smutzer, G., and Wang, J. H. (1984). A highly active oxygen-evolving Photosystem II preparation from *Synechococcus lividus*. *Biochim. Biophys. Acta* **766**, 240–244.

Sotiropoulou, G., Lagoyanni, T., and Papageorgiou, G. C. (1984). Effects of Ca^{2+} ions on the light-induced electron transport activities of *Anacystis nidulans* permeaplasts and spheroplasts. *In* "Advances in Photosynthesis Research" (C. Sybesma, ed.), Vol. 2, pp. 663–666. Nijhoff, The Hague.

Sparrow, R. W., and England, R. R. (1984). Isolation of a calcium-binding protein from an oxygen-evolving Photosystem II preparation. *FEBS Lett.* **177**, 95–98.

Stewart, A. C., and Bendall, D. S. (1979). Preparation of an active oxygen-evolving photosystem II particle from a blue-green alga. *FEBS Lett.* **107**, 308–312.

Stocking, C. R., and Ongun, A. (1962). The intracellular distribution of some metallic elements in leaves. *Am. J. Bot.* **49**, 284–289.

Stosic, V., Penel, C., Marmé, D., and Greppin, H. (1983). Distribution of calmodulin-stimulated Ca^{2+} transport into membrane vesicles from green spinach leaves. *Plant Physiol.* **72**, 1136–1138.

Susor, W. A., and Krogmann, D. W. (1964). Hill activity in cell-free preparations of a blue-green alga. *Biochim. Biophys. Acta* **88**, 11–19.

Sybesma, C., ed. (1984). "Advances in Photosynthesis Research," Vol. 1 (various articles). Nijhoff, The Hague.

Tanaka, A., and Tsuji, H. (1980). Effects of calcium on chlorophyll synthesis and stability in the early phase of greening in cucumber cotyledons. *Plant Physiol.* **65**, 1211–1215.

Tanaka, A., and Tsuji, H. (1981). Changes in chlorophyll *a* and *b* content in dark-incubated cotyledons excised from illuminated seedlings. *Plant Physiol.* **68**, 567–570.

Ward, B., and Myers, J. (1972). Photosynthetic properties of permeaplasts of *Anacystis. Plant Physiol.* **50**, 547–550.

Wavare, R. A., and Mohanty, P. (1983). Cations stimulate electron transport associated with Photosystem II and inhibit electron flow linked with Photosystem I in spheroplasts of the cyanobacterium *Synechococcus cedrorum. Photobiochem. Photobiophys.* **6**, 189–199.

Wensink, J., Dekker, J. P., and van Gorkom, H. J. (1984). Reconstitution of Photosynthetic water splitting after salt-washing of oxygen-evolving photosystem II particles. *Biochim. Biophys. Acta* **765**, 147–155.

Williamson, R. E., and Ashley, C. C. (1982). Free Ca^{2+} and cytoplasmic streaming in the alga *Chara. Nature (London)* **296**, 647–651.

Yamagishi, A., Satoh, K., and Katoh, S. (1981). Fluorescence induction in chloroplasts isolated from the green alga, *Bryopsis maxima IV. Biochim. Biophys. Acta* **637**, 252–263.

Yamashita, T., and Butler, W. L. (1968). Photoreduction and photophosphorylation with triswashed chloroplasts. *Plant Physiol.* **43**, 1978–1986.

Yamashita, T., and Tomita, G. (1974). Effects of manganese, calcium, dithiothreitol and bovine serum albumin on the light-reactivation of Tris-acetone-washed chloroplasts. *Plant Cell Physiol.* **15**, 69–82.

Yamashita, T., and Tomita, G. (1975). Comparative study of the reactivation of oxygen evolution in chloroplasts inhibited by various treatments. *Plant Cell Physiol* **16**, 283 296.

Yamashita, T., and Tomita, G. (1976). Light reactivation of (Tris-washed)-DPIP-treated chloroplasts: Manganese incorporation, chlorophyll fluorescence, action spectrum and oxygen evolution. *Plant Cell Physiol.* **17**, 571–582.

Yerkes, C. T., and Babcock, G. T. (1981). Surface charge asymmetry and specific calcium ion effect in chloroplast photosystem II. *Biochim. Biophys. Acta* **634**, 19–29.

Yu, C. M.-C., and Brand, J. J. (1980). Role of divalent cations and ascorbate in photochemical activities of *Anacystis* membranes. *Biochim. Biophys. Acta* **591**, 483–487.

Index